The Future of Road Transportation

The Future of Road Transportation presents rapidly growing research towards electrified and automated vehicles. It explains the workings and drawbacks of a conventional vehicle's powertrain, braking, and steering systems before exploring ADAS equipment and driverless car technologies.

Emphasizing the necessary changes in conventional transport systems towards sustainable and smart mobility, this book discusses advanced future mobility technologies and the challenges and considerations for developing sustainable vehicle designs. It overviews the construction details and the research-level contents of the power train, battery, charging infrastructure, and other control systems of the electrical vehicles.

This book is intended for automotive and electrical engineers and researchers working on electric vehicle technology, autonomous and automated vehicles, and automotive sustainability. It will also be useful for mechanical and electrical engineering students taking courses in Automotive/Vehicle Engineering and Automotive Systems and Design.

The Future of Road Transportation

Electrification and Automation

Edited by
Jeyaprakash Natarajan, Mahendra Babu Kantipudi,
Che-Hua Yang, and Yaojung Shiao

CRC Press
Taylor & Francis Group
Boca Raton London New York

CRC Press is an imprint of the
Taylor & Francis Group, an **informa** business

Designed cover image: Shutterstock

First edition published 2024
by CRC Press
2385 NW Executive Center Drive, Suite 320, Boca Raton FL 33431

and by CRC Press
4 Park Square, Milton Park, Abingdon, Oxon, OX14 4RN

CRC Press is an imprint of Taylor & Francis Group, LLC

Library of Congress Cataloging-in-Publication Data
Names: Natarajan, Jeyaprakash, editor. | Kantipudi, Mahendra Babu, editor.|
Yang, Che-Hua, editor. | Shiao, Yaojung, editor.
Title: The future of road transportation : electrification and automation /
edited by Jeyaprakash Natarajan, Mahendra Babu Kantipudi, Che-Hua Yang, and Yaojung Shiao.
Description: First edition. | Boca Raton, FL : CRC Press, 2024. |
Includes bibliographical references and index.
Identifiers: LCCN 2023024365 (print) | LCCN 2023024366 (ebook) |
ISBN 9781032408330 (hardback) | ISBN 9781032408347 (paperback) |
ISBN 9781003354901 (ebook)
Subjects: LCSH: Electric automobiles—Design and construction. |
Electric automobiles—Batteries. | Automobiles—Automatic control. |
Transportation—Technological innovations.
Classification: LCC TL220 .F88 2024 (print) | LCC TL220 (ebook) |
DDC 629.22/93—dc23/eng/20230525
LC record available at https://lccn.loc.gov/2023024365
LC ebook record available at https://lccn.loc.gov/2023024366

ISBN: 978-1-032-40833-0 (hbk)
ISBN: 978-1-032-40834-7 (pbk)
ISBN: 978-1-003-35490-1 (ebk)

DOI: 10.1201/9781003354901

Typeset in Times
by codeMantra

Contents

List of Contributors

G. Suresh Babu
Chaitanya Bharathi Institute of Technology (A)
India

M. Baskaran
K.S. Rangasamy College of Technology
India

B. Mouli Chandra
QIS College of Engineering and Technology
India

Muralimohan Cheepu
Super-TIG Welding Company Limited
Republic of Korea

Milon Selvam Dennison
Kampala International University
Uganda

Ashok Kumar Dewangan
National Institute of Technology, Delhi
India

T. Ganapathy
S Veerasamy Chettiar College of Engineering and Technology
India

Suresh Goka
National Institute of Technology
Warangal
India

R. Gopalakrishnan
K.S. Rangasamy College of Technology
India

Xiaoqiang Guo
Yanshan University
China

N. Jeyaprakash
China University of Mining and Technology
China

Venkata Charan Kantumuchu
Bradley University, IL
USA

Sundara Subramanian Karuppasamy
National Taipei University of Technology
Taiwan

Bollepelly Manichandra
National Taipei University of Technology
Taiwan

Syed Quadir Moinuddin
College of Mechanical Engineering
King Faisal University,
Al Hofuf - 31982
Kingdom of Saudi Arabia

Dora Nagaraju
GITAM Deemed to be University
India

S. Hari Prasadh
K.S. Rangasamy College of Technology
India

Koganti Radhika
Dhanalakshmi Srinivasan College of Engineering and Technology
India

R. Rajasekaran
Dhanalakshmi Srinivasan College of Engineering and Technology
India

M. Ravichandran
K. Ramakrishnan College of Engineering
India

Tara Kalyani Sandipamu
Jawaharlal Nehru Technological University
India

S. Moses Santhakumar
National Institute of Technology
Tiruchirappalli
India

S. Sathish
Madras Institute of Technology
India

S. Senthilraja
K.S. Rangasamy College of Technology
India

Yaojung Shiao
National Taipei University of Technology
Taiwan

Sandeep Singh
National Institute of Technology, Puducherry
India

Freddy Tan Kheng Suan
University of Nottingham
Malaysia

V. Sundararaju
K.S. Rangasamy College of Technology
Inida

Ahmad Syed
Chaitanya Bharathi Institute of Technology (A)
India

A. Cyril Thomas
SASTRA Deemed to be University
India

R. Vidya
SASTRA Deemed to be University
India

C. Vijayakumar
K.S. Rangasamy College of Technology
India

S. Yamini
National Institute of Technology Tiruchirappalli
India

Che-Hua Yang
National Taipei University of Technology
Taiwan

Preface

Transportation is the heart of human evolution in this modern, connected global market era. In recent years, vehicles have been tuned to be much more usable, efficient, and smarter. However, global warming, safety, and other traffic issues are giant challenges for automobile engineers. Chapters 1–3 elucidate the working and drawbacks of the conventional vehicle's powertrain, braking, and steering systems. The objective of this book is to study future mobility that is going to be happening—the concept of the *go greens and go safe*. This book emphasizes the essential and required changes in the conventional transport systems that civilization should make to move towards sustainable and smart mobility. Fourteen chapters cover advanced future mobility technologies. This overview guides future automobile engineers to develop sustainable designs.

First, this book is planned for engineers and researchers who need detailed knowledge of electric vehicle technology. Electrical vehicles have the advantages of zero emissions in the area of the vehicles and less dependency on fossil fuels. This technology is not new for transportation, since electrical trains are widely used. But the electric vehicles have not attained commercial success in road transportation due to limitations. Battery technology has today advanced to the level where electric vehicles can stand commercially successful. Chapters 4–10 thoroughly overview the construction details and the research-level contents of the power train, battery, charging infrastructure, and other control systems of the electrical vehicles.

Second, ADAS are arrangements in a vehicle that back the driver in a variety of functions. The ADAS can give significant information about traffic flow like signals, obstruction, and closings and can advise restored routes to avoid the above glitches. It can also notice any tiredness or distraction from the driver, alarm the driver, and even take control of the vehicle automatically. These ADAS equipment and driverless car technologies can solve the man-made road accident problems. Chapters 11–13 explain the working and construction details of sensing and control systems for ADAS and driverless cars. Chapter 14 discusses the present situation and future projections of these electrical and automated vehicles.

This book aims to develop a world of incredible diversity and environmentally intent societies with inventive and advanced mobility architecture. The reader can not only understand the technical and trendy information but also observe rapidly growing research towards electrified and automated mobility.

About the Editors

Jeyaprakash Natarajan is an Associate Professor at the School of Mechanical and Electrical Engineering, China University of Mining and Technology, Xuzhou, China. He received his bachelor's degree in Mechanical Engineering from Anna University, Chennai, India in 2006. Dr. Jeyaprakash Natarajan received his master's degree in Manufacturing Engineering in 2009 from Anna University, Chennai, India and his PhD in Production Engineering in 2018 from the National Institute of Technology, Tiruchirappalli, India. In 2007, he was appointed to the position of Lecturer at the Department of Automobile Engineering, Thanthai Roever Institute of Polytechnic College, Perambalur, India. In 2009, he was appointed as an Assistant Professor in the Department of Mechanical Engineering at Periyar Maniammai University, Tanjore, India. In 2011, he was appointed to the position of Assistant Professor in the Department of Engineering, Hamelmalo Agricultural College, Keren, Eritrea, North East Africa. In 2013, he was appointed as Assistant Professor in the Department of Production Engineering, Defence University, Debre Zeit, Ethiopia, Africa. From 9, 2018 to 12, 2018 he was a Research Assistant in Additive Manufacturing Center for Mass Customization Production, National Taipei University of Technology, Taiwan. From 2019 to 2020, he was a Postdoctoral Research Associate in Additive Manufacturing Center for Mass Customization Production, Graduate Institute of Manufacturing Technology, National Taipei University of Technology, Taiwan. In 2021, he was promoted as Research Assistant Professor in Graduate Institute of Manufacturing Technology, National Taipei University of Technology, Taiwan. Dr. Jeyaprakash Natarajan has published more than 70 peer-reviewed research articles, book chapters, and two books titled Laser Surface Treatments for Tribological Application and Advances in Additive Manufacturing Processes in 2021. He teaches various postgraduate courses in automotive, mechanical, manufacturing and laser ultrasound, and conducts academic research and consultancy. Dr. Jeyaprakash Natarajan is serving as a reviewer for many reputed international journals. Also, Dr. Jeyaprakash Natarajan has completed four research projects in the field of additive manufacturing, surface engineering, tribology, laser cladding, corrosion, etc.

Mahendra Babu Kantipudi received his B.Tech. in Mechanical Engineering from Pondicherry University, India, in 2007 and his master's degree in Automotive Engineering from the University of Hertfordshire, England, in 2009. Then, he worked as an Assistant Professor of Mechanical Engineering at reputed engineering colleges in India for 7 years. He got his PhD from the College of Mechanical and Electrical Engineering, National Taipei University of Technology, Taiwan in 2020. Currently, he is a senior researcher in the National Taipei University of Technology and has been involved in the project of the Ministry of Science and Technology, Taiwan. He is the author and co-author of several research papers in SCI-indexed journals. He has presented the paper at national and international conferences. His areas of research interest include the areas of magnetorheological fluid (MR)-based actuators, anti-lock braking systems (ABS), smart vehicle control systems, IC engine performance study, electromagnetic analysis, future mobility, thermal studies, and multi-physics analysis.

Che-Hua Yang is a Professor at the Graduate Institute of Manufacturing Technology, National Taipei University of Technology, Taiwan. He received his bachelor's degree in Mechanical Engineering from National Taiwan University, Taiwan in 1984. Prof. Che-Hua Yang received his Masters and PhD in Mechanical Engineering in 1989 and 1991, respectively, from Johns Hopkins University, USA. In 1991, he was appointed as a research associate at Johns Hopkins NDE Center, Johns Hopkins University USA, where he worked on the structural analysis and Opto-mechanical integration and implementation. In 1993, he was appointed as an Associate Professor in the Department of Mechanical Engineering at Chang Gung University, Taiwan. In 2005, he was promoted as Professor in the Department of Mechanical Engineering, Chang Gung University, Taiwan. In 2007, Prof. Che-Hua Yang was appointed as Professor in the Department of Mechanical and Electrical Engineering, National Taipei University of Technology, Taiwan and he has been the Deputy Director in National Taipei University of Technology, Taiwan, from 2007 to 2008. From 2007 to 2008, he was also an Adjust Professor at Chang Gung University, Taiwan. From 2008 to 2011, he was a Professor and Director of the Institute of Engineering, National Taipei University of Technology, Taiwan. From 2011 to 2017, he was a Professor and Dean of School of Mechanical and Electrical Engineering, National Taipei University of Technology, Taiwan. From 2017 to 2018, he was a Professor and Vice President of National Taipei University of Technology, Taiwan. From 2018 to 2021, he was a Professor and Director of Additive Manufacturing Center for Mass Customization Production, National Taipei University of Technology, Taiwan. Prof. Che-Hua Yang has published more than 120 peer-reviewed research articles, book chapters and two books titled Laser Surface Treatments for Tribological Application and Advances in Additive Manufacturing Processes in 2021. He teaches various postgraduate courses in automotive, mechanical, manufacturing and laser ultrasound, supervises Master and PhD students, and conducts academic research and consultancy. Prof. Che-Hua Yang is serving as a reviewer for many reputed international journals.

Yaojung Shiao received his bachelor's degree from the Department of Mechanical Engineering, National Cheng Kung University, Taiwan. He received his master's degree and PhD from the Department of Mechanical Engineering, University of Wisconsin, Madison, USA. He has worked as an Assistant Professor of Heat Transfer in the Mechanical Engineering Department, University of Wisconsin, Madison, USA. He worked as an Associate Professor in the Department of vehicle engineering, National Pingtung University of Science and Technology, Taiwan from 1996 to 2001. He was the head of the Department of vehicle engineering at the National Taipei University of Technology, Taiwan from 2014 to 2019. He is presently a Professor in the Department of vehicle engineering, National Taipei University of Technology, Taiwan. He is the author of over 100 publications in smart material structures, actuators, intelligent control systems, connected vehicles, vehicle control systems, and gait control systems. He is working on rail vehicles, electric vehicle systems, advanced low-energy engines, road-friendly headlight systems, advanced braking systems and suspension systems, and locomotive ABS/TCS systems. His research has obtained several invention patents.

1 Fundamentals and Challenges of Conventional Road Transportation

Dora Nagaraju
GITAM Deemed to be University

1.1 INTRODUCTION

1.1.1 History

Automobile engines are helpful for the transportation of goods and passengers. The automobile engines involve light-, medium-, and heavy-duty vehicles; these are self-propelled running through internal combustion engine principles. The automobile engines in the modern scenario are made of an efficient propulsion system and excellent frame structure to assist the body and transmitting systems. Through tyres, wheels, and proper suspension, the power generated at the combustion chamber is utilized for the desired purpose. Figure 1.1 presents a block diagram of the history of vehicles and vehicle components.

The historical development of engines started in 1769 in France and was designed by implementing the steam engine principle. Indeed, the unique merits of automobiles were explored simultaneously in various locations of the world. In 1885, German engineers built a gasoline engine, and from the initiation of thorough investigations, an Otto cycle was designed in 1914 [1]. Later, American scientist Henry Ford built a two-cylinder engine to power bicycle-type wheels. Therefore, the significance of the concept of internal combustion engines has come into the limelight over the next four decades.

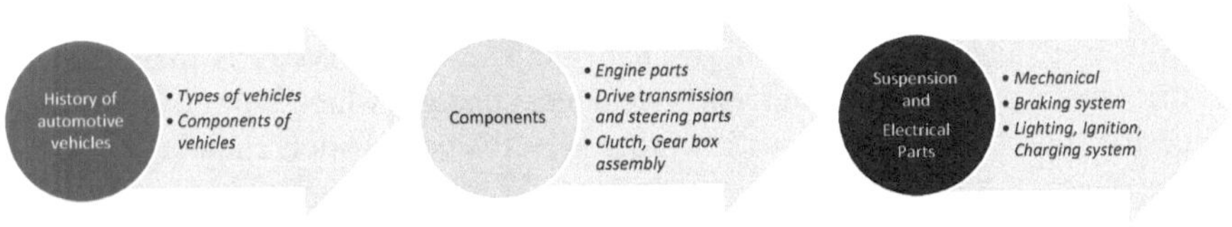

FIGURE 1.1 History of vehicles and vehicle components.

DOI: 10.1201/9781003354901-1

Moreover, automobiles have played a vital role in providing mobility to transport in the progress of modern society to safeguard and satisfy the needs of humankind purpose in everyday life. Eventually, the rapid population growth led to speedy automobile industry growth since the dependence on automobiles increased for advanced humankind. Furthermore, revenue generation and economic development growth have served as a backbone.

However, the continuous usage of automobiles by the population causes unavoidable environmental damage and severe health hazards to human life as automobiles release serious emissions into the atmosphere and subsequently increase global warming and air pollution. Moreover, the depletion of fossil resources on the earth leads to stringent issues; alternative fuel sources should be paramount if it continues. In this regard, several researchers have carried out enormous investigations through research and development activities. It is perceived that efficient combustion, clean transportation, and safety are much needed. Researchers have suggested various alternative sources such as biofuels, electric-hybrid vehicles, electric vehicles, dual-model operative vehicles, and fuel cell vehicles.

1.1.2 Types of Vehicles

Automobile vehicles are categorized based on load-carrying capacities, such as heavy-, medium-, and light-duty vehicles. Examples of these vehicles are heavy- and light-duty motor vehicles. Also, based on wheels, the vehicles are classified as two-, three-, and four-wheeler vehicles (e.g., scooter, autorickshaws, cars, etc.).

Based on the fuel usage in the engine, the vehicle can be gasoline (petrol), diesel, gas (CNG), battery-driven or steam vehicles. Also, there are other types namely dual-mode engines. Any conventional engine can be modified as dual-mode by adopting a suitable advancement technique. The body style of vehicles is different; on this basis, Sedan hatchback cars, vans, and cars have a particular purpose. On the other hand, the vehicle can be called either manual or automatic based on the transmission system. In the automatic model, the gears will be switched automatically, whereas they must be performed manually in the former model.

1.1.3 Components of Vehicle

The components of automobiles are divided into five parts, which are presented in the following sections.

1.1.3.1 Engine Parts

The engine is the heart of the automobile, consisting of a combustion chamber, piston, valves, and valve actuating mechanism. The engine converts the available chemical energy of fuel into valuable mechanical energy, which will be helpful for the desired purpose. The combustion process in the combustion chamber involves various stages. In this context, the vehicle can be classified as an external or internal combustion engine based on the fuel combustion types. If the mixture of fuel and air is prepared externally to the engine, it is called an external combustion engine (EC);

on the other hand, if the combustion happens inside the engine, it is named an internal combustion engine (IC).

1.1.3.2 Drive Transmission and Steering Parts

The desired purpose of the vehicle will be served by an efficient transmission system and its parts. The function of the transmission system is to supply the output of the IC engine to drive the wheels. The transmission system will monitor the vehicle speed by altering the torque according to the load applied to the engine. In practical scenarios, the transmission system is also used in pedal bicycles and some fixed-bed machines to control the rotational speed. The components of the transmission system are explained in the following sections.

1.1.3.2.1 Clutch

The clutch engages and disengages the developed power from the drive shaft to the driven shaft. The mechanism enables the rotary motion shaft when desired, and both drive and driven axes coincide.

1.1.3.2.2 Gearbox Assembly

The gearbox is the transmission system that moves the vehicle at different speeds. The gear ratio selection depends on the torque available at the flywheel. The vehicle bears high inertia forces at a lower speed. Eventually, as the speed increases, the torque on the vehicle is reduced, and subsequently, the vehicle will maintain the desired speed at higher gear ratios.

1.1.3.2.3 Propeller Shaft

The main purpose of the propeller shaft is to transmit the developed brake power from the engine to the wheels. The propeller shaft consists of three parts, namely, shaft, universal joint, and slip joint. First, the shaft bears the typical torsional stresses produced during the twisting. Later, the universal joint is useful to handle the up and down vehicle movements. The slip joint's purpose is to adjust the propeller shaft when required.

1.1.3.2.4 Steering Mechanism

A steering mechanism can do the vehicle's turning movements and control vehicle directions in different paths. It consists of a steering wheel, steering gear that converts rotary motion into a straight path, and steering linkages. The power steering system is manually controlled or uses electric or hydraulic power drives.

1.1.4 Suspension and Brake Parts

The system consists of a spring system and, more importantly, shock absorbers connected with various linkages to the vehicle wheel. The system is mainly helpful for holding the vehicle wheel with the road while applying the brake and avoiding slippages, providing passengers with comfortable driving conditions. Moreover, the suspension system must isolate the vehicle from the vibrations caused during running and other noises such as bums. The components of the mechanical suspension system are explained in detail.

1.1.4.1 Mechanical Suspension

The components are various springs such as leaf, coil, and rubber to handle the vehicle dynamics. Torsion bars are helpful to keep the tyres in contact with the road surface and support the vehicle's weight. Eventually, the absorbed forces during vehicle movement will be balanced. As part of the hydraulic suspension, it consists of a hydraulic shock absorber and a telescopic fork absorber. These systems serve the vehicle to handle the pressure exerted when the vehicle travels in different directions and up/down distances. The hydraulic fluid will be moved out of the tube by force exerted by the piston. Another important system is air suspension which consists of air springs.

1.1.4.2 Brakes

The braking system consists of various links and mechanical linkages or brake line used to stop the vehicle, and hence kinetic energy is converted into heat energy; as a result, the vehicle decelerates or sometimes stops. The heat generated will be dissipated into the atmosphere, and the braking system can act as a precaution against accidents and prevent the vehicle from severe damage.

1.1.5 Electrical Parts

Electrical and electronic systems play a vital role in the functioning of automobiles. The following are the systems used in automobiles.

- *Starting system*: It is driven by electrical power taken from the battery.
- *Ignition system*: The homogenous charge prepared in the pre-mix chamber will be ignited using an ignition system. The fuel-air mixture will be sent to the combustion chamber to ignite at the end of the compression stroke.
- *Charging system*: It regulates, generates, and supplies power to charge the vehicle's battery.
- *Lighting system*: It consists of various essential lights used in vehicles, such as panel board lights, headlights, fog lights, etc.

1.2 IC ENGINE

The internal combustion engine ignites and burns the fuel-air mixture inside the engine cylinder. The working principle is that the fuel can burn under extreme pressure generation conditions to generate power. The entire combustion process occurred in a closed system, consisting of an engine cylinder, piston, crank, valves, and valve actuating mechanism. The charge will be ignited in the cylinder, generating high pressure and temperature. Further, the piston will exert a pressure force resulting in rectilinear motion of the piston. Later, it converts into rotational speed by employing a crank connected through a connecting rod. The fuel used in this engine type is either gaseous fuel like CNG or LPG or volatile fuels such as petrol, diesel,

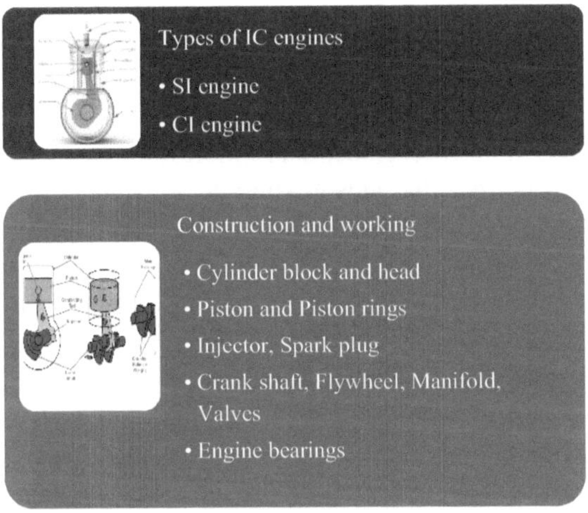

FIGURE 1.2 Schematic representation of cylinder bore and stroke.

and other alternative fuels. The applications are various industries, automobiles, and electric power generators [2]. The typical workflow of types of engines and their construction is presented in Figure 1.2.

- *Advantages of the IC engine*: The unique operating principle of the IC engine has found several advantages over EC engines.
  - High overall efficiency
  - It needs less space and is a compact type
  - Less initial cost and easy cold start since it uses volatile fuels

In the following section, the types of IC engines are explained in detail as well as their working principles.

1.2.1 SI Engine

SI engines use gasoline fuel, a highly volatile hydrocarbon liquid. This fuel mixes with pure outside air to prepare a homogeneous mixture before it enters the combustion chamber. The ignition system will ignite this mixture by means of spark plug. Further, the generated flame will be travelled all over the engine cylinder to complete the combustion. The combustible gases will be burned and expanded in the cylinder, which exerts a force on the piston, resulting piston moving in a downward direction to generate power.

1.2.2 CI Engine

The compression ignition (CI) engine uses diesel as fuel during the suction stroke, the fresh air enters the combustion chamber through the orifice, and piston compresses it. It moves from the bottom dead centre (BDC) to the top dead centre (TDC). At the end of the compression stroke, as piston moves towards the TDC of the cylinder, the fuel injection system will supply the metering diesel fuel at a particular pressure into a highly pressurized air environment. Then the sprayed fuel particles in the vicinity of air get vaporized; at this stage, the generated temperature would be approximately 538°C. The fuel exhibits both chemical and physical delay phases by atomization resulting in the prepared heterogenous mixture being ignited, and further combustion will be done.

1.2.3 Construction and Working

1.2.3.1 Cylinder Block and Cylinder Head

The cylinder is the main body part of the IC engine; it is part of the combustion chamber where the fuel-air mixture will be burned. The function of the cylinder is to guide the piston while delivering the power through connecting the rod to the crankshaft. The engine cylinder must be cooled to avoid excess heat generation during the combustion since it has direct contact with the combustion products and gases. Therefore, a cooling jacket will be provided around the cylinder. A liquid cooling system will be preferred in the case of four-wheeler vehicles.

On the other hand, two-wheeler fins are provided on the cylinder surface to increase the heat transfer surface area, and atmospheric air is allowed to flow through the equipped surface area to remove excess heat from the cylinder. Coming to constructional details of the cylinder, the cylinder head is situated at the upper end, and the crankcase is bolted at its bottom end. The cylinder upper portion, i.e., a cylinder head, is used to execute the combustion process, and during combustion, the gases are produced at higher temperatures and pressure. Hence, the material selection for the cylinder has a vital role, and it should withstand these extreme conditions, i.e., high compressive strength and stability with induced thermal stresses. In general, cast iron material is chosen for the cylinder and made using the casting process.

The engine cylinder top portion is closed with the cylinder head; it has two ports to permit intake charge, and another is useful to escape exhaust gases. Valves and valve actuating mechanisms operate both these valves. Moreover, other essential parts, such as spark plug or fuel injector, are bolted on the cylinder head. Aluminium or cast iron is used to fabricate the cylinder head, and the primary function of the cylinder head is to seal the engine cylinder.

1.2.3.2 Piston

The piston transmits the thrust forces offered by combustible gases to the connecting rod and is an essential prime mover in an automobile engine. Besides, it seals the cylinder throughout the bore by maintaining enough clearance between the cylinder wall and the piston surface. Since the prime mover must be efficient while transmitting the generated thrust forces, it should be light in weight and continuously experience hot gas temperatures; therefore, the material should withstand the induced thermal stresses. In general, aluminium material is chosen to fabricate pistons, and sometimes cast iron is preferable to avoid cylinder clearance issues because aluminium, like light alloy materials, expands more than cast iron.

1.2.3.3 Piston Rings

While delivering the generated thrust forces to the connecting rod, the piston must work against frictional forces. The piston is fitted with a loose fit, making it easy to transmit forces with the desired speed. Moreover, the piston surface has two rings for lubrication and sealing. These piston rings are fitted in the grooves, and according to the type of application, the number of piston rings required will be decided. The two-stroke engine needs two piston rings and four-stroke machines equipped with extra oil rings. The piston rings are cast iron or steel and should be highly elastic to work against high temperatures.

1.2.3.4 Connecting Rod

The connecting rod transmits the trust forces received from piston to the crankshaft with fewer losses and works against frictional forces. The small end of the connecting rod is connected to piston, and the other end is connected to the piston using a piston pin. The connecting rod materials are nickel, chrome vanadium steel, and chrome. Also, aluminium is used for small engines.

1.2.3.5 Crankshaft

The crank converts the rectilinear motion of the piston into rotary movement; it mounts on the bearing and moves freely while transferring the desired motion. The design involves the shape and size of the piston, and it depends on the arrangement of cylinders and their number. In general, spheroidal graphite or nickel-based alloys are used to make crankshafts to provide good service life; sometimes, steel forging is preferable.

1.2.3.6 Engine Bearing

The bearing's purpose is to support the rotary/linear motion of various engine parts, and the bearings support the crankshaft, connecting rod end attached to the piston pin. Also, it serves to reduce friction in moving parts. In the IC engine, both sliding and rolling bearings rotate freely, and the bush pin, namely the sliding direction, is used to connect the connecting rod, piston, and crankshaft. Typically, it is made of steel.

1.2.3.7 Crankcase

The engine cylinder parts, crankshaft, bearings, and other main body parts are mounted on the crankcase. It is also called an oil sump that provides lubrication for the engine and serves as a lubrication system where all oils are placed.

1.2.3.8 Valves

The charge inlet and generated exhaust emissions from the engine to the atmosphere and vice-versa will be operated by valves. Two valves, namely the inlet and outlet, will be operated using a valve actuating mechanism; during suction, fuel-air/air charge will be taken through the inlet valve, and after expansion of hot gases, the unwanted emissions will be escaped through the outlet valve. The valve actuating mechanism holds the valve either in an open or in a closed position according to the requirement.

1.2.3.9 Spark Plug

The spark function is to ignite the charge taken into the engine cylinder; the high voltage potential is supplied from the ignition system, and then a flame is generated immediately. Subsequently, the generated flame propagated throughout the chamber, and thus the combustion was completed. The spark plug consists of a metal shell with two electrodes insulated using a sufficient air gap. The SI engine uses a spark plug to get combustion done.

1.2.3.10 Injector

The CI engine uses the fuel injector to spray diesel at pressure with metering fuel quantity at the end of the compression stroke. When the injector sprays fuel into the combustion chamber, the fuel chooses a spherical shape. The practical pressure requirements, spray cone angle, and other fuel spray characteristics must be optimized for effective combustion efficiency.

1.2.3.11 Manifold

It supplies the fuel-air mixture into the combustion chamber and collects the exhaust emission from the cylinder. IC engine uses two types of manifolds, inlet and outlet manifolds, which are made of aluminium.

1.2.3.12 Camshaft

The valve mechanism operates the valve at exact timing through the camshaft, the valve/port timings are managed by an oval shape cam, and it exerts pressure force on the valve to open or close the valve at an instance. The engine's combustion efficiency depends on the valve timing while executing the intake and exhaust strokes. The cam is placed at the bottom of the cylinder, and a belt drive drives it.

1.2.3.13 Gudgeon Pin or Piston Pin and Pushrods

It is lightweight and has hardened steel fitted parallel to the piston bosses, and the small end eye is connected to connecting rod.

The valve receives the camshaft motion through pushrods, and when the camshaft is situated at the cylinder's bottom end, the push rod is used to carry the desired motion to the cylinder head.

1.2.3.14 Flywheel

It is a heavy steel wheel, and its size depends on the number of cylinders and the engine construction. It is attached to the crank shaft's rear end and secured on the crankshaft. The flywheel allows additional levelling-off of the generated power when power impulses are required since the engine's power is not smooth.

1.3 PERFORMANCE CURVES

1.3.1 Performance Parameters

The essential parameters required to assess the performance of the IC engine are explained below.

- *Bore and stroke*: The engine cylinder diameter, also known as the bore and the cylinder, has two extreme heights, namely, TDC and BDC, as shown in Figure 1.3. The distance between TDC and BDC is known as stroke. It is the distance travelled by piston inside the bore. The bigger the bore size than the stroke, it is called over the square, and vice-versa is said to be under the square. For tractors and trucks, under-square types of engines are preferred.
- *Crank throw*: The distance between the axis of the crank pin and crankshaft bearing is known as crank throw, and generally, the twice of throw is said to be the engine's stroke.
- *Displacement*: The cylinder volume between the TDC and BDC is called displacement, represented in cubic centimetres. The approximate power out of the engine can be defined as the sum of the cylinder's displacement.
- *Compression ratio*: The ratio between cylinder volume, i.e., when Piston rests at BDC, to the clearance volume, i.e., the Piston rests at TDC. The power increases with the compression ratio, so the less octane rating fuels are burned quickly, and the charge may explode, causing a sudden rise in heat instead of steady burning when the compression ratio is high. It results in pre-ignition and causes abnormal combustion.
- *Torque*: The trust force times the velocity resulting in torque is a twisting force. The exerted thrust force is transmitted to the crank pin through a connecting rod, and the transmission system is responsible for delivering the torque as per the load applied to the engine.
- *Power*: The power generated by the IC engine is known as brake power or output power. It is measured at the crankshaft, and the power inside the engine is more than the power available at the crankshaft because while transmitting, the power may be significantly lost due to friction, known as frictional power. The power available at the piston crown indicates power (IP) and is some of the brake and frictional power. The power can also be

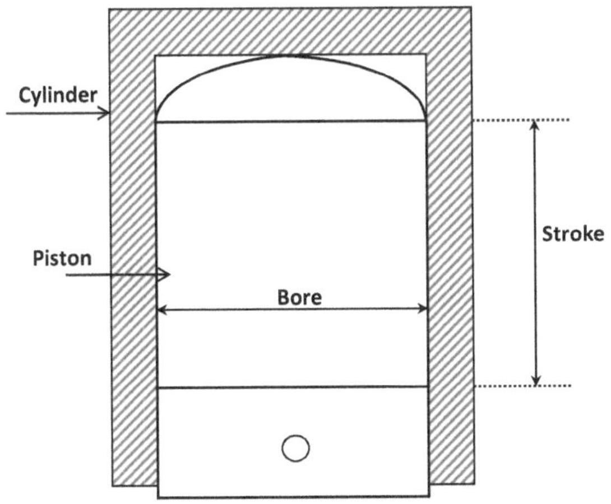

FIGURE 1.3 Schematic representation of cylinder bore and stroke.

represented as the mean adequate pressure generated during the combustion depending on the engine construction type.

- *Engine efficiency*: IC engine efficiency can be assessed in various ways, such as brake thermal efficiency, indicated thermal efficiency, mechanical efficiency, and volumetric efficiency. The available chemical energy is converted into heat energy by combustion, and the generated gases work on the piston, and then mechanical energy will be delivered at the end. The brake thermal efficiency is the ratio of brake power to input fuel energy, which is certainly 30%–45% for an IC engine. Indicated power is generated at the piston crown and converted into brake thermal efficiency. On the other hand, mechanical efficiency is the brake power ratio to indicated power. The lower the mechanical efficiency, the more generated power is consumed to overcome the friction. Moreover, the volumetric efficiency of the engine resembles the breathing capacity of the engine. It should be high, and generally, it is in the range of 95%–98% [3].

1.3.2 Performance Curves

The mean effective pressure (MEP) significantly changes with vehicle speed, and the frictional power decreases with speed resulting in higher mechanical efficiency. MEP value is less at a lower speed than its maximum, which is corroborated by carburation effects and mismatching of valve timings. In most cases, the valve timing is designed at a particular speed; MEP shows less value if the speed drops below the designed speed. Moreover, if the engine speed exceeds the designed MEP again decreases due to the less volumetric efficiency. Also, when the vehicle maintains high speed, stresses and bearing loads become high due to high inertia forces, which may cause fracture or bearing seizure. The power out from the engine remains constant when MEP decreases with an increase in engine speed, whereas if the fall of MEP is rapid, the power output drastically reduces with engine speed. On the other hand, the brake mean effective pressure (BMEP) is diminished significantly at higher engine speeds, as shown in Figure 1.4, and high frictional losses corroborate the reason. It can be perceived from the figure that the BP curve departs from the ideal straight drastically than the IP curve, i.e., points A, B, C, and D. Also, the maximum mean IP obtained at speed is two times of BP, which occurs at point G.

Figure 1.5 depicts the variation of torque against engine speed. BMEP can be expressed as TV × 4π where V is stroke volume, and others are constant. So, the torque variation is like BMEP, and these plots are drawn for full-throttle conditions. The variation of brake-specific fuel consumption, i.e., the quantity of fuel required to produce 1 kW power, is shown for full-load engine operation and the whole speed range [4].

Supercharging is the option to enhance the power output, and these improvements depend on the degree of supercharging. In this scenario,

a. Supercharging is sufficient to maintain the optimum volumetric efficiency, and then BMEP and power will be enhanced in the high-speed range.

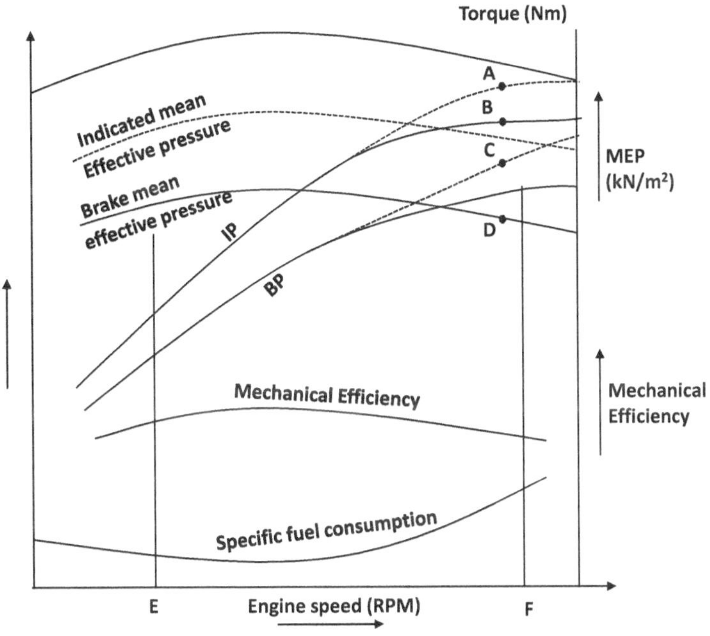

FIGURE 1.4 Performance curves of IC engine with speed.

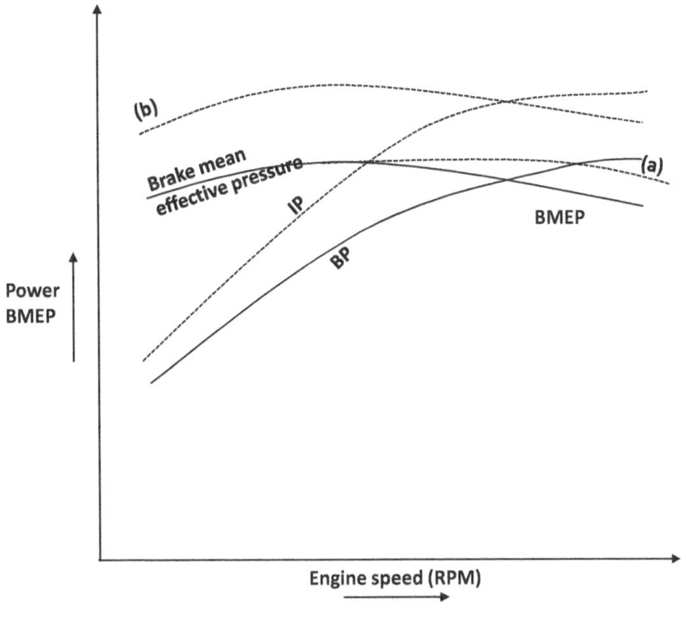

FIGURE 1.5 Variation of BMEP with speed.

b. In case of a high degree of supercharging, power will be enhanced in the entire speed range, and crankshaft torque will increase. It has an adverse impact on SFC, which cause abnormal heating and waste heat disposal problem [5].

1.4 EMISSIONS OF IC ENGINES

Automobile vehicles utilize hydrocarbon fuels and derive energy through proper combustion to execute desired necessary motion. Combustion is the reaction process between hydrocarbon components and air, releasing heat and combustion products. The generated heat is converted to power by the engine, and exhaust emission or combustion products will be released into the atmosphere. The hydrocarbon (HC) is made of carbon and hydrogen chains, and upon complete combustion process, in an ideal scenario, carbon monoxide (CO_2) and water (H_2O) will be produced. Indeed, plant on earth utilizes CO_2 for their photosynthesis; on the other hand, animals do not suffer from CO_2 emissions unless the oxygen (O_2) content is almost absent. In a realistic scenario, due to various reasons such as design aspects, availability of fuel-air mixture, etc., the combustion of HC is not complete, always resulting in the emissions involving nitrogen oxides (NO_x), carbon monoxide (CO), sulphur oxides (SO_x), soot, particle matter (PM), and unburned HCs. These emissions stringently cause harmful human health hazards.

1.4.1 Carbon Monoxide

The incomplete combustion, i.e., HCs unable to participate in combustion due to the unavailability of oxygen, thereby formation of CO instead of CO_2. These emissions are harmful to human and animal health, and their effects, like high dizziness, reduce mental abilities and sometimes cause death. Carbon monoxide has more vital binds with blood haemoglobin than oxygen, and the average body could not break the chain propagation. Moreover, high pressure is required to break the haemoglobin-CO chains, so the human infected with heavy CO breathing must be treated in a pressurized chamber.

1.4.2 Nitrogen Oxide

During the combustion process, very high temperatures and pressures will be generated. At this instant, the nitrogen can efficiently react with oxygen, and NO_x will be formulated in exhaust emissions. Various possible formulations are nitrogen oxide (NO), which is commonly found, nitrogen dioxide (NO_2) in small quantities, and nitrous oxide (N_2O). These emissions are responsible for acid rain, brownish colour smog, and deforestation's significant effect on acid rain [6].

1.4.3 Unburned Hydrocarbons

The incomplete combustion results in unburned HCs, and depending on the type of HCs, it may harm living human beings. Some of them may show direct position effects, increasing the ultraviolet radiation from the Sun, and reacting with NO to increase ozone depletion. Owing to severe consequences of ozone attack on membranes of living cells, asthmatics and toddlers suffer from high ozone concentrations. Moreover, in developed cities, it is reported that a significant density of people died due to exposure to elevated ozone concentration.

1.4.4 Other Pollutants

Sometimes the unwanted impurities in fuel cause pollutants like PM, soot, and sulphur oxides. Sulphur in concentration is present in diesel fuel, jet fuel, and less in gasoline fuels. The combustion products of sulphur are sulphur oxides (SO_x); these emissions are significantly released from transportation and burning coal. Steel and power industries generate sulphur dioxide emissions into the atmosphere, causing severe harmful human health problems. Petroleum companies use chemical compounds as an additive to enhance engine performance. These additives are tetraethyl lead, also called lead, to improve knock resistance characteristics and engine efficiency [7]. However, the combustion results in the generation of lead metal and causes severe neurological issues. In most countries, it is forbidden and replaced with other eco-friendly chemicals.

1.4.5 Global Warming Statistics

It results from accumulating CO_2 and other harmful gases in the atmosphere. Dangerous gas like methane captures the Sun's IR (infrared radiation) reflected from the ground; thus, the heat energy remains retained in the atmosphere, resulting in increased temperature.

The drastic rise in temperature influences the sustainability of humans on earth in several ways, such as ecological damage, dynamic climate conditions, heat waves, and cold waves. It also causes natural disasters and damages natural resources, more importantly, causing the disappearance of some species because it threatens the natural resources that feed populations. It also observed the migration of indigenous species from one warm condition to the former cold northern conditions [8]. Global warming stringently has meteorological phenomena that further disturb the South Pacific region and cause floods, droughts, and tornadoes. The melting of polar ice caps is another major problem because global warming and rising sea levels cause permanent tidal waves inundation of coastal regions. The variation in CO_2 emissions from various sectors is presented in Figure 1.6. The transportation sector is responsible for 32% and the residential and commercial sectors for 19% and 15%, respectively [9].

1.4.6 Emission Control Techniques

The combustion process in a four-stroke engine starts only 25–50 ms, depending on the vehicle's operating conditions. As the combustion ends, the components like unburned HCs continue to react with other constituents during work done by gases through expansion stroke. Later, these pollutants will be escaped into the atmosphere through the port/valve.

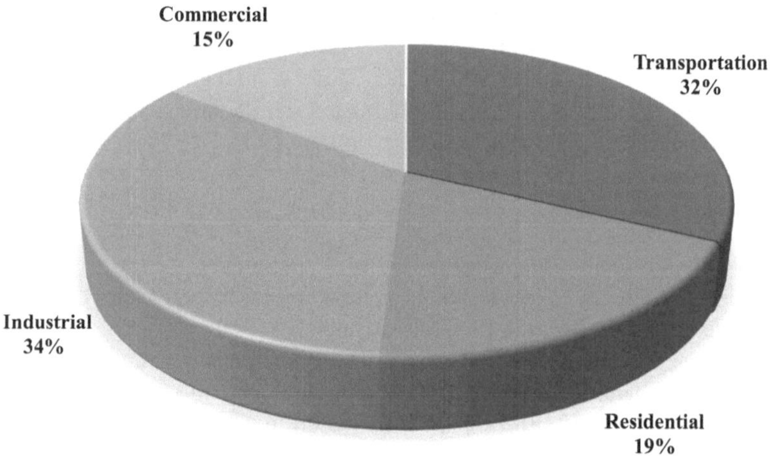

FIGURE 1.6 Contribution towards the global from various sectors.

1.4.6.1 Catalytic Converters

The emissions generated in the engine can be controlled and adopted by several automobiles using a catalytic converter. These have chambers mounted on the gas flow system, and emission pollutants are allowed to flow through the system. The chambers contain catalytic prompts for oxidation, meaning carbon monoxide and water will be oxidized to CO_2 and H_2O. If thermal converters are present, the temperature will be held in the range of 600°C–700°C [10]. Sometimes specific catalyst presents the temperature required to sustain and reduce to 250°–300°C. Aluminium oxide is a ceramic material used for the catalytic converter; mainly other materials are palladium and platinum, which promote CO and HC oxidation, and rhodium, which promotes NO_x emissions.

$$\left.\begin{array}{c} NO+CO \rightarrow \frac{1}{2}N_2+5CO_2 \\ 2NO+5CO+3H_2O \rightarrow 2NH_3+5CO_2 \\ 2NO+CO \rightarrow N_2O+CO_2 \\ NO+H_2 \rightarrow \frac{1}{2}N_2+H_2O \\ 2NO+5H_2 \rightarrow 2NH_3+2H_2O \\ CO+H_2 \rightarrow CO_2+H_2 \end{array}\right\} \text{Reactions catalytic converters}$$

1.4.6.2 Reducing Emissions by Chemical Methods

Cyanuric acid has been adopted to reduce NO_x emissions; it is the less expensive solid material that sublimes in the exhaust flow system. The gas produces isocyanides and dissociates by maintaining an operating temperature of about 500°C and reacts with

NO_x to form nitrogen, hydrogen, and carbon dioxide. This method achieves 95% of NO_x reduction without affecting the engine performance. This technique is not implemented in the present automobile sector since it has limitations like bulk weight and complexity. Chemical systems are developed to capture H_2S emissions; when the engine operates at a rich mixture, it converts into SO_2 when it receives lean or excess air conditions.

$$H_2S + O_2 \rightarrow SO_2 + O_2$$

1.4.6.3 Ammonia Injection System

An ammonia injection system is implanted in marine and large stationary engines to reduce NO_x emissions. The system sprays NH_3 into exhaust gas flow, and the following reactions result in emissions reduction.

$$4NH_3 + 4NO + O_2 \rightarrow N_3 + 6H_2O$$

$$6NO_2 + 8NH_3 \rightarrow 7N_2 + 12H_2O$$

1.4.6.4 Exhaust Gas Recirculation

NO_x can be reduced by holding the combustion temperature. It is quite critical since it affects the brake thermal efficiency of the engine. Exhaust gas recirculation (EGR) is a remarkable technique to recover waste heat from exhaust gases; recirculating hot gases can do it. The flow can be as high as possible, i.e., 30% of the total intake charge. EGR technique diminishes the peak temperature rise by combining the fresh intake charge with the previous cycle residual left in the cylinder. Engine management system (EMS) can monitor the EGR flow rate. EGR is the defined ratio of the mass flow rate of EGR to the total cycle mass flow rate where $\dot{m}_{cyl}$ is the total mass flow into the cylinder.

$$EGR = \frac{m_{EGR}}{\dot{m}_{cyl}} \times 100 \quad (1.1)$$

EGR reduces overall combustion efficiency, and high EGR results in partial burns and extreme accumulation, causing misfires during combustion. Moreover, EGR significantly decreases the NO_x but should bear the cost of HC emissions and have less brake thermal efficiency.

1.5 FUEL SHORTAGES

Population growth and usage of automobiles, on-road vehicles, stationary engines, and other industries increase the demand for fuel. The oil reservoirs serve as the supplier as per demand. The oil resources support the demand and depend on discovering new oil reservoirs. Based on statistics, oil consumption will increase and cross cumulative numbers by 2050. Moreover, the rate of finding new oil reservoir

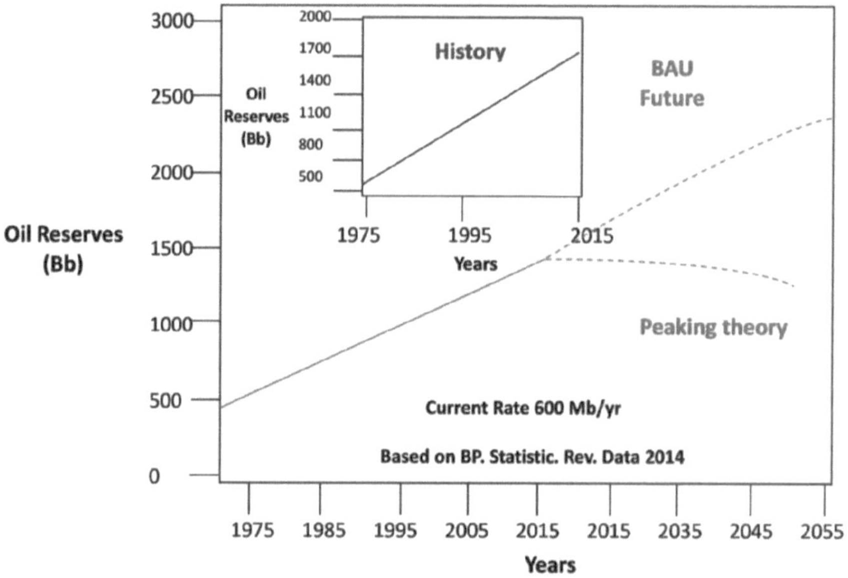

FIGURE 1.7 Fuel shortage and depletion of oil reservoirs consumption [11].

resources decreases slowly and parallelly, increasing the demand for oil consumption. The year-wise variation of oil is shown in Figure 1.7. It is perceived that the same trend of renting oil reservoirs will become a severe problem in future, and the oil will be more expensive. It is observed that oil consumption must be conserved, and alternative fuels for sustainability should exist.

1.6 VEHICLE DYNAMICS

1.6.1 Vehicle Resistance

Vehicle resistance faces issues while attempting to move from stationary to acceleration condition. The inertia forces, including the vehicle's weight, should be overcome by developing sufficient power by the engine. To sustain the vehicle's motion, the engine should be designed to overcome the inertia forces and vehicle resistance. The vehicle resistance is more when it tries to move uphill or travel against heavy wind blow. The vehicle will groan and slow down because of poorly inflated tyres. Broadly, vehicle resistance can be categorized as,

a. Drag force due to aerodynamics
b. Linear gradient resistance
c. Rolling resistance
d. Inertia resistance

To maintain sustainable motion, the tractive force should be greater than the resistive force. The resistance forces mentioned the aforementioned result in tractive force and can be represented as equation 1.2.

$$F = F_{req} = F_A + F_G + F_R + F_I \tag{1.2}$$

where

- F_A = force due to air resistance
- F_G = force due to the gradient of a slope
- F_G = force due to rolling resistance
- F_I = force due to moving or static inertia

When a vehicle accelerates or decelerates, the last force, FI, will be significant, whereas other forces need to be considered when moving or at a constant speed.

1.6.2 Air resistance/Aerodynamic Drag

The denser fluid may resist vehicle speed, the air molecules collide with the vehicle surface, and then some energy will be absorbed, resulting in decreased vehicle speed. The resistance force exerted by air depends on density, velocity, and temperature. Air resistance increases proportionally with vehicle speed. The drag force exerted by the air on the vehicle surface can be represented as equation 1.3.

$$F_A = -\tfrac{1}{2} \times C_d \times P \times V^2 \tag{1.3}$$

where

- C_d = coefficient of discharge
- P = pressure
- V = velocity of the vehicle

1.6.3 Gradient Resistance

The gradient resistance force experienced by the vehicle when it travels in uphill and downward directions. The vehicle's weight is resolved into two components, one in the direction opposite to its motion. The vehicle would slow down or roll in the backward direction if no sufficient energy was supplied.

Let's consider the vehicle's weight is W, and the slope uphill is θ. The weight component along the vehicle speed direction (W Sin θ), i.e., along the road surface, tries to reduce the vehicle speed, and the component perpendicular to the road surface is (W Cos θ). Then the gradient resistance is expressed as $F_G = W$ Sin θ (refer Figure 1.8).

1.6.4 Rolling Resistance

Rolling resistance will be experienced by the vehicle when the vehicle's tyre moves on the road, and then the relative motion between two surfaces exerts friction.

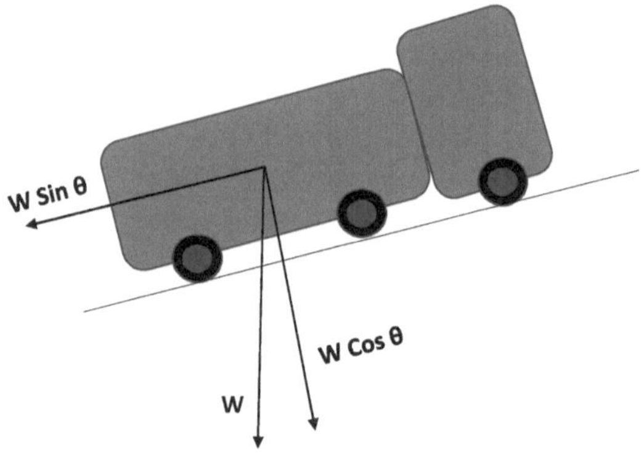

FIGURE 1.8 Truck moving in an uphill direction and resolution of forces.

Road surfaces and the tyre should not be rigid and must provide slightly relative motion. The contact between the road surface and the tyre is dynamic while rolling. The gradual deformation changes abruptly; the deformation has attained the highest value at the bottommost point and the least at the entry and exit. The slip of the vehicle tyre on the road surface produces energy loss, further resulting in resistance. The rolling resistance consists of components such as tyre rolling, road rolling, tyre slip angle, bearing friction, and residual braking. The rolling resistance can be expressed as equation 1.4.

$$F_R = F_{R,T} + F_{R,Tr} + F_{R,\alpha} + F_{R,fr} \tag{1.4}$$

Here the tyre rolling resistance can be estimated as equation 1.5; it is due to the flexure of the tyre, air resistance on the tyre, and friction.

$$F_{R,T} = F_{R,T,flex} + F_{R,T,A} + F_{R,T,fr} \tag{1.5}$$

Moreover, the coefficient of rolling friction can be expressed as equation 1.6.

$$k_R = F_R / F_{Z,w} \tag{1.6}$$

1.7 BRAKING SYSTEM

The brakes of the automobile engine absorb the energy generated by the engine and the momentum of the vehicle. The absorbed energy from the friction is converted into heat and dissipated to the atmosphere. When the brake is applied on the vehicle wheels, the speed reduces, also known as retardation. The brake must pull the vehicle and bring it to rest position slowly. In general, the vehicle speed is reduced or retarded by drum or disc brakes. In some transmission systems, the applied brake is due to friction, known as a friction brake.

1.7.1 Brake Actuating Mechanism

The hydraulic brake actuating mechanism working principle is based on Pascal's law that pressure applied to the enclosed fluid in any area is transferred in all directions to every interior surface of the vessel. Hydraulic brakes are applied by the vehicle driver using a brake actuating mechanism system, where the master cylinder and wheel cylinder, along with connecting lines and hoses, are enclosed in the closed vessel, as presented in Figure 1.9. In this regard, any pressure applied on the master cylinder is transmitted undiminished to every cylinder.

1.7.2 Drum and Disc Brakes

Brakes are used to stop the vehicle or slow the vehicle by absorbing the kinetic energy, and thereby absorbed energy is converted into heat. The principle is the same as peddling a bicycle. In general, drum brakes are used in automobiles. However, advanced cars replace these with disc brakes due to their remarkable advantages. Drum brakes are mounted on the rear wheels, and disc brakes are mounted on the car's front wheels. In the design aspect of drum brake, two leading shoes and trailing shoes or one leading and trailing shoe are preferable. Figure 1.10 depicts the leading and trailing shoe brake; the type of shoe and its parts have been shown by pointing the show anchor location concerning forward wheel rotation. In drum brake, the anchor brake shoe is known as the heel, and the brake actuating force is applied on the toe of the shoe. If the brake drum rotates in the direction of the shoe toe towards the heel, it is known as the leading shoe. On the other hand, the rotation from the heel towards the shoe toe is called a trailing shoe.

1.7.3 Disc Brakes

The disc brake's design aspect is quite different from the drum brake. The circular plate and calliper replace the drum and brake shoe. The friction pads support the calliper, and these pads are forced inwardly by applying the force, and the retardation

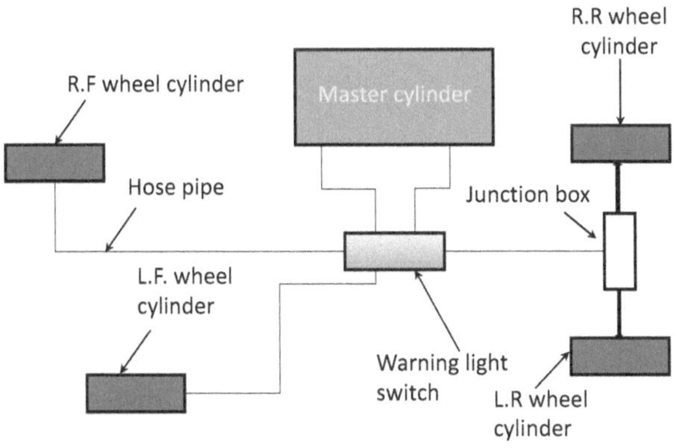

FIGURE 1.9 Hydraulic brake actuating mechanism.

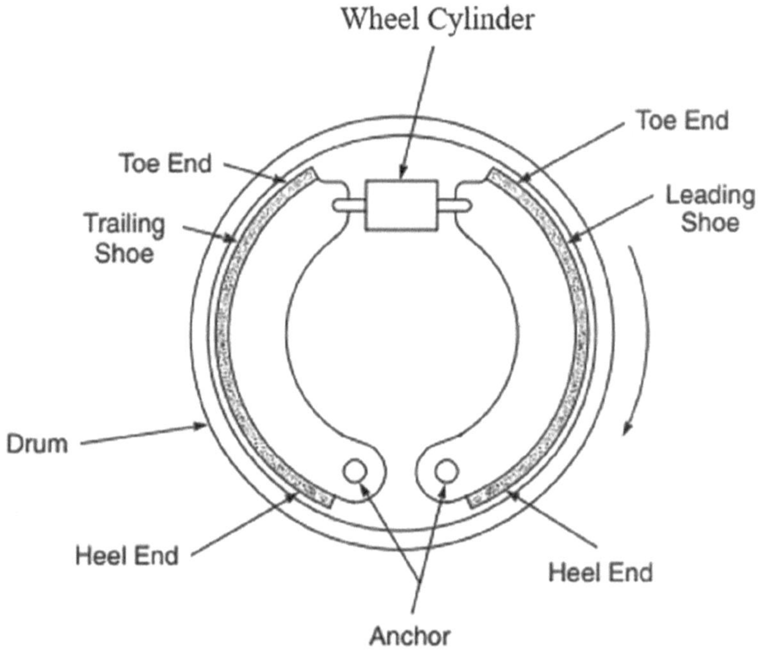

FIGURE 1.10 Leading and trailing shoe brake [12].

results from the applied disc. The benefits are lightweight and better cooling provision. Because, unlike drum brakes, the disc brake surface is exposed directly to the air. Disc brakes are best in resistance force than drum type. Disc brakes support the uniform pressure distribution, whereas drum brakes do not support uniform pressure distribution. Disc brakes do not have the self-servo effect, and the brake pads can be replaced easily, which means there is no need to remove the tyre like in the drum brake. Moreover, disc brakes design involves the self-adjustment of drum brakes. The limitations are that it is challenging to install parking brake attachment, and high pressure is required on peddle to stop the vehicle.

1.7.4 Hydraulic Brakes

In hydraulic brakes, the fluid exerts pressure in all directions equally in a closed circuit and does not require mechanical linkages. Frictional losses can also be significantly decreased; these brakes are used in passenger and medium-capacity vehicles. Figure 1.11 depicts the lock head hydraulic brake system, and it is used in the Ambassador car, which is manufactured in India; it consists of a fluid tank, a master cylinder where hydraulic pressure is produced, and a wheel cylinder. The exerted pressure from the master cylinder is transmitted to each cylinder utilizing pipelines and flexible hoses. A hydraulic brake works when an external force presses the brake pedal, and the rod applies pressure force on the master cylinder piston; thus, hydraulic pressure is applied. The fluid forces apart from the cylinder and the piston of the

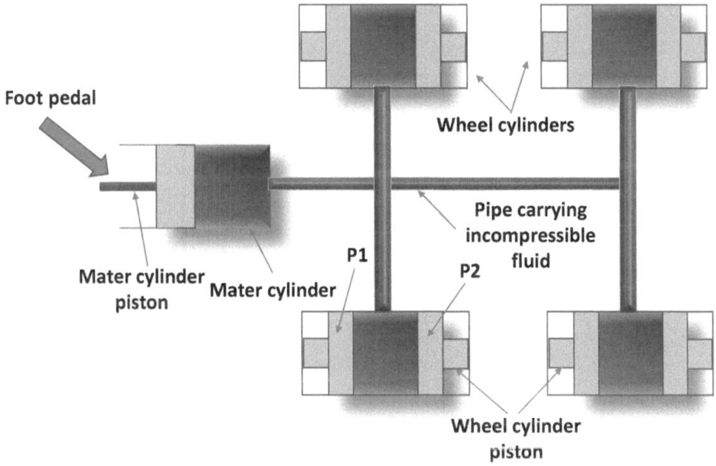

FIGURE 1.11 Lock head hydraulic brake actuating mechanism.

master cylinder returns to the former state upon releasing the pedal. Eventually, the fluid in the vessel flows bask to the tank and brake master cylinder thereby, the brake ceases.

1.7.5 Anti-lock Braking System

In a regular braking system, the brake pedal will be pressed with external force, and the brake pads are then tightly pressed against the wheel disc to stop the vehicle instantaneously. When vehicle rotation is stopped, it cannot be steered, which means the driver loses control over the steering. Sometimes the skids cause severe accidents, and the ABS is preferable to avoid this situation. The components of ABS are explained below.

- *Speed sensors*: To monitor the vehicle speed.
- *Valves*: The functions of valves are to block and release the pressurized fluid when necessary.
- *Pumps*: When brakes are applied, the pump supplies fluid under pressure to brakes and callipers.
- *ECU*: Electronic control unit (ECU) is the heart of the ABS; it receives signals from various sensors and vice-versa. In ABS, when the brake is applied, the speed sensor tracks the wheel rotation and its magnitude, and when brakes are about to stop the vehicle rotation, a signal will be sent to ECU. ECU releases the brake pads, valves, and pump, and the wheel will be rotated continuously. The unique advantages of ABS are that wheel allows maintain control over the speed in heavy braking situations.

In the absence of ABS, the vehicle sometimes stops and skids due to the wheel's translational velocity. If the vehicle wheel is subjected to skidding, it must travel more considerable distances and lose control over the vehicle. Moreover, the vehicle will experience traction surfaces when both left and right wheels are subjected to frictional forces, which may cause spinning uncontrollably.

- *Advantages of anti-lock braking system*: It reduces the braking distance by applying the appropriate pressure on the wheel to stop the vehicle. It avoids uneven tyre wear since it doesn't lock the brake while trying to stop the vehicle. ABS is supplied with a traction control system, which reduces the wearing of brake pads and disc brakes.
- *Disadvantages of anti-lock braking system*: Variable braking system causes intricate arrangement of ECU, sensors, and complex mechanisms may charge high cost.

1.8 STEERING SYSTEM

The steering system controls the vehicle accurately and enables the driver to operate the automobile in the desired direction. It consists of steering gears and steering linkages; steering gears multiply the driver's effort, and steering linkages connect the gearbox to the front wheels. Alignments in the front wheel and directional control, along with comfortable steering, influence the functionality of the steering system.

1.8.1 Axles

The shaft for rotating the vehicle wheel is known as axle. Axle's one end connects to the differential gear (sun gear), and another end will be connected to the wheel. The axle is fixed at the wheel and allowed to rotate around; sometimes, wheels are only allowed to rotate the axle that remains fixed to the vehicle. This can be achieved by mounting the bearing at proper locations on the axle and vehicle wheel. In general, manual and automatic transmission systems use synchromesh and epicyclic gearboxes. Axle types are rear, front, and stub axles. The rear axle transmits power from the differential to the driving wheel. It is to be noted from the construction of the rear axle that it is not a single piece, but the two halves of the axle connect by differential and another part of the axle as half shaft. The differential box's sun gear is connected by the shaft's inner end and the driving wheel's other outer end.

The rear wheels are driving wheels in the case of rear-wheel drive vehicles, whereas front wheels act as driving wheels in front wheel drive vehicles. In modern passenger vehicles, live axles are preferred since they revolve around vehicle wheels. On the other hand, dead axles are said to be fixed or stationary vehicles because they do not move with wheels. In this structure, a housing encloses completely with rear axles differential box, protecting from moisture contact and foreign particles. It will be mounted with bearings, which will be provided with proper lubricants. The front axle system will carry the weight of the vehicle's front part and facilitates driving, steering, and absorbing shocks due to vibrations. Therefore, the construction of the axle must be robust in design, and the design should withstand vehicle bending loads.

The components of the front axle are stub, swivel pin, track rod, and beam. Stub axles minted on the front wheel connected to the front wheel employing kingpins. These are forged materials with 3% Ni steel, steel made of alloy with Cr, and molybdenum. The stub axle is a soft drive fit in the glowing eye, which the cotter pin will lock. Also, a bronze bush is used to fit the forked ends of the axle provided on the kingpin surface.

1.8.2 Components of Steering

1.8.2.1 Steering Wheel

The driver's steering wheel will control the vehicle, consisting of various systems such as traffic indication light, vehicle lighting control, wiper switch, etc.

It is also known as the hand wheel and is used in most vehicles like light, medium, and heavy-duty types. The schematic diagram of steering components is shown in Figure 1.12.

1.8.2.2 Steering Column or Shaft

The steering column is a shaft fitted into the hollow steering column's inner portion. The vehicle turns, and the steering column also rotates; the motion is transmitted to the steering box. The shaft is located at the top portion of the steering system and is directly attached to the steering wheel. It will be connected to the concerned part utilizing the universal joint and the immediate shaft.

1.8.2.3 Steering Gearbox

In the rocker arm of the steering box, one end is splined through the Pitman's arm, and another end is linked to the drag link utilizing a ballpoint joint. It contains the gears to transmit the input from the driver's steering linkages, further, it turns the wheel, thereby multiplying the steering changes from the driver. Therefore, the front wheels of a vehicle are observed with more motion than the steering wheel.

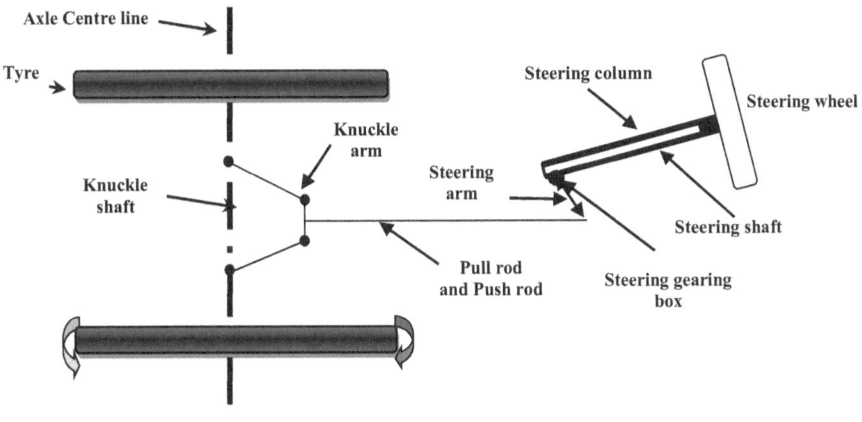

FIGURE 1.12 Steering mechanism and components.

1.8.2.4 Drop Arm or Pitman's Arm

As the steering wheel takes a turn, either right or left, the power is transmitted by the Pitman, which is received from the steering gearbox. This will be further sent to the tie rod; Pitman's arm is used to make necessary corrections in the steering whenever the vehicle needs suspension.

1.8.2.5 Ball Joints

It prevents the steering parts from forming dirt and protects them from encasing the complete assembly. The ball joints are spherical bearings that connect the steering knuckles vis control arm. It is a tapered shape threaded direction to fit into the steering knuckle hole.

1.8.2.6 Drag Link

It converts the sweeping arc to a linear direction when the vehicle takes turns using the steering system. The drag links connect the steering arm vis Pitman, and some applications are directly related to a tie rod assembly.

1.8.2.7 Steering Arm

The steering arm transmits the vehicle's turning momentum from the steering gear to the drag linkage, particularly in automotive vehicles. The steering arm provides enough driving comfort and reduces driver efforts for safe driving.

1.8.2.8 Stub Axle

As the steering wheel rotates, the rotary motion will be sent to the gearbox vis Pitman's arm and transmitted to the drag linkage system. Further, the drag linkage transfers the vehicle moment to the stub axle, where it rotates about kingpin, resulting vehicle will turn in the exact direction.

1.8.2.9 Left Tie Rod Arm

Both left and right tie rod arms are connected through the centre link and are mounted on Pitman's arm. The entire structure is connected to the steering gear and idler arm near the passenger side. The structure involves three phenomena: correcting the steering, power steering, and cornering the force. Most heavy-duty vehicles prefer power steering, as shown in Figure 1.13. This system needs a separate pump to supply the hydraulic fluid at a flow rate. The pump is mounted on the front side of the vehicle and is driven by the belt drive, which is connected through hydraulic linkages on the steering box. Also, other components include hydraulic valves, piston assembly and gears. They are coming to the functionality; when the driver takes a turn, hydraulic pressure is routed through the valve to the cylinder. It pushes them towards the driver's turn and makes easy handling of automobiles safe. If the system fails to execute the task correctly, the driver does not feel discomfort driving. The power steering mechanism is the best in maintaining the speed range and reduces the driver's efforts to control the vehicle on the road. Moreover, while taking the reverse direction of the vehicle, the power steering handles easily, safely, and safely parking the vehicle.

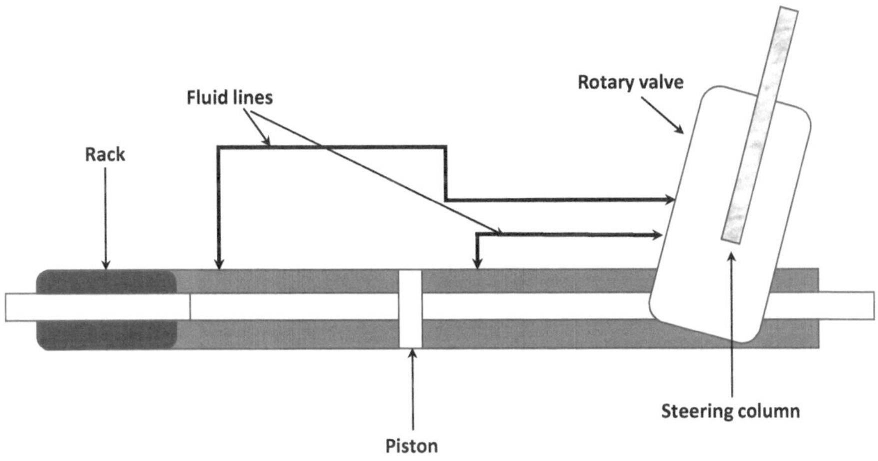

FIGURE 1.13 Schematic of the power steering mechanism.

1.8.3 Steering Geometry

The study of various supporting angles in the structure formed employing different supporting arms connected is known as steering geometry. The steering geometry angles are caster, camber, kingpin, and toe in/out. The top axle towards the vehicle's rear axle makes the caster angle. If the vehicle turns towards the front, then it is positive "+"; if it turns back represented as unfavourable. As shown in Figure 1.14, the caster angle gives a trailing effect on the vehicle's front wheels; when the vehicle trails, the wheel aligns in the right direction and the line of weight moves in the same vehicle direction. Therefore, it is advantageous and easy to steer the vehicle in a straight line. Camber angle is the inward/outward tilt angle at the top of the wheel. If the vehicle tilts outward, it is represented by positive; if it tilts inward, it will be negative camber. The chamber's purpose is to exact contact the vehicle wheel with

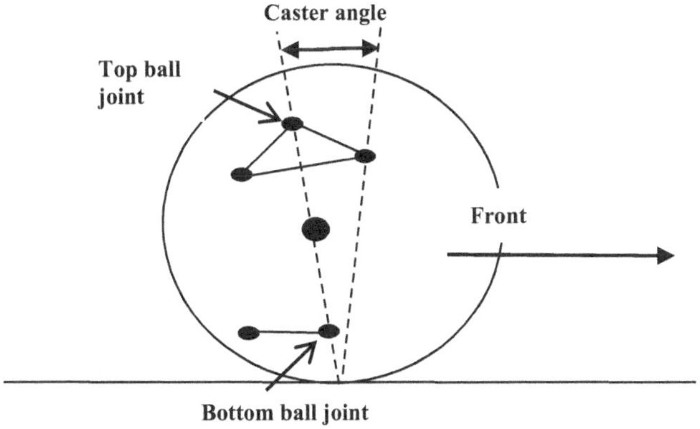

FIGURE 1.14 Caster angle.

the road and near the point of the load to throw the weight either on the inner wheel side or outer side, whichever is larger in the direction point of view.

Kingpin angle is the inclination angle tilted at the top of the wheel towards the vehicle to handle the vehicle's weight, and it will throw the tyre centre. The angle made by the vehicle line and kingpin centre is known as kingpin inclination angle. Toe in/out angles help spread the vehicle front wheel and handle excessive turns. These angles' primary purpose is to reduce the tyre's wear and tear and prevent excessive material loss due to friction. Therefore, the steering angles made by various steering system components are adjustable as per the manufacturer's specifications. The wheel's alignment must be verified without affecting power generation.

1.9 ROAD SAFETY STATISTICS

As per the statistics, every year, it is perceived that 1.3 million people are cut short due to road traffic, and approximately 20–50 million people suffer from injuries. Indeed, most of the population became disabled due to several accidental injuries. Injuries in road traffic cause economic loss to families, individuals, and nation as a whole. The financial losses involve hospital treatment charges, productivity loss and suffer from significant injuries. In most countries, 3% of gross domestic product costs are due to road traffic crashes. It is perceived that most young people are involved in road traffic crashes, in which most are males than females. Approximately 73% of road traffic deaths are caused by people who are under 25 years, and it is almost three times more than compared with female deaths. The reasons for unexpected deaths due to drastic changes in vehicle speed are directly related to increased severity. The consequences of the road crash are likely to be uncontrolled vehicle speed ranges. It is found that a 1% increase in average speed results in a 4% rise in fatal road crash risks, eventually causing 3% of severe traffic crashes [11,12].

Moreover, the paediatric death risks are abruptly increased from 3 to 4.5 times due to sudden speed rise beyond 60 km/h. Also, impacts from car to car are approximately 85% when the vehicle crosses beyond 65 km/h speed. Other reasons influencing situations like driving while consuming alcohol and drug or psychoactive incident increase the risk of accidents and sometimes agonizing death. The general indication of risk is assessed based on the blood alcohol concentration (BAC), which increases abruptly when BAC exceeds 0.04 g/dl. Driving while talking through mobile phones is likely to cause accidents approximately four times the normal driving situation [13].

1.10 NEED FOR EV AND SELF-DRIVING VEHICLES

According to data, it is found that 94% of crashes occur due to driver behaviour, which is treated as an error, and self-driving vehicles have the scope to reduce the uncertainty. The advancements and adoption of automation in vehicle design prompt and show the potential to reduce dangerous as well as risky behaviour of the driver. The damage to the vehicle and painful death can be reduced by the proper occupant seat belt, safety, speed control, not encouraging driving while consuming drugs

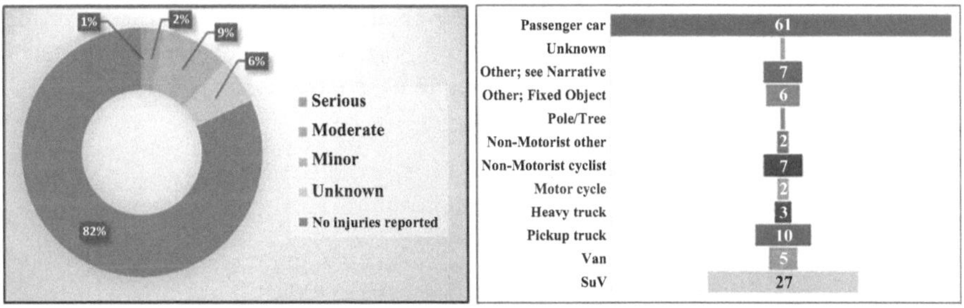

FIGURE 1.15 ADS report on (a) highest injury severity and (b) collisions with other vehicles [17].

or alcohol and pairing appropriate driver behaviour [14]. Self-driving cars' unique advantage is the potential use of artificial intelligence and machine learning algorithms. Further, the person with disabilities can use them and enhance road safety. Automated vehicles (AVs) can reduce personal transportation costs and increase mobility. In the future, AVs could provide convenient driving with all safety principles. In addition, it reduces the traffic issues in cities, which further helps to reach or drop the person at the right time. National Highway Traffic Safety Administration (NHTSA) has released a standing GO (general order) requirement to identify SAE level 2 manufacturers which are equipped with advanced and automated driving assisting systems and also generate a report to certain agencies for crash reports involving SAE level 3 vehicles through five automated driving systems (ADS) [15,16].

ADS equipped vehicles reported on various crashes, which are classified based on the type of automated driving system used in vehicles [17]. Figure 1.15 depicts the injury rates that happened and collisions with other vehicles; it is observed that serious injuries as 82%, and most of the time, passenger cars collide with other vehicles. It is intuited that once self-driving cars are fully developed, there will be potentially substantial positives in personal and professional lifestyles. The technology and high-tech vision systems enhance human ergonomics while driving vehicles and protect from dangerous accidental situations. Moreover, algorithms developed to find faster and safest routes to destinations improve fuel economy and efficiency.

1.11 CONCLUSION

The fundamentals of internal combustion engines, the importance of vehicle resistance and dynamics have been discussed. The demand for transportation energy will increase, and petroleum fuels must supply this. It is observed that the discovery of new oil reservoirs is complex, and the share of oil for industries and transportation might reduce to 80%–90% by 2040 compared to the current percentage, i.e., 95%. The importance of necessary modification in both SI and CI engines is identified without affecting the performance. The effects of automobile emissions are discussed, and the importance of antiknock quality fuels or fuel additives is registered significantly. Indeed, emissions regulation of passenger vehicles and mitigation of road traffic crashes are mandatory. Moreover, the future of vehicles is striving towards self-driving and electric vehicles. In this regard, the development of technologies in various aspects, such as cooling systems, control systems, and implementation of machine learning, is much needed.

REFERENCES

[1] U.S. World shale resource assessments. U.S. Energy Information Administration. https://www.eia.gov/analysis/studies/worldshalegas/&gt (2013).

[2] C.K. Westbrook, Chemical kinetics of hydrocarbon ignition in practical combustion systems. *Combust. Inst.* 28 (2000) 1563–1577.

[3] J.C. Livengood, P.C. Wu, Correlation of autoignition phenomena in internal combustion engines and rapid compression machines. *Combust. Inst.* 3 (1955) 347.

[4] A.M.K.P. Taylor, Science review of internal combustion engines, *Energ. Policy* 36 (2008) 4657–4667.

[5] A. Alagumalai, Internal combustion engines: progress and prospects, *Renew. Sustain. Energ. Rev.* 38 (2014) 561–571.

[6] A. Abdel-Rahman, On the emissions from internal-combustion engines: a review, *Int. J. Energ. Res.* 22 (1998) 483–513.

[7] G. Kalghatgi, Is it really the end of internal combustion engines and petroleum in transport? *Appl. Energ.* 225 (2018) 965–974.

[8] G.T. Kalghatgi, Developments in internal combustion engines and implications for combustion science and future transport fuels, *Proc. Combus. Inst.* 35 (2015) 101–115.

[9] N. Abas, A. Kalair, N. Khan, Review of fossil fuels and future energy technologies, *Futures* 69 (2015) 31–49.

[10] N. Dora, T.J.S. Jothi, Emission studies in CI engine using LPG and palm kernel methyl ester as fuels and di-ethyl ether as an additive, *J. Inst. Eng. India Ser. C* 100 (2019) 627–634.

[11] Y. Shiao, P. Gadde and M.B. Kantipudi, Mode strategy for engine efficiency enhancement by using a magneto-rheological variable valve train, *ASME. J. Energy Resour. Technol.* 143 (2021) 062307.

[12] Y. Shiao, M. Babu Kantipudi, High torque density magnetorheological brake with multipole dual disc construction, *Smart Mater. Struct.* 31 (2022) 045022.

[13] R. Saidur, M. Rezaei, W.K. Muzammil, M.H. Hassan, S. Paria, M. Hasanuzzaman, Technologies to recover exhaust heat from internal combustion engines, *Renew. Sustain. Energy Rev.* 16 (2012) 5649–5659.

[14] X. Sun, Z. Li, X. Wang, C. Li, Technology development of electric vehicles: a review, *Energies* 13 (2020) 90.

[15] G. T. Kalghatgi, The outlook for fuels for internal combustion engines, *Int. J. Eng. Res.* 15 (2014) 383–398.

[16] G.H. Abd-Alla, Using exhaust gas recirculation in internal combustion engines: a review, *Energ. Convers. Manag.* 43 (2002) 1027–1042.

[17] NHTSA, Summary report: standing general order on crash reporting for automated driving systems. https://www.nhtsa.gov/.

2 Challenges of Present Transportation

Milon Selvam Dennison
Kampala International University

R. Rajasekaran and Koganti Radhika
Dhanalakshmi Srinivasan College of Engineering and Technology

T. Ganapathy
S Veerasamy Chettiar College of Engineering and Technology

2.1 INTRODUCTION

Due to the exponential growth of population and globalization, modern society highly relies on different modes of transportation [1, 2]. In recent decades, automobiles have given greater freedom and convenience to society and nowadays, it is impossible to imagine our routine lifestyle without automobiles. For everyday transportation, people are increasingly choosing a motorcycle, personal car, public bus, or train [3]. A few decades ago, large streets in cities were free, and the city residents did not know what a traffic jam was. Nowadays there are more and more automobiles in real-time practice. Every family has a motorcycle, car and someone even has several. The modern automotive industry offers a huge variety of vehicles from exclusive to economic options [4].

The automobile sector plays a vital role in global economics, as said the routine lifestyle of society highly depends on the transportation system [1, 4]. The present transportation system highly relies on fossil-fuelled internal combustion (IC) engines, and the IC engines are considered as the heart of automobiles [5]. But the dark fact of the IC engine-based automotive system is the emission of toxic gases such as oxides of carbon, oxides of nitrogen, and unburnt hydrocarbons (HCs) in a relatively larger amount that causes severe environmental pollution, which leads to global warming [6]. In addition, vehicles have a serious impact on the environment not only during their use but also during production and disposal [7]. For IC-engine-based vehicles, it is estimated that the CO_2 emissions produced over the life of a vehicle, 10% comes from manufacturing, 5% from disposal, and the remaining 85% from use. On the other side, these IC engines use fossil fuel supplements such as petrol and diesel as the fuel which is non-renewable, and it causes depletion of the natural resource [8].

DOI: 10.1201/9781003354901-2

This made the engineers to look for the alternative and automated solution in the transportation sector.

Automated electric vehicles (AEV) possess more advantages over human-driven electric vehicles, such as low-frequency accidents, higher traffic efficiency, and 360-degree vision while driving the vehicles. The hardware with technologies such as deep learning and Light Detection and Ranging (LiDAR) are recently used to develop autonomous vehicles (AV). The energy of the EVs is saved by automation due to the accomplishment of smooth driving, optimal routing design, and flocking. However, EV rage is frequently affected due to the higher weight (sensor weight + computing load) and improved drag. According to certain studies, the AEV's range is reduced for both suburban and city driving by 5%–15% [9]. The improved range for the AEVs can be achieved by launching sensors with aerodynamic designs in the future.

2.2 NEED FOR SUSTAINABLE TRANSPORTATION

The existing IC engine-based transportation system is the backbone of this modern society, whereas it has several issues such as greenhouse gas emissions (GHEs), air pollution, rapid depletion of natural resources, and thus a raise in the oil prices that affects the routine lifestyle of humans [5, 8].

In the world, over 95% of energy in transport comes from fossil fuels, mainly gasoline and diesel [10]. Exhaust gases combine several hundred chemical compounds. It is difficult to accurately calculate the contribution of vehicle exhaust gases to global air pollution since the data vary greatly by region and city. A report by CE Delft [11], commissioned by a consortium of European non-profit organizations, found that on average the European transport sector is responsible for 40%–50% of oxides of nitrogen (NO_x) emissions and 10%–15% of particulate matter (PM) emissions.

'The IC engine-based transportation system that accounts for more than a quarter of GHEs globally. It is mandatory to decarbonize all means of transport to reach net zero emissions by 2050 worldwide' and decarbonizing the present transportation system requires all nations to address the emissions from road transportation systems to aviation [12, 13]. Priorities in this regard include phasing out the production of IC engine-based vehicles by 2040. The World Bank reports that shifting to a sustainable transportation system could save 33 trillion United States Dollars (USD) by 2050 [14]. The transportation system's impact on the atmosphere can be reduced by improving the environmentally friendly transportation system. Therefore, it is essential to work towards developing an environmentally friendly transportation system. This section details the existing transportation system's problems.

2.2.1 Air Pollution and Transportation Emissions

Nowadays, most of the large cities in the world use automobiles that run on IC engines, which is one of the significant causes of air pollution. The fossil fuel-powered IC engine-based transportation system damages nature and human health more. Its harmful emissions worsen the problem of global warming, making people sick, particularly in the respiratory and nervous systems. Good combustion in an IC engine can be well understood from equation 2.1.

$$HC + O_2 + N_2 = H_2O + CO_2 + N_2 \tag{2.1}$$

These IC-engine-based vehicles pollute the air mainly in the process of movement. The emissions consist of gases and dust, which are formed during the combustion of fossil fuels, as well as PM from the friction of tyres on the road surface. As a result, air quality is noticeably reduced in the immediate vicinity of busy roads.

The substances that are released during the operation of IC engines are carbon monoxide (CO), oxides of nitrogen (NO_x), sulphur dioxide (SO_2), ammonia (NH_3), methane (CH_4), PMs, and GHEs [15]. The side effects of these emissions are detailed in this section.

Carbon monoxide (CO) is a poisonous gas; if inhaled, the human might be susceptible to headache, shortness of breath, and nausea. Large concentrations of CO can even lead to death. The NO_x in high concentrations in the atmospheric air can lead to irritation of the mucous membranes and chronic respiratory diseases [16].

Sulphur dioxide (SO_2) and ammonia (NH_3) are dangerous substances that cause coughing, runny nose, vomiting, and choking in humans. But, comparatively methane (CH_4) causes less harm to human health because it does not accumulate in the body and does not irritate the respiratory tract when inhaled. However, very high concentrations of CH_4 in the air in the range of 20%–30% cause suffocation [16, 17].

The World Health Organization (WHO) has recognized PM as the most dangerous air pollutant. The vehicles throw them out when burning fuel, rubbing tyres on asphalt, and destroying the road surface. Researchers isolated these particles with diameters less than 0.01 mm as PM_{10} and less than 0.00 25 mm as $PM_{2.5}$, which are 40 times thinner than a human hair. These suspended PMs are dangerous as they can adsorb and carry toxic substances or harmful micro-organisms. $PM_{2.5}$ particles can enter the bloodstream and accumulate in the human body, which leads to cardiovascular disease [18].

Another important type of pollution for which transport is responsible is noise. Noise pollution comes from engine sounds, tyres rubbing against the road surface, and metal wheels rubbing against rails. Constant noise worsens the quality of sleep and contributes to rapid fatigue in humans. A strong noise for many years can lead to the development of dangerous cardiovascular diseases.

In addition, transportation is one of the major sources of GHEs, the main driver of climate change. In the UK, for example, road transport accounts for 22% of the country's total carbon dioxide (CO_2) emissions [11].

2.2.1.1 Greenhouse Gas Emissions

The IC engine-based transportation systems such as light vehicles, heavy-duty vehicles, railways, airways, and shipping are major reasons for GHEs [19]. The by-product of burning fossil fuel in an IC engine is carbon dioxide (CO_2) and it is one of the major GHEs that adversely influence the climatic conditions. Half of the world's aviation CO_2 emissions come from just 1% of passengers. In early 2021, the European Union (EU) launched the destination 2050 plan to achieve carbon neutrality in aviation by 2050.

Let's discuss the phenomenon of the 'greenhouse' effect and its consequences. The greenhouse effect regulates Earth's temperature. This phenomenon is important for life on Earth because without it, the average temperature would be −18°C instead of 15°C. Pollution caused by GHEs in the atmosphere is the main issue. The greenhouse effect plays a crucial role in maintaining Earth's temperate climate [20].

GHEs are atmospheric gases that occur often. They take in some of the sun's energy and then release it in a different form. Its increasing atmospheric concentration is a major contributor to 'global warming,' which is caused by human activities like transportation, industry, etc. These gases are the primary contributors to the greenhouse effect, as the names imply. Carbon dioxide (CO_2), water vapour (H_2O), methane (CH_4), nitrous oxide (N_2O), and ozone (O_3) are the gases responsible for the greenhouse effect [20, 21].

A greater concentration of GHEs exists in the atmosphere. Human activities, such as growing cattle that produce methane (CH_4) or driving vehicles powered by fossil fuels, emit huge quantities of GHEs, resulting in the concentration of additional greenhouse effect that raises the average global temperature [20].

The greenhouse effect, induced by the intensity of fossil fuel consumption, has already raised temperatures by 1°C due to the industrial era. There are already some changes in the climate that can be seen, but their effects are still small. This is why we need to act fast and cut down on GHEs [20, 21].

Figure 2.1 explicits the global GHE data, showing that CO_2 is the dominant emittance which is very harmful to the universe followed by N_2O and other GHEs [22]. Figure 2.2 gives the details of CO_2 in the United States, and the increasing trend is an alarming factor for the environment [23].

In 2014, Intergovernmental Panel on Climate Change (IPCC) released data on global GHEs by economic sector and in this, the transportation sector is ranked in the fourth position and as shown in Figure 2.3 [24]. Whereas, in 2020, the Environmental Protection Agency (EPA) of the United States released a report on GHEs in various sectors and pollution by various levels of vehicles in the transportation sector. The report reveals that the transportation sector is the highly polluting segment and it is evident from Figure 2.4 [25]. This might be due to the increasing population and people's lifestyles. Among the transportation sector, light-duty vehicles (motorcycles and cars) are a highly used transportation system in the United States, and it ranks top in the GHEs, which is evident from Figure 2.5 [26].

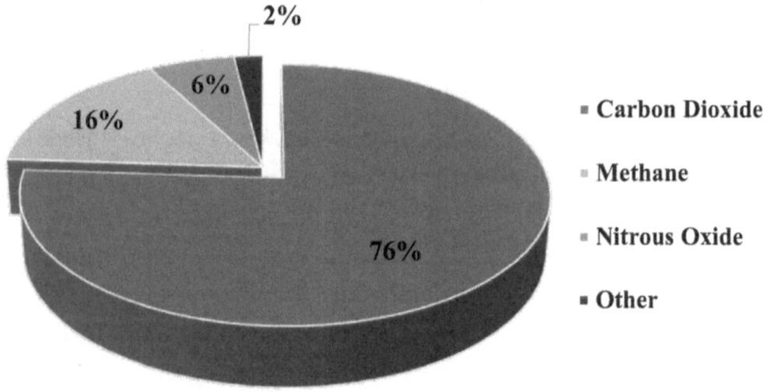

FIGURE 2.1 Global GHE by gas.

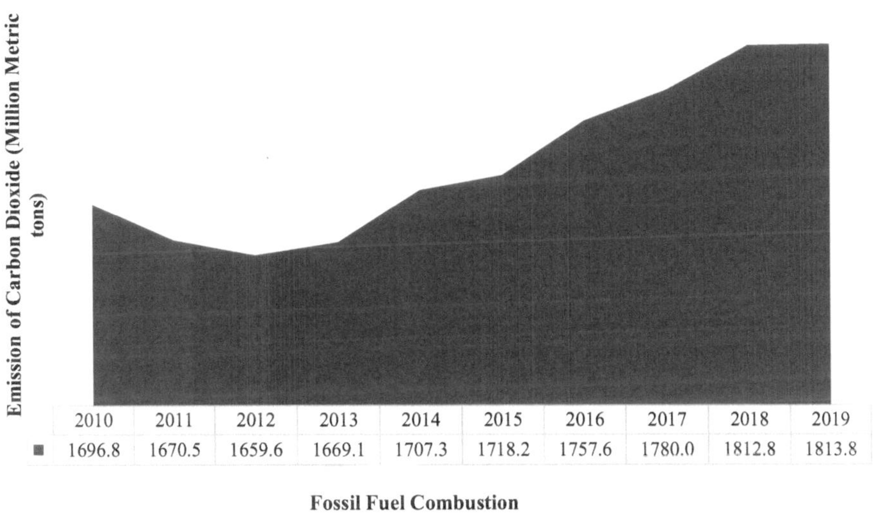

FIGURE 2.2 CO_2 emissions from the transportation sector in the US.

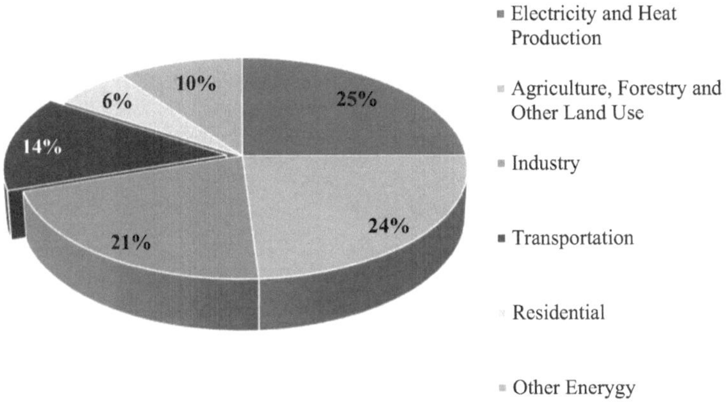

FIGURE 2.3 Global GHEs data by economic sector.

The balance of our ecosystems will be at risk more and more as climate change gets worse. So, a rise in the average temperature of the Earth of more than 1.5°C would cause extreme weather events that would have a direct effect on things like melting ice, rising sea levels that would flood coastal cities, more powerful hurricanes, and the need for some populations and species to move.

In 2018, the IPCC issued a separate report describing the details of the consequences of global warming of 1.5°C, which may lead to heavy rains, devastating hurricanes, intense droughts, etc. [27].

Humans can only mitigate the greenhouse effect by reducing GHEs. Thus, in 2015, 192 countries mobilized at the *21st Conference of the Parties* (COP21) by

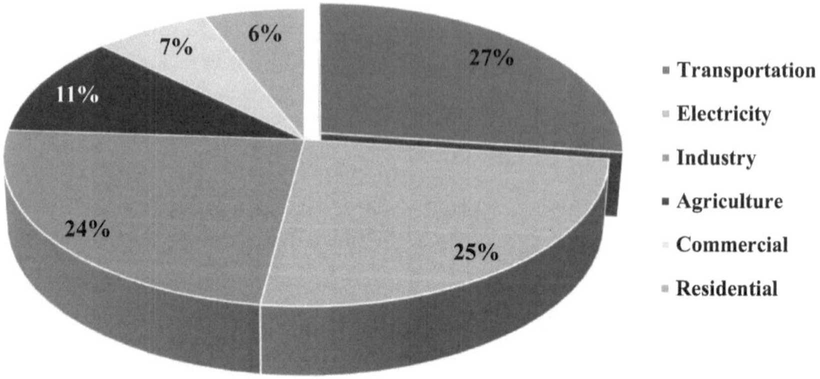

FIGURE 2.4 GHEs in the US by economic sector.

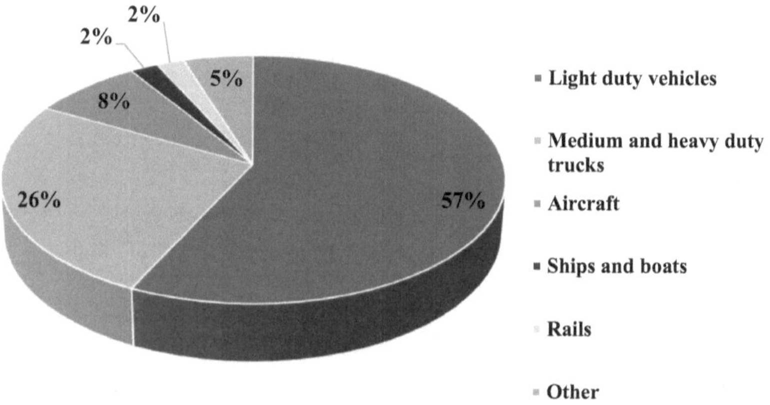

FIGURE 2.5 GHEs of the transportation sector in the US.

signing the Paris Agreement and establishing the goals to keep global warming below 2°C by 2100 [28].

Figure 2.6 shows the emission factors (g/km) of a light-duty passenger car (LDPC) and a light-duty diesel truck (LDDT). From the graph, it is evident that the LDDT emits higher quantity of NOx when compared to LDPC which negatively affects the environment. However, there is not much difference in the HC emission [29].

2.2.2 Fossil Fuel Depletion

As long as the current way of life does not change significantly, humanity will continue to depend heavily on oil, gas, and coal. And in the specific case of oil, this dependency will continue to be inevitable as long as most automobiles run on petrol, diesel, coal, and natural gas [30]. This scenario will not change significantly as long as other alternative propulsion systems such as electric or hybrid vehicles are not

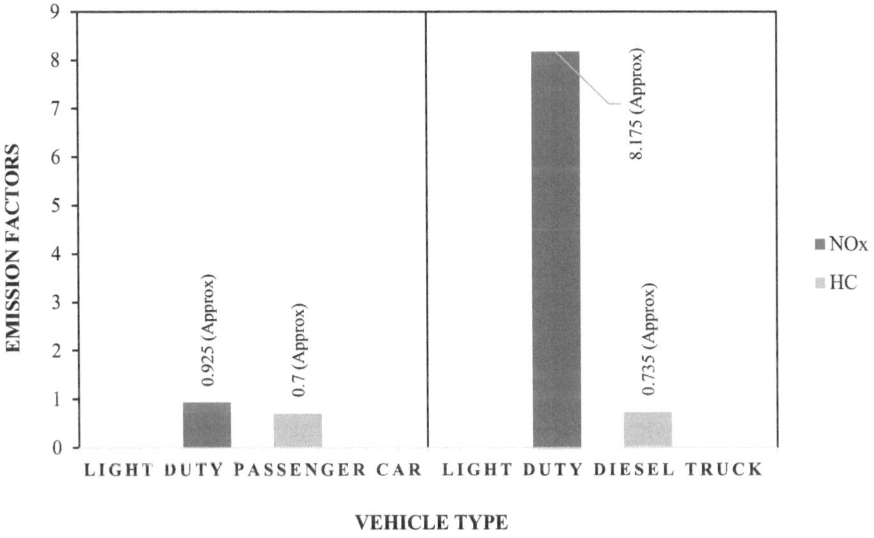

FIGURE 2.6 Comparison graph of emission factors.

sufficiently developed to compete on equal terms with the former. Depending on the fuel, the depletion forecasts for each of the fossil fuels are as follows:

2.2.2.1 Oil

It is the fuel for which it is most difficult to estimate the duration of currently known reserves. It has been used significantly in the modern era since approximately 1850, although it was not until the arrival of the IC engine in the early years of the 19th century that its consumption began to skyrocket. It is the scarcest fossil fuel due to the resource's intensive utilization, which has decreased its depletion estimates to between 40 and 50 years.

2.2.2.2 Coal

Even though coal has been used a lot for more than 200 years, there is still a lot of it, and its deposits are not all in one place. This means that there are no supply problems, and none are expected in the short or medium term. There is enough coal for more than a century at the current rate of extraction.

2.2.2.3 Natural Gas

It is the most recently used fossil fuel. In the US, its consumption became widespread in the 1950s, in Europe from the 1970s, and in Spain from the mid-1990s. The technology for making energy from this type of fuel is very advanced, which makes it possible for its use to skyrocket. This has led to a big drop in the known reserves, so it is estimated that they will be gone in 60–80 years.

The fossil fuel depletion (FFD) is calculated using the following equation 2.2:

$$\text{Fossil fuel depletion (FFD)} = \text{FFE}_t * \text{OILEQ}_t \tag{2.2}$$

where FFE_t = fossil fuel extraction by type '*t*' in either MJ/unit or kg/unit output, $OILEQ_t$ = oil equivalent characterisation factor by type '*t*' in either kg*oil-eq/MJ or kg*oil-eq/kg, type '*t*' represents the fossil fuel type such as natural gas, crude oil and coal and finally, FFD is measured as kg*oil-eq/unit output [31].

2.3 ALTERNATIVE SOLUTIONS

The most popular technology in the current transportation industry that burns fuel made of oil and natural gas is the IC engine. The fundamental support for human survival regarding transportation in the modern period is provided by conventional energy sources such as coal, oil, and natural gases. In this current era, 72%–78% of the world's population depends on conventional fuels [32]. Several detrimental effects and hazards including noise, air pollution, and global warming are caused by conventional energy sources to the environment. Another notable point in the transportation sector is the continuous decline of petroleum product availability. Hence researchers and business people have tended to favour alternate effective energy sources like batteries and fuel cells (FC) as a result of these benefits. The alternative methods for decarbonizing the transportation industry include solar photovoltaics (PVs), electric and hybrid electric vehicles (EVs and HEVs). Furthermore, these vehicles are less polluting and may reduce the GHEs produced by traditional vehicles. In recent years, industrial advancements have increased interest in FC and clean energy sources, among these energy units' kinds [33–35].

2.3.1 Solar Mobility/Photovoltaics (PVs)

The solar electric vehicle (SEV) is an electrically powered vehicle that significantly uses direct solar energy for its operation [36]. SEVs employ photovoltaic (PV) cells on their solar panels to convert sunlight into electricity that may be stored in a battery or used to power the vehicle's electric motor. SEVs have limited autonomy due to solar panel efficiency and aerodynamic resistance.

In addition, the most experimental prototypes of SEVs are designed and manufactured with very low-weight materials to facilitate the energy efficiency of the whole vehicle. On the other hand, there is a dependence on the sun to recharge the batteries of SEVs. It is obvious, but depending on the country and the latitude, this will be a challenging issue.

When talking about the operation of the SEVs, there has been no choice but to recount their disadvantages or rather the problems that engineers face to create truly efficient vehicles. But if they can bring the models they are working on to the market, it will also have plenty of advantages such as zero emissions and it's a pure 'green technology' [37].

2.3.2 Battery Electric Vehicles

The idea behind battery electric vehicles (BEVs) is straightforward: instead of an internal combustion engine vehicle (ICEV) and a fuel tank, a battery-powered electric motor is used, and when not in use, the car is hooked into a charging station.

They have many advantages, such as high efficiency, zero exhaust emissions that benefit the environment, swift acceleration, and the capacity to be charged overnight with cheap electricity from any kind of power plant, including one powered by renewable energy. The existing IC engine-powered vehicles are the dominant propulsion method for more than ten decades but the depleting fossil fuel is encouraging engineers to invent an alternative transportation system such as electric vehicle (EV) or BEV. The electric motors serve as the heart of the EVs, which may be one or more. The major battery technologies used in the EVs are given in Figure 2.7 [38]. The lead-acid technology is used for small vehicles with lower performance, and it is a familiar technology for all industries. The Ni-MH technology is widely used for hybrid vehicles since it is higher potential with economic for the industry needs. Even though Li-ion is more potential the probability of the fire is more due to the higher electrochemical potential of lithium. The $Na/NiCl_2$ battery systems possess multiple advantages such as higher safety, higher duty cycle, and lower cost as compared to the other technologies. However, due to its inferior specific energy always supercapacitors are needed with the $Na/NiCl_2$ battery systems

Nowadays, the electricity-powered automobile system includes, light and heavy moving vehicles, rail vehicles, drones, electric spacecraft, etc. The EVs arrived late back in the 19th century but the technical difficulties such as the development of batteries, transmission systems, limited speed, etc., restrained the real-time practice of the electric-powered vehicles. Whereas in the 21st century, EVs are making a comeback due to technological developments and the widespread use of renewable energies. The depletion of fossil fuels increased the demand for EVs, and the automobile research community has started sharing the technical details for electric-powered vehicle conversions [39, 40]. On the other hand, the United States and the EU have created government incentives to increase the adoption of EVs. Data shows that EVs are expected to increase from 2% in 2016 to 22% in 2030 [41].

The development of the metal oxide semiconductor (MOS) technology led to the progress of modern EVs and the MOS—field effect transistor (MOSFET), and the microcontroller, were significant advances in the development of electric vehicle technology [42]. And another important technology was lithium-ion batteries which allowed the innovation of electric-powered vehicles capable of travelling long distances.

Many nations now provide financial incentives for the purchase and use of EVs in an effort to mitigate the adverse environmental and economic effects of these vehicles. Several incentives aim to boost the purchase of electric vehicles by offsetting

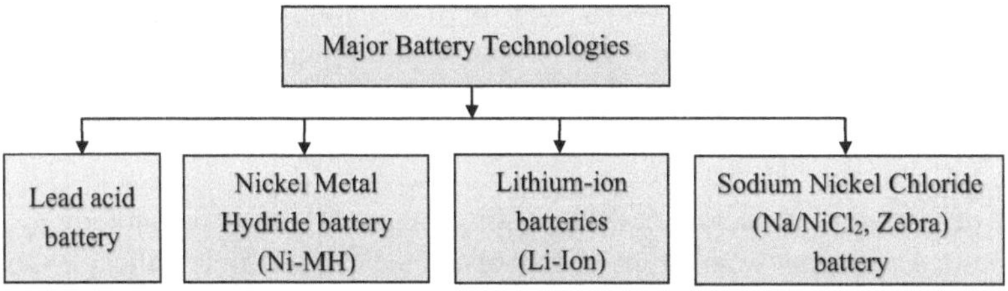

FIGURE 2.7 Major four types of batter technologies.

the cost of the vehicle with a subsidy. Investing in electric charging infrastructure and lowering tax rates are two further motivations.

The major global brands in the automotive sector are immersed in research and development projects to exponentially increase the sustainability of electric-powered vehicles.

There exist certain challenges to promote the sale and use of EVs. It should be noted that, in many localities, it is still very difficult to recharge these electric-powered vehicles, which forces the public authorities to design ambitious strategies for the installation of charging points. On the other hand, the rapid recharging of these EVs has attracted consumers to purchase and use these vehicles. Also, the cost of EVs is still high which even discourages many users to buy these electric-powered vehicles.

2.3.3 Fuel Cell Electric Vehicle

A fuel cell electric vehicle (FCEV) is powered by electricity that is generated when hydrogen stored in the vehicle unites with air [43]. This type of technology does not produce any type of emissions except pure water (H_2O), and thus it can be categorized as 'green technology'. It is also called a 'roadside air purifier' as it is capable of removing 99.9% of ultrafine $PM_{2.5}$ from the atmosphere [44, 45].

For an FCEV, the tank that stores hydrogen can have the same capacity as those that stores fuel in an existing IC engine-powered vehicle and its recharge time is similar to filling up with gasoline at a service station. It might drastically reduce fossil fuel use and climate change emissions [46].

The lack of adequate hydrogen charging infrastructure has prevented FCEVs from being successful in the marketplace. While the market is constantly growing, FCEVs are still classified as a type of green vehicle of the future. There are a few problems that need to be fixed before these vehicles can compete with those with IC engines, but the benefits of this technology could be huge. And on the other hand, the FC technology is not as durable as IC engines, particularly in certain temperature and humidity ranges. At the moment, fuel cell batteries only last half as long as they need to for commercial use. In the past few years, durability has improved a lot, going from 29,000 to 75,000 miles. However, experts say that an FCEV needs to last 150,000 miles to compete with gasoline vehicles [47, 48].

2.3.4 Hybrid Electric Vehicle

When a vehicle is powered by both an IC engine and an electric motor that vehicle is referred to as a 'hybrid vehicle (HV)' [49]. With this configuration, a hybrid vehicle can take advantage of both sources to obtain maximum energy and move economically with excellent performance [50]. These vehicles are partially categorized as 'green vehicles' because it emits fewer GHEs when compared to the existing IC engine-powered vehicles.

One of the main characteristics that distinguish an HV is its technology to intelligently manage the power and type of propulsion to be used. An HV includes several important components that properly regulate its function such as an IC engine, generator, battery, electric motor, and computer management system.

In most HVs, the IC engine, also known as a thermal engine, is either gasoline or diesel. These engines function normally like the existing IC engine-powered vehicle. However, the fuel consumption of the HVs is lower than that of any common combustion vehicle. While accelerating the vehicle, the surplus energy supplied by the IC engine is recovered with the aid of a generator and accumulated in the battery, and this accumulated electric power is fed to the electric motor for the hybrid function. The batteries provide the energy for the electric motor to function. They are usually placed on the rear side of the vehicle or the floor under the rear seats and are supplemented by the 12 V battery that the vehicle carries to start it. These batteries can be made of different materials such as nickel-metal hydride, lead-acid, nickel-cadmium, or lithium-ion. Depending on the application of the HV, there may be one or two electric motors in the vehicle. In either case, the electric motor is always connected to the transmission or directly to the wheel axle. And finally, the computer management system or the information system that enables the driver's preference to determine the work of the HV, for example, if maximum power is required, the fuel motor will come into operation, while, for medium speed, the electric motor may be more than enough. The very interesting fact about the HV is that its technology allows the battery to be recharged along the way through regenerative braking.

Choosing an HV gives us fuel-saving benefits as well as being conducive to caring for the environment. The other benefits include;

- Due to the combination of technology, an HV emits fewer polluting gases since it not only uses IC engines to drive the vehicle [51].
- It consumes less fossil fuel thus lowering the GHEs since it has mechanisms to make fuel consumption more efficient [52].
- The HVs are ideal for large cities and they allow us to travel long distances while taking advantage of their electrical operation [53].

2.4 CHALLENGES IN ROAD SAFETY AND TRAFFIC CONGESTION

Nowadays the government has the primary focus on road safety due to the need for health, humanitarian and economic factors. 1.25 million people are killed in motor vehicle collisions annually, 3,400 people are injured in such collisions daily, and 30–50 million people are hurt yearly, as stated by the WHO [54]. Notably, 90% of traffic injuries are accounted for in low- and middle-income countries. Also, worldwide, the mortality rate per one lakh population ranges from approximately 3–40. However, in high-income countries, the rate is less than 9 and almost 20 in low- and middle-income countries. Throughout the previous few decades, road safety has improved in high-income countries, whereas in low- and middle-income countries, the trend has been less encouraging. By 2020, almost 2 million people a year are expected to die on the roads.

Several countries have considered this a global problem and adopted many resolutions on road safety. One such example, that the United Nations has come up with is the Decade of action for road safety 2011–2022, which has addressed worldwide concerns over the last couple of years. However, the future of road safety is uncertain and it mostly depends on where an individual lives. Countries with a rational approach to road safety and a desire to achieve further progress are intended to adopt a proactive approach: the safe system approach. A study on road safety reported that several low and middle-income countries are extremely intended to design and implement road safety strategies.

Traffic congestion is an important parameter that influences the occurrence of a road accident. It results from the high population density, increase in the number of motor vehicles, and development of rideshare and delivery services [55]. Recurring congestion occurs when there is an interruption to the normal flow of traffic by a higher number of vehicles and results in excess travel time. Unpredictable situations like weather, festivals, incidents, and peak time zones are the main causes of non-recurring congestion [56, 57]. The Federal Highway Administration of the United States Department of Transportation (DOT-FHWA) reports that only around 40% of traffic congestion is due to regular occurrences, while the remaining 60% is caused by irregular events. It is essential to have a comprehensive understanding of the effects of congestion on traffic accidents in order to work towards reducing accident rates.

2.4.1 Road Safety

The measurement of road safety is the first task where one should have an understanding of the elements to be included in the definition. The number of road crashes and the associated casualties, and their negative consequences are the main elements that define road safety [58]. However, the data related to these elements are not properly collected to account for road safety measurements. Also, the reported number of road crashes in many countries is unmatching the real-time scenario. For many reasons, the personnel from the department of the police are underreporting the incidents. It is reported that cycle crashes are mostly underreported. Scientific knowledge and road safety data statistics play an important role in managing evidence-based and data-driven road safety management [59]. However, these methods proved to be more complicated and non-realistic compared with traditional methods.

2.4.1.1 Advanced Driver Assistance System

The majority of road accidents are due to human factors such as drowsiness, using mobile phones while driving, careless driving, etc. This made the automobile manufacturers show intense interest in the advanced driver assistance system (ADAS), which makes road trips safer and more comfortable. The schema of the ADAS is shown in Figure 2.8.

The ADAS can significantly improve vehicle safety by integrating the key factors for fatal accidents such as cruise control and by maintaining a safer distance between the vehicles. Indeed, the ADAS also minimizes human errors including the pedestrian and cyclist collision risk, capable of alerting the driver or braking automatically, which could prevent several accidents on road [9, 58, 59].

The installation of black boxes in the vehicle can help us to understand the reason for vehicle crashes and also the alcohol countermeasure system would prevent the drivers who have consumed alcohol from driving the vehicles. Although the ADAS assists us to drive safer on roads, it is also necessary to consider some of its limitations such as varying weather conditions, the state of the roads, crooked routes, etc.

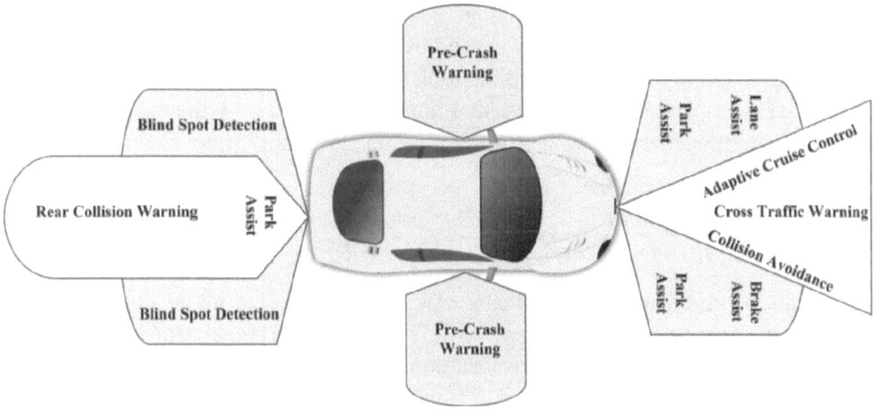

FIGURE 2.8 Advanced driver assistance system (ADAS).

2.4.2 Relation Between Congestion and Road Accidents

The impact of traffic congestion on accident rates is well-studied. In the range of 7,000 vehicles and higher, the number of accidents per million vehicle miles is proportional to the average daily traffic (ADT). However, it decreases gradually with the increase in congestion due to the decreased speed. The correlation between accidents, annual average daily traffic (AADT), and congestion index indicate that the increase in traffic volumes is the main attribute related to the increased accident rates.

In 2012, the International Organization for Standardization (ISO) published the first version of the International Standard for the Road Traffic Safety (RTS) Management Systems (ISO 39001), which contains requirements and recommendations for good practices that can be adopted by the public and private organizations, from companies, government agencies to civil organizations. Some of the good practices in road safety that can be adopted by society are as follows;

- *Road culture and education*: The main and most profitable investment that can be made in safety matters is related to people, with a focus on the development of skills, knowledge, and awareness towards prevention.
- *Lead by example*: For any initiative, business strategy, or public policy to be effective and achieve the desired results, the leaders are the ones who must first be convinced, committed, and set an example. Both directors and managers in companies, officials and traffic agents in public agencies as well as ourselves from our family and social environment must set the example by complying with the laws and good practices of road safety.
- *Being participatory and an agent of change*: Some unduly accepted practices represent a serious safety problem, such as taxis and passenger trucks that fail to comply with elementary safety standards such as not having seat belts or having trained drivers. With simple actions such as reporting improper acts or not using a service that does not comply with safety regulations, we will be making a difference and positively influencing society.

Even when there is a significant gap to achieve the purpose of reversing the negative impacts on public health and safety caused by road accidents, the various initiatives led at the global and local levels and the tools such as the international standard ISO 39001, give us a perspective positive that must be supported by the participation of all to achieve the generation of a culture of road safety that improves our quality of life [60].

2.4.3 Smart Transportation System

Smart transportation is a revolutionary new way of thinking about how we move in a cleaner, safer, and more efficient way. The solutions to break traffic congestion include a wide variety of new transportation alternatives such as bicycles, electric skateboards, subways, AV, etc. Another alternative way of moving smart is sharing a vehicle with other citizens, either as the owner or as a passenger [61–63]. However, the concept of smart transportation is summarized in the following principles:

- *Efficiency*: It is about the traveller reaching their destination in the shortest possible time and with the minimum number of interruptions.
- *Integration*: The entire route is planned from end to end, no matter what means of transport is used.
- *Flexibility*: As there is a great diversity of transport modes, users can choose which one suits them best according to their needs.
- *Clean technology*: Polluting transport is increasingly being left aside to be replaced by zero-emission transport.
- *Accessibility*: This new way of life must be accessible to all citizens, without hierarchies.
- *Security*: Fatalities, injuries, and accidents are drastically reduced by greater control.
- *Social benefit*: One of the reasons for smart transportation is to provide general well-being and an improvement in the quality of life for everyone.

2.5 CONCLUDING REMARKS AND FUTURE SCOPE

We are at a crucial moment where the global population is continuing to grow, yet economic activity cannot be stopped, and fossil fuel decreases inevitably every year. One of the most significant contributors to air pollution around the world is the transportation sector. To mitigate transportation's negative effects on the environment, however, it is important to adopt more sustainable modes of transportation. However, plenty of challenges are prevailing in the alternative system. Although the electric-powered vehicles are said to be 'green vehicles', we are only just beginning to realize the damage to the environment caused by the extraction of lithium, from which batteries are made. In this regard, a question may arise whether EVs are 100% 'green'. But when compared to the existing IC engine-powered vehicles the EVs are much more environmentally friendly. For example, replacing one diesel bus with an electric bus reduces 60 tonnes of CO_2 per year.

In order to ensure road safety and traffic congestion, the effective way is to move towards a smart transportation system, and it can be initiated through data collection. In this system, the security issues can be identified and addressed before the problem arise. These data can range from the identification of collision points on city streets to the detection of possible signalling problems. In addition to the collection of data as a basis, initiatives can be used in which the mobilization of citizens is promoted by bicycles, public transport, on foot, etc. On the other hand, there are also more innovative options to establish solutions to traffic congestion. The use of sensor networks that are located at different strategic points in the city helps the users to avoid congested streets.

Another technological solution in the smart transportation system is to allow different types of vehicles to be connected to the internet that obtain data from it. In this way, it is possible to contribute to the problems such as traffic congestion, high fuel consumption, or CO_2 emissions by vehicles. And another objective of this system is to increase the safety and efficiency of land mobility.

REFERENCES

[1] Banister, D., Anderton, K., Bonilla, D., Givoni, M. and Schwanen, T., 2011. Transportation and the environment. *Annual Review of Environment and Resources*, *36*, pp. 247–270.

[2] Pietzcker, R.C., Longden, T., Chen, W., Fu, S., Kriegler, E., Kyle, P. and Luderer, G., 2014. Long-term transport energy demand and climate policy: Alternative visions on transport decarbonization in energy-economy models. *Energy*, *64*, pp. 95–108.

[3] Berg, J., Henriksson, M. and Ihlström, J., 2019. Comfort first! Vehicle-sharing systems in urban residential areas: The importance for everyday mobility and reduction of car use among pilot users. *Sustainability*, *11*(9), p. 2521.

[4] Mattioli, G., Roberts, C., Steinberger, J.K. and Brown, A., 2020. The political economy of car dependence: A systems of provision approach. *Energy Research & Social Science*, *66*, p. 101486.

[5] Fantin Irudaya Raj, E. and Appadurai, M., 2021. The hybrid electric vehicle (HEV)-an overview. *Emerging Solutions for e-Mobility and Smart Grids*, *1*, pp. 25–36.

[6] Lee, S. and Park, S., 2017. Optimization of the piston bowl geometry and the operating conditions of a gasoline-diesel dual-fuel engine based on a compression ignition engine. *Energy*, *121*, pp. 433–448.

[7] Brown, S., Pyke, D. and Steenhof, P., 2010. Electric vehicles: The role and importance of standards in an emerging market. *Energy Policy*, *38*(7), pp. 3797–3806.

[8] Ma, H., Balthasar, F., Tait, N., Riera-Palou, X. and Harrison, A., 2012. A new comparison between the life cycle greenhouse gas emissions of battery electric vehicles and internal combustion vehicles. *Energy Policy*, *44*, pp. 160–173.

[9] Mohan, A., Sripad, S., Vaishnav, P. and Viswanathan, V., 2020. Trade-offs between automation and light vehicle electrification. *Nature Energy*, 5(7), pp. 543–549.

[10] Vohra, K., Vodonos, A., Schwartz, J., Marais, E.A., Sulprizio, M.P. and Mickley, L.J., 2021. Global mortality from outdoor fine particle pollution generated by fossil fuel combustion: Results from GEOS-Chem. *Environmental Research*, *195*, p. 110754.

[11] Sander de Bruyn and Joukje de Vries, 2020. Health costs of air pollution in European cities and the linkage with transport, *Delft, C.E,* 1, pp. 1–86.

[12] https://blogs.worldbank.org/transport/why-sustainable-urban-transport-great-investment (accessed on 1 August 2002).

[13] https://www.ipcc.ch/site/assets/uploads/2018/02/ipcc_wg3_ar5_chapter8.pdf (accessed on 15 August 2022).

[14] https://press.un.org/en/2021/sgsm20971.doc.htm (accessed on 15 August 2022).

[15] Aziz, M.A., Kassim, K.A., Bakar, W.A.W.A., Jakarmi, F.M., Ahsan, A.A., Rosid, S.J.M. and Toemen, S., 2019. Traffic pollution: Perspective overview toward carbon dioxide capture and separation method. *Fossil Free Fuels*, *1*, pp. 149–186.

[16] Das, S., 2020. Toxic gases. In: *Toxicology Cases for the Clinical and Forensic Laboratory* (pp. 387–396). Academic Press.

[17] Mitchell, R. and Sweeney, F., 2018. *Air Pollution*. Scientific e-Resources.

[18] World Health Organization, 2021. *WHO Global Air Quality Guidelines: Particulate Matter (PM2. 5 and PM10), Ozone, Nitrogen Dioxide, Sulfur Dioxide, and Carbon Monoxide*. World Health Organization.

[19] Seto, K.C., Churkina, G., Hsu, A., Keller, M., Newman, P.W., Qin, B. and Ramaswami, A., 2021. From low-to net-zero carbon cities: The next global agenda. *Annual Review of Environment and Resources*, *46*(1), pp. 377–415.

[20] Kweku, D.W., Bismark, O., Maxwell, A., Desmond, K.A., Danso, K.B., Oti-Mensah, E.A., Quachie, A.T. and Adormaa, B.B., 2018. Greenhouse effect: Greenhouse gases and their impact on global warming. *Journal of Scientific Research and Reports*, *17*(6), pp. 1–9.

[21] Mikhaylov, A., Moiseev, N., Aleshin, K. and Burkhardt, T., 2020. Global climate change and greenhouse effect. *Entrepreneurship and Sustainability Issues*, *7*(4), p. 2897.

[22] https://www.epa.gov/ghgemissions/global-greenhouse-gas-emissions-data (accessed on 17 August 2022).

[23] https://cfpub.epa.gov/ghgdata/inventoryexplorer/#transportation/entiresector/allgas/select/all (accessed on 17 August 2022).

[24] https://www.epa.gov/ghgemissions/global-greenhouse-gas-emissions-data (accessed on 17 August 2022).

[25] https://www.epa.gov/ghgemissions/sources-greenhouse-gas-emissions (accessed on 17 August 2022).

[26] https://www.epa.gov/ghgemissions/sources-greenhouse-gas-emissions (accessed on 17 August 2022).

[27] Hoegh-Guldberg, O., Jacob, D., Bindi, M., Brown, S., Camilloni, I., Diedhiou, A., Djalante, R., Ebi, K., Engelbrecht, F., Guiot, J. and Hijioka, Y., 2018. Impacts of 1.5°C global warming on natural and human systems. In: *Global Warming of 1.5°C. Intergovernmental Panel on Climate Change*, *1*, pp. 175–311.

[28] https://unfccc.int/most-requested/key-aspects-of-the-paris-agreement (accessed on 18 August 2022).

[29] Mei, H., Wang, L., Wang, M., Zhu, R., Wang, Y., Li, Y., Zhang, R., Wang, B. and Bao, X., 2021. Characterization of exhaust CO, HC and NOx emissions from light-duty vehicles under real driving conditions. *Atmosphere*, *12*(9), p. 1125.

[30] Johnsson, F., Kjärstad, J. and Rootzén, J., 2019. The threat to climate change mitigation posed by the abundance of fossil fuels. *Climate Policy*, *19*(2), pp. 258–274.

[31] Huijbregts, M.A., Steinmann, Z.J., Elshout, P.M., Stam, G., Verones, F., Vieira, M. and van Zelm, R., 2016. *A Harmonized Life Cycle Impact Assessment Method at Midpoint and Endpoint Level Report I: Characterization* (pp. 2016–0104). National Institute for Public Health and the Environment.

[32] Waseem, M., Sherwani, A.F. and Suhaib, M., 2019. Integration of solar energy in electrical, hybrid, autonomous vehicles: A technological review. *SN Applied Sciences*, *1*(11), pp. 1–14.

[33] Hong, B. K. and Kim, S. H., 2018. Recent advances in fuel cell electric vehicle technologies of Hyundai. *ECS Transactions*, *86*(13), p. 3.

[34] Turoń, K., 2020. Hydrogen-powered vehicles in urban transport systems-current state and development. *Transportation Research Procedia, 45*, pp. 835–841.
[35] Zeynali, S., Nasiri, N., Marzband, M. and Ravadanegh, S.N., 2021. A hybrid robust-stochastic framework for strategic scheduling of integrated wind farm and plug-in hybrid electric vehicle fleets. *Applied Energy, 300*, p. 117432.
[36] Bhatti, A.R., Salam, Z., Abdul, M.J.B. and Yee, K.P., 2016. A comprehensive overview of electric vehicle charging using renewable energy. *International Journal of Power Electronics and Drive Systems, 7*(1), p. 114.
[37] Boxwell, M., 2010. *Solar Electricity Handbook: A Simple, Practical Guide to Solar Energy-Designing and Installing Photovoltaic Solar Electric Systems.* Greenstream Publishing.
[38] Andwari, A.M., Pesiridis, A., Rajoo, S., Martinez-Botas, R. and Esfahanian, V., 2017. A review of Battery Electric Vehicle technology and readiness levels. *Renewable and Sustainable Energy Reviews, 78*, pp. 414–430.
[39] Kumar, R.R. and Alok, K., 2020. Adoption of electric vehicle: A literature review and prospects for sustainability. *Journal of Cleaner Production, 253*, p. 119911.
[40] Goel, S., Sharma, R. and Rathore, A.K., 2021. A review on barriers and challenges of electric vehicle in India and vehicle to grid optimisation. *Transportation Engineering, 4*, p. 100057.
[41] Nogueira, T., Sousa, E. and Alves, G.R., 2022. Electric vehicles growth until 2030: Impact on the distribution network power. *Energy Reports, 8*, pp. 145–152.
[42] Mihet-Popa, L. and Saponara, S., 2018. Toward green vehicles digitalization for the next generation of connected and electrified transport systems. *Energies, 11*(11), p. 3124.
[43] Thomas, C.E., 2009. Fuel cell and battery electric vehicles compared. *International Journal of Hydrogen Energy, 34*(15), pp. 6005–6020.
[44] Mac Kinnon, M., Shaffer, B., Carreras-Sospedra, M., Dabdub, D., Samuelsen, G.S. and Brouwer, J., 2016. Air quality impacts of fuel cell electric hydrogen vehicles with high levels of renewable power generation. *International Journal of Hydrogen Energy, 41*(38), pp. 16592–16603.
[45] Cox, J., Isiugo, K., Ryan, P., Grinshpun, S.A., Yermakov, M., Desmond, C., Jandarov, R., Vesper, S., Ross, J., Chillrud, S. and Dannemiller, K., 2018. Effectiveness of a portable air cleaner in removing aerosol particles in homes close to highways. *Indoor Air, 28*(6), pp. 818–827.
[46] Sulaiman, N., Hannan, M.A., Mohamed, A., Majlan, E.H. and Daud, W.W., 2015. A review on energy management system for fuel cell hybrid electric vehicle: Issues and challenges. *Renewable and Sustainable Energy Reviews, 52*, pp. 802–814.
[47] İnci, M., Büyük, M., Demir, M.H. and İlbey, G., 2021. A review and research on fuel cell electric vehicles: Topologies, power electronic converters, energy management methods, technical challenges, marketing and future aspects. *Renewable and Sustainable Energy Reviews, 137*, p. 110648.
[48] Wilberforce, T., El-Hassan, Z., Khatib, F.N., Al Makky, A., Baroutaji, A., Carton, J.G. and Olabi, A.G., 2017. Developments of electric cars and fuel cell hydrogen electric cars. *International Journal of Hydrogen Energy, 42*(40), pp. 25695–25734.
[49] Fuhs, A., 2008. *Hybrid Vehicles: and the Future of Personal Transportation.* CRC press.
[50] Xing, Y., Ma, E.W., Tsui, K.L. and Pecht, M., 2011. Battery management systems in electric and hybrid vehicles. *Energies, 4*(11), pp. 1840–1857.
[51] De Souza, L.L.P., Lora, E.E.S., Palacio, J.C.E., Rocha, M.H., Renó, M.L.G. and Venturini, O.J., 2018. Comparative environmental life cycle assessment of conventional vehicles with different fuel options, plug-in hybrid and electric vehicles for a sustainable transportation system in Brazil. *Journal of Cleaner Production, 203*, pp. 444–468.

[52] Demirdoven, N. and Deutch, J., 2004. Hybrid cars now, fuel cell cars later. *Science, 305*(5686), pp. 974–976.

[53] Liu, J., Kockelman, K.M., Boesch, P.M. and Ciari, F., 2017. Tracking a system of shared autonomous vehicles across the Austin, Texas network using agent-based simulation. *Transportation, 44*(6), pp. 1261–1278.

[54] WHO, 2015. *Informe Sobre la Situación Mundial de la Seguridad Vial 2015* (pp. 1–12). World Health Organization

[55] Reed, T.; Kidd, J,. 2019. *Global Traffic Scorecard.* INRIX Research.

[56] Ghosh, B. 2019. Predicting the Duration and Impact of the Nonrecurring Road Incidents on the Transportation Network. Ph.D. Thesis, Nanyang Technological University, Singapore.

[57] Falcocchio, J.C.; Levinson, H.S., 2015. Managing nonrecurring congestion. In: *Road Traffic Congestion: A Concise Guide* (vol. 7, pp. 197–211). Springer Tracts on Transportation and Traffic. https://doi.org/10.1007/978-3-319-15165-6_15.

[58] Wegman, F., 2013. Traffic safety. In: B. van Wee, J.-A. Annema, D. Banister (Eds.), The *Transport System and Transport Policy. An Introduction* (pp. 254–280). Edward Elgar.

[59] Wegman, F., Berg, H.-Y., Cameron, I., Thompson, C., Siegrist, S., & Weijermars, W. 2015. Evidence-based and data-driven road safety management. *IATSS Research, 39*(1), 19–25. doi: 10.1016/j.iatssr.2015.04.001.

[60] Norge, S., 2012. *Road Traffic Safety (RTS) Management Systems Requirements with Guidance for Use: NS-ISO39001:2012.* Standard Norge.

[61] Karami, Z. and Kashef, R., 2020. *Smart Transportation Planning: Data, Models, and Algorithms. Transportation Engineering, 2*, p. 100013.

[62] Dogra, A.K. and Kaur, J., 2022. Moving towards smart transportation with machine learning and Internet of Things (IoT): A review. *Journal of Smart Environments and Green Computing, 2*(1), pp. 3–18.

[63] Agarwal, S., Mustavee, S., Contreras-Castillo, J. and Guerrero-Ibañez, J., 2022. Sensing and monitoring of smart transportation systems. In: *The Rise of Smart Cities* (vol. 1, pp. 495–522). Butterworth-Heinemann.

3 Alternative Propulsion Systems

Sundara Subramanian Karuppasamy
National Taipei University of Technology

N. Jeyaprakash
China University of Mining and Technology

Che-Hua Yang
National Taipei University of Technology

3.1 INTRODUCTION

Electric vehicle does not rely on any non-renewable energy sources. These vehicles are propelled with the help of electricity generated from batteries. By means of chemical reactions happening in the battery, the electricity is produced, and this electricity will actuate the electric motor which in turn accelerates the wheels for movement. Electric vehicles have many advantages compared to traditional internal combustion engine vehicles [1,2]. Some of the major advantages are (i) noise-free operation, (ii) compactness, (iii) emission-free or emission-less transportation, (iv) cost-efficient since only electric energy is needed, whereas the IC engines need petrol or diesel for operation, (v) less maintenance, (vi) ease of drive, and so on. Thus, electric vehicles are far much better than the traditional internal combustion engine vehicle [3–5].

In the past, electric vehicles are entirely dependent on the batteries for actuating the electric motors installed in them. The battery electric vehicle holds good for short distance transportation [6]. A single battery pack cannot meet the demand for long distance transportation. Moreover, for long distance, there is a need for large and heavy battery packs to deliver the nominal voltage required for transportation. Besides this drawback, there are other disadvantages faced by battery electric vehicles such as limited battery range, short lifespan, long charging times, need for high-end battery charging systems, low top speeds, and the battery packs are quite expensive. Hence there is a need for alternate sources that could overcome these drawbacks with emission-free/emission-less transportation [7].

The electric vehicles which need two or more energy sources for movement are termed as hybrid electric vehicles (HEVs). It consists of a battery pack and an IC engine as the primary energy source to actuate the vehicle. The combined effect of these two energy sources will provide the required energy to move the vehicles at high top speeds [8]. In most cases, the battery pack is made up of lead-acid and

DOI: 10.1201/9781003354901-3

lithium-ion batteries, whereas internal combustion engines are propelled by traditional fossil fuels like petrol and diesel. Limited capacity, minimal cycle life, slow charging, greater self-discharge rate, and high maintenance cost are the major drawbacks of Pb batteries. On the other hand, overheating could not withstand high voltage, aging, need for a battery management system, and so on are reported as the limitations of the Li^+ ion batteries. These batteries (lithium ion and lead acid) limitations are unavoidable and could be controlled to some extent. On the other hand, the fuels used in the IC engine have pollution as the major limitation. Hence, researchers are working on advanced batteries and alternative fuels that could be implemented in future HEVs. The major objective of this chapter is to describe the cell chemistries of advanced batteries and alternative sources that could be adopted in future HEVs considering emission-free/emission-less transportation as the primary key aspect. It also describes the supercapacitors and fuel cell technology which serves as alternative propulsion systems in the near future.

3.2 HYBRID ELECTRIC VEHICLES

In major cases, HEV employs two or more energy sources for transportation. Mostly, IC engines and battery packs are installed. For some distance, the IC engine will propel the vehicle, and the installed battery pack is charged through a regenerative braking mechanism, and the power supplied from the battery pack will drive the electric vehicle when the IC engine propulsion is not in use. Some of the examples of HEVs are as follows: Toyota Yaris Hybrid and Ford Mondeo Hybrid. It was reported that the global sales of HEVs will hit above 100 million by the year 2050 due to the demand for emission-less transportation [9]. The major unit of a HEV contains an energy storage device, a controller unit, and a drive train. The combination of these three components paved the way for different configurations as follows:

i. Series configured HEV.
ii. Parallel configured HEV.
iii. Series-parallel configured HEV.
iv. Complex configured HEV.

3.2.1 Series Configured HEV

The necessary trust needed to actuate the vehicle is supplied by the electric motor in the series configuration whereas the battery pack is charged by means of the power obtained from the IC engine through a generator. Without charging or discharging the battery, this type of configuration can actuate the electric vehicle based on the combined effect from the IC engine generator. Due to this effect, the user can electrically separate the IC engine from the shaft as in diesel-electric locomotives. Further, this type has six types of operation modes: (i) energy storage device alone, (ii) internal combustion engine alone, (iii) combined mode (both battery and engine provide the necessary power to actuate the vehicle), (iv) power split mode (engine-generator pair powers the battery and actuate the vehicle), (v) regenerative braking mode, and (vi) stationary battery charging mode [10].

3.2.2 Parallel Configured HEV

In this type, the motor and engine can work together to drive the wheels. By means of clutches, the engine and electric motor are coupled with the driving shaft. Engines, electric motors, or both can provide traction power [11]. Here, the electric motor is reported to work as a generator by means of regenerative braking. The major advantage of this type of architecture is the less amount of electromechanical power loss.

3.2.3 Series-Parallel Configured HEV

This type of architecture combines the pros of the series and parallel configuration. Moreover, this type holds good at low and high speeds followed by achieving high top speed in most cases. But the limitation is that there is a need for planetary gear and also the degree of system's complexity is high [12]. This configuration has a greater number of operation modes than the previous two architectures.

3.2.4 Complex Configured HEV

Complex configured architecture is reported to be the most complicated configuration than the other types. This type employs a bidirectional converter which makes it different from the rest of the configurations. The bidirectional converter enables the electric motor's power to flow in both directions. Versatile operating modes can be seen in this type where the bidirectional power flows through both the electric motor and the IC engine. Many researchers reported that this architecture is very complex and expensive compared to the other configurations [13].

3.3 ADVANCED BATTERIES

Since the traditional battery packs used in the HEVs are made up of Pb and Li^+ ion batteries which have limitations as discussed above, researchers work on different materials that could be adapted for producing electricity with minimal limitations. This section deals with alternative (advanced) batteries, which could be incorporated for storing energy in future HEVs.

3.3.1 Nickel Metal Hydride Batteries

Most researchers consider nickel as an effective replacement for traditional batteries due to their enormous advantages. Nickel metal hydride batteries are rechargeable batteries where any metal hydrides (hydrogen absorbing material) are used as the anode materials. Mostly, nickel oxyhydroxide is preferred as the cathode material (Figure 3.1). The metal hydride is oxidized to give protons and electrons, whereas KOH acts as the electrolyte. These protons travel through potassium hydroxide solution and reach the nickel oxyhydroxide electrode. In the nickel oxyhydroxide electrode, the protons and electrons from the anode combine with the nickel oxyhydroxide where it is reduced to nickel hydroxide. By using the external circuit, the e^- ions flow from anode to cathode, thereby producing electricity to run the

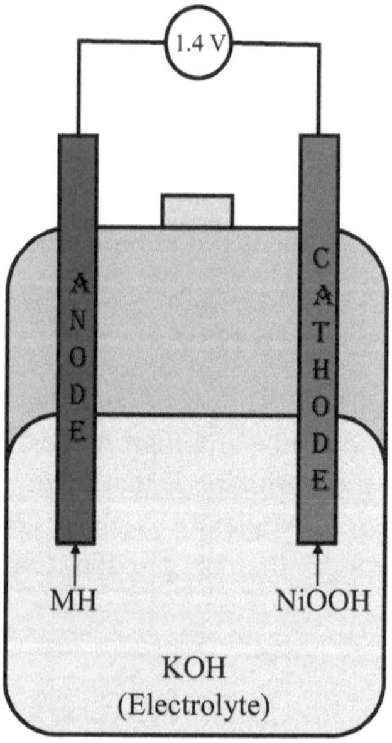

FIGURE 3.1 Schematic illustration of nickel metal hydride battery.

vehicle [14,15]. Thus, by means of redox reactions happening in the metal hydride and nickel oxyhydroxide electrodes, electricity is produced. Some of the key characteristics of these batteries are greater energy density, better discharge rate, high specific power, effective alternative for alkaline batteries, greater capability, longer life cycle, and so on [16].

Since any form of metal hydride that absorbs hydrogen is used as the anode, many materials can be adapted as the anode materials. The three basic forms of alloy families (materials) which could be implemented as anode materials are as follows:

1. AB_2: LaCePrNdNiCoMnAl alloy
2. AB_5: LaCePrNdSmMgNiCoMnAlZr alloy
3. A_2B_7: VTiZrNiCrCoMnAlSn alloy

Out of these three alloys, the AB_5 (LaCePrNdSmMgNiCoMnAlZr) alloy type is more capable of storing or absorbing hydrogen than the AB_2 and A_2B_7 alloy [17]. For improving the charge transfer rate, the electrolyte KOH (potassium hydroxide solution) could be combined with sodium hydroxide or lithium hydroxide solution which gives rapid charge transfer between the electrodes. The cell chemistries behind the nickel metal hydride batteries are given below:

- *Anode reaction*: $MH + OH^- \leftrightarrow M + H_2O + e^-$
- *Cathode reaction*: $NiOOH + H_2O + e^- \leftrightarrow Ni(OH)_2 + OH^-$
- *Overall reaction*: $MH + NiOOH \leftrightarrow M + Ni(OH)_2$

3.3.2 Nickel-Zinc Batteries

Nickel-zinc battery (Figure 3.2) belongs to the class of nickel-based battery and has better characteristics compared to the rest of nickel-based batteries. It is a type of rechargeable battery which involves zinc (metallic zinc) as the anode material. The cathode material (nickel oxyhydroxide) used for the nickel-zinc battery is the same as the cathode used in NiMH battery. By using electrolyte solution (potassium hydroxide), the anodic zinc is oxidized to form Zn^+ ions and electrons. These Zn^+ ions react with the OH^- ions which results in the formation of zinc hydroxide. The anodic zinc is oxidized as zinc hydroxide with the release of electrons. These electrons pass the external circuit where electricity is produced to run the electric vehicle and reaches the cathode (nickel oxyhydroxide). In cathode, these electrons combine with the nickel oxyhydroxide to form nickel hydroxide. Thus, as the end result, nickel hydroxide and zinc hydroxide were produced [18,19]. The overall cell reactions in the nickel-zinc batteries are given below:

- *Anode reaction*: $Zn + 2OH^- \leftrightarrow Zn(OH)_2 + 2e^-$
- *Cathode reaction*: $NiOOH + H_2O + e^- \leftrightarrow Ni(OH)_2 + OH^-$
- *Overall reaction*: $Zn + 2NiOOH + 2H_2O \leftrightarrow Zn(OH)_2 + 2Ni(OH)_2$

Some of the notable features in the nickel-zinc batteries include fast rechargeability, better specific energy, higher cycle life, compactness, cost-efficient electricity production, better nominal voltage compared to the other nickel-based battery types. Like the nickel metal hydride battery, the electrolyte can be mixed with sodium hydroxide and lithium hydroxide solution to improve the charge transfer rate. Both the NiMH and NiZn batteries can be implemented in electric vehicles in different configurations. The most commonly used configurations are button configuration, cylindrical

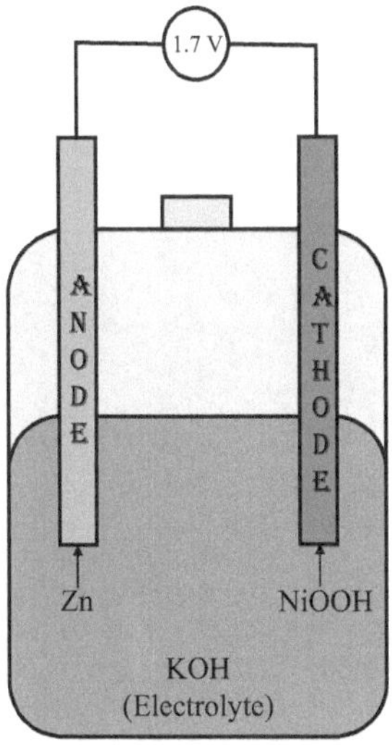

FIGURE 3.2 Schematic illustration of nickel-zinc battery.

configuration, large prismatic configuration, and monoblock configuration. These batteries can be implemented in BEVs and HEVs depending on the capacity [20,21].

3.3.3 Lithium-Sulfur Batteries

Since lithium-ion batteries provide very efficient nominal voltage, their limitations are unavoidable. Hence researchers think about integrating lithium with other materials. One of the best combinations of this kind is lithium-sulfur batteries. These batteries are rechargeable and considered as the future batteries. Here, the metallic lithium or lithium-based alloy is preferred as the anode material and the sulfur or sulfur-based alloy is used as the cathode material. The solution made up of lithium salts serves as the electrolyte for this battery. During the electrochemical reaction, the anodic lithium gets oxidized into Li^+ ions with the release of electrons. The electrons produced as a result of oxidation flow through the circuit and reach the other end (cathode). In cathode, the sulfur gets reduced to sulfide ions (S^{2-}) and combines with the lithium ions (Li^+) to form lithium sulfide [22,23]. The overall cell chemistries behind the electricity generation in the lithium-sulfur batteries (Figure 3.3) are given below:

- *Anode reaction*: $2Li \leftrightarrow 2Li^+ + 2e^-$
- *Cathode reaction*: $S + 2e^- \leftrightarrow S^{2-}$
- *Overall reaction*: $2Li + S \leftrightarrow Li_2S$

The overall output is the formation of lithium sulfide, this sulfide is formed in the order as given below.

$$Li_2S_8 \rightarrow Li_2S_6 \rightarrow Li_2S_4 \rightarrow Li_2S_2 \rightarrow Li_2S$$

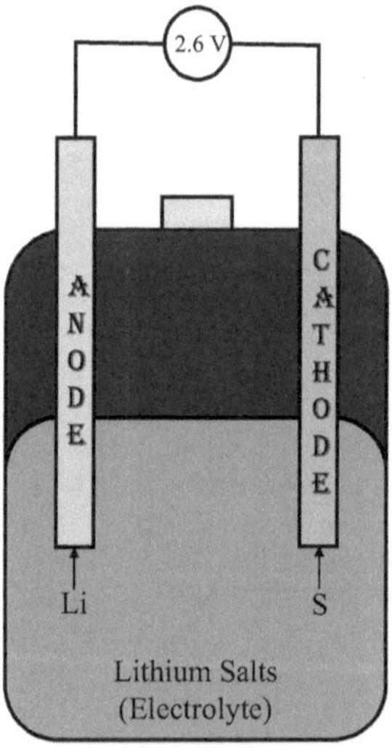

FIGURE 3.3 Schematic illustration of lithium-sulfur battery.

Since lithium alloys and sulfur alloys are used as electrode materials, a variety of materials hold good to get implemented as anode and cathode. Some of the commonly used anode materials are metallic lithium and lithium tin carbon alloys whereas sulfur, Li_2S-C composites, sulfur in a PAN network, sulfur compounds obtained from petroleum processing, and combination of carbon, PVDF (polyvinylidene fluoride), NMP (N-methyl pyrrolidinone) are preferred as the cathode materials. Depending on the anode-cathode combinations, different electrolytes such as composite gel polymer membrane, solid lithium-conducting polysulfidophosphates, polyethers in dioxolane, ion liquids, and polysulphide electrolytes are adopted in these batteries. The major advantages of this kind are (i) eco-friendly compared to the Li-ion batteries, (ii) greater energy storage, (iii) well suits for low-temperature operation, (iv) enhanced gravimetric energy density, and (v) longer sustainability [24,25].

3.3.4 Lithium-Air Batteries

These batteries are termed as future batteries for electric vehicles. They are well known for their storage capacity (same as petrol engine vehicles) and produce a nominal voltage which is nearer to the lithium-ion batteries. Lithium-air battery (Figure 3.4) uses metallic lithium as the oxidizing electrode and atmospheric oxygen is preferred as the reducing electrode material. In anode, the metallic lithium is reported to split into Li^+ ions and e^- ions. The e^- ions reach the reducing electrode (air cathode) through a circuit, whereas the anodic lithium ions pass through the electrolyte in order to reach the reducing electrode. In the reducing electrode, O_2 is reduced to O^{2-} ions and combines with the lithium ions to form lithium peroxide (Li_2O_2) [26,27]. The cell chemistries behind the lithium-air batteries are discussed below:

- *Anode reaction*: $2Li \leftrightarrow 2Li^+ + 2e^-$
- *Cathode reaction*: $O_2 + 2Li^+ + 2e^- \leftrightarrow Li_2O_2$
- *Overall reaction*: $2Li + O_2 \leftrightarrow Li_2O_2$

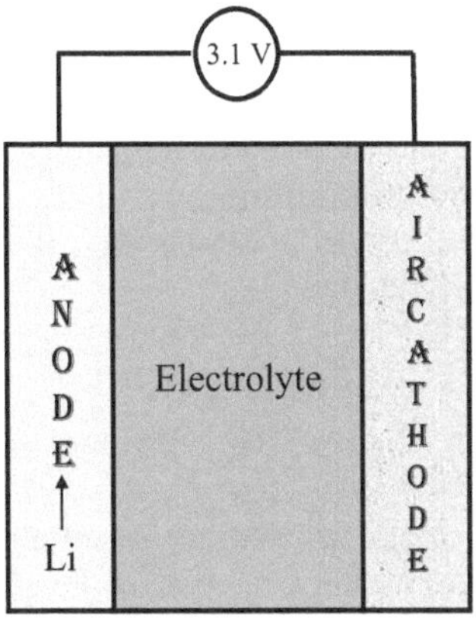

FIGURE 3.4 Schematic illustration of lithium-air battery.

Since these are air batteries, a range of electrolytes can be used as given below [28,29]:

i. *Aqueous electrolytes*: The electrolytic solution is prepared by dissolving any lithium salt in H_2O. The aqueous electrolytes are more beneficial since it restricts the electrode clogging in the lithium-air battery because the by-products produced through aqueous electrolytes are water-soluble. Also, this electrolyte is reported to have a better discharge rate.
ii. *Acidic electrolytes*: These electrolytes contain acid solution to transfer the lithium ions from anode to cathode.
iii. *Alkaline aqueous solutions*: As the name suggests, alkaline aqueous electrolyte contains a mixture of water and some bases.
iv. *Aprotic electrolytes*: These electrolytes are termed as nonaqueous electrolytes. In the presence of this electrolyte, oxide anions react with more lithium ions to form two oxides namely lithium oxide and lithium peroxide. The mixture containing carbonates of ethylene and propylene along with lithium hexafluorophosphate is used as the aprotic electrolyte.
v. *Aqueous-aprotic electrolytes*: These electrolytes are a mixed combination of both aqueous and aprotic electrolytes. Here, each electrode uses a distinct electrolyte. For example, the aqueous electrolytes serve for the oxidizing electrode, whereas the aprotic electrolyte is used in the reducing electrode or vice-versa.
vi. *Solid-state electrolytes*: This type of electrolyte is preferred considering the safety of the battery pack. Polymer ceramic composites and glass ceramic composites are used as the solid-state electrolytes for the lithium-air battery.

3.3.5 Sodium-Ion Batteries

This kind (Figure 3.5) is similar to the lithium-ion batteries. These batteries produce electricity due to the flow of Na^+ ions between the two electrodes. These batteries are considered as fast rowing batteries, and they may be an alternative to lithium-ion batteries in the near future. The production cost is very cheap compared to the Li^+ ion batteries, and it has higher energy density. Metallic sodium or sodium alloys can be used as the anode materials and the sodium-based alloys are used as the cathode materials. The metallic sodium is reported to get oxidized in the presence of electrolytes into Na^+ ions and e^- ions. The transmission of Na^+ ions from anode to cathode produces electricity required to drive the vehicle [30,31]. The cell chemistries behind the sodium-ion battery are given below:

- *Anode reaction*: $\mathrm{Na(C)} \leftrightarrow \mathrm{C} + \mathrm{Na}^+ + \mathrm{e}^-$
- *Cathode reaction*: $\mathrm{Host} + \mathrm{Na}^+ + \mathrm{e}^- \leftrightarrow \mathrm{Na(Host)}$

Since, both the cathode and anode use sodium-based alloy as the electrode materials, a wide range of alloys can be implemented as the electrode materials. Some of the most commonly used electrode materials are metallic sodium, carbon materials (graphite), tin-doped hard carbon, metal phosphates (NASICON-type $NaTi_2(PO_4)_3$),

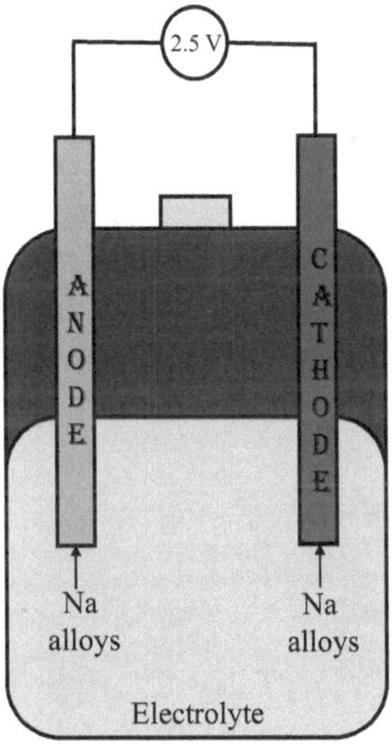

FIGURE 3.5 Schematic illustration of sodium-ion battery.

metal oxides (amorphous TiO_2, $Na_2Ti_3O_7$), and other alloys ($Na_{15}Sn_4$ and Na_4Si_4 alloy). On the other hand, the various cathode materials are manganese oxides, sodium layered oxide phases, sodium-ion conductors (NASICON), olivines ($FEPO_4^-$ amorphous structure), sodium vanadium fluorophosphate, layered sodium iron fluorophosphate, tavorite sodium iron fluorosulfate, and so on. Based on the electrode material, various types of electrolytes (aqueous and nonaqueous electrolytes) and electrolyte additives are used in sodium-ion battery. The most commonly used electrolytes: $C_3H_4O_3$, $C_4H_6O_3$, $C_3H_6O_3$, $C_5H_{10}O_3$, and $C_4H_8O_3$, respectively. Moreover, the sodium-ion batteries have the following advantages, (i) abundant electrode materials, (ii) cost-effective, (iii) good safety behaviors, and (iv) better charge and discharge rate [32,33].

3.3.6 Sodium Air Batteries

Sodium air batteries (Figure 3.6) work similarly to the lithium-air batteries. These batteries are emission-less batteries since one of the electrodes is air-based electrode (oxygen as electrode). Sodium air batteries are well known for their energy storage and energy density. This battery employs metallic sodium as the oxidizing electrode whereas atmospheric air is reported as the reducing electrode. Both aqueous and nonaqueous materials are preferred as electrolytes depending upon the requirements. The anodic (metallic) sodium is reported to get oxidized to Na^+ and e^- ions. In the presence of electrolyte, the Na^+ ions reach the air cathode. The anodic Na^+ combines with the cathodic oxygen (O^{2-}) ions to form sodium superoxide as the product [34,35]. The cell chemistries based on the aqueous and nonaqueous electrolytic medium is explained as below:

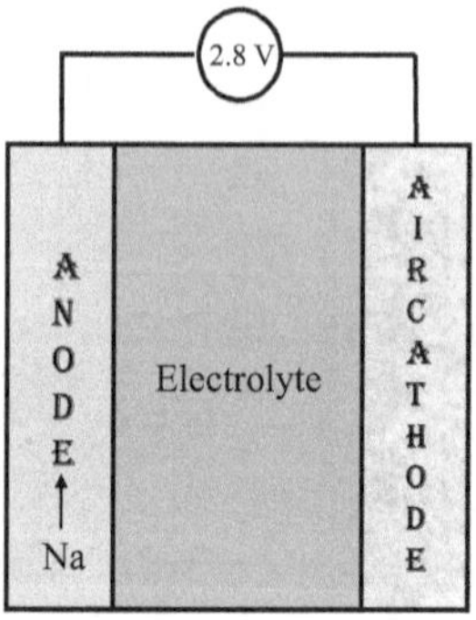

FIGURE 3.6 Schematic illustration of sodium air battery.

i. Aqueous electrolytic medium:
   - *Anode reaction*: $Na \leftrightarrow Na^{+} + e^{-}$
   - *Cathode reaction*: $O_2 + 2H_2O + 4e^{-} \leftrightarrow 4OH^{-}$
   - *Overall reaction*: $4Na + O_2 + 2H_2O \leftrightarrow 4NaOH$

ii. Nonaqueous electrolytic medium:
   - *Anode reaction*: $Na \leftrightarrow Na^{+} + e^{-}$
   - *Cathode reaction*: $O_2 + e^{-} \leftrightarrow O_2^{-}$
   - *Another form of cathode reaction*: $O_2 + 2e^{-} \leftrightarrow O_2^{2-}$
   - *Overall reaction*: $Na + O_2 \leftrightarrow NaO_2$
   - *Another form of overall reaction*: $2Na + O_2 \leftrightarrow Na_2O_2$

The aqueous electrolytes can be prepared from any sodium salts dissolved in water. For aqueous electrolytes, sodium hydroxide is formed whereas in the case of nonaqueous electrolytes, the resulting product is the formation of sodium peroxide (superoxide). Some of the most commonly used aqueous electrolytes are solutions of sodium tetrafluoroborate, sodium hexafluorophosphate, sodium bis(fluorosulfonyl)imide, and sodium bis(trifluoromethylsulfonyl)imide. These electrolytes can be used along with additives in order to enhance the charge transfer rate (flow of ions). Separators are used in an aqueous electrolytic medium to avoid the mixing of electrolytes. These separators not only restrict the flow of electrolytes into the cells but also act as a path of ions movement. NASICON and ceramic oxide plates (sodium-ion conductor) are used as the separators for the sodium-ion batteries [36,37].

3.3.7 Magnesium Batteries

Magnesium battery (Figure 3.7) belongs to the new age of batteries which have two definite forms (primary and secondary battery). The former form are non-rechargeable which are implemented in small-scale applications like watches, clocks, etc., whereas the latter form are rechargeable and are applicable in electric vehicles. The magnesium battery uses metallic magnesium as the oxidizing electrode and molybdenum sulfide as the reducing electrode. The electrolytic solution is prepared by mixing a certain amount of magnesium organochloro aluminate with polyethers.

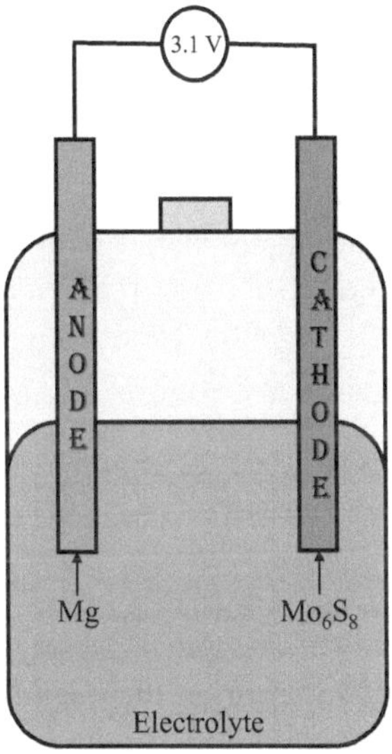

FIGURE 3.7 Schematic illustration of magnesium battery.

In the presence of this electrolytic solution, the anodic magnesium is oxidized to form magnesium ions (Mg^{2+}) and electrons. These electrons pass through the outer terminal where the electricity required for the acceleration of an electric vehicle is produced. The magnesium ion reaches the cathode and combines with the electrons to form $Mg_xMo_6S_8$ where x can be 1 or 2 [38,39]. The cell chemistries behind the magnesium battery are given as follows:

- *Anode reaction*: $Mg \leftrightarrow Mg^{2+} + 2e^-$
- *Cathode reaction*: $Mg^{2+} + 2e^- + Mo_6S_8 \leftrightarrow Mg_xMo_6S_8$ ($x = 1$ or 2)

The double-headed arrow denotes that these batteries are rechargeable. Depending on the requirement, the anode, cathode, and electrolyte materials may vary.

i. *Anode materials*: Metallic magnesium, nano magnesium, metallic bismuth, aluminum zinc magnesium alloys, silicone magnesium alloys, tin magnesium alloys, and bismuth magnesium alloys are mostly used as the anode materials.
ii. *Cathode materials*: Chevrel phase, mixed chevrel phase, vanadium oxide, titanium diboride, molybdenum diboride, titanium disulfide, ruthenium dioxide, zirconium sulfide, manganese oxide, tungsten trioxide, cobalt tetraoxide, lead oxide, lead dioxide, and other metal sulfides/borides/oxides were used as the cathode materials.
iii. *Electrolyte materials*: Polar aprotic electrolytes (amides, esters, and acetonitrile), nonaqueous electrolytes (propylene carbonate, dimethylformamide, sodium perchlorate with CH_3NO and C_2H_3N), grignard reagents

(magnesium tetrabutyl borate, magnesium organochloro aluminates, lewis acids), polymer electrolytes with magnesium ions conductivity, and gel electrolytes could be suitable for magnesium batteries.

Moreover, recently magnesium air batteries are used in electric vehicles. This battery uses metallic magnesium as anode, atmospheric air (oxygen) as cathode and seawater as the electrolyte [40,41]. The cell chemistries of the magnesium air batteries are as follows:

- *Anode reaction*: $Mg \leftrightarrow Mg^{2+} + 2e^-$
- *Cathode reaction*: $O_2 + 2H_2O + 4e^- \leftrightarrow 4OH^-$
- *Overall reaction*: $Mg + 1/2\ O_2 + H_2O \leftrightarrow Mg(OH)_2$

3.3.8 Fluoride Batteries

In general, fluoride ions are well known for their aggressiveness which means that they are highly reactive in any conditions. Implementing these fluoride ions as electrode material will have rapid charge transfer. Hence, fluoride battery (Figure 3.8) belongs to the class of new age batteries. The anode is responsible for the formation of metal fluoride whereas the fluoride ions are released by the cathode. By combining Li, Ca, and Ln with metal fluorides, these batteries can deliver a nominal cell voltage of up to 3 V [42,43]. The cell chemistries behind the fluoride battery are represented below:

- *Anode reaction*: $M + xF^- \leftrightarrow MF_x + xe^-$
- *Cathode reaction*: $M'F_x + xe^- \leftrightarrow M' + xF^-$
- *Cell reaction*: $M'F_x + M \leftrightarrow M' + MF_x$

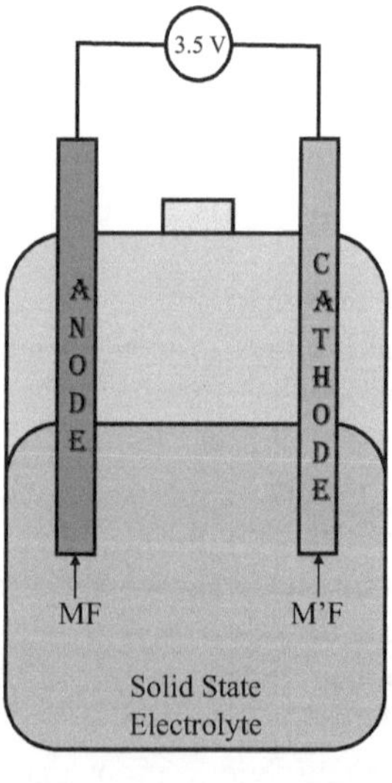

FIGURE 3.8 Schematic illustration of fluoride battery.

The various anode, cathode, and electrolyte materials are described below [44,45]:

i. *Anode materials*: Lithium, calcium, lanthanum, and cerium in their metallic forms are preferred as the anode materials.
ii. *Cathode materials*: Manganese (II) fluoride, cobalt (III) fluoride, cupric fluoride, bismuth fluoride, potassium tetrafluoroborate, and tin (II) fluoride are used as cathode materials in fluoride battery.
iii. *Electrolytes*: Solid-state electrolytes such as calcium fluoride, strontium fluoride, barium fluoride, lanthanum fluoride, and cerium fluoride have been used as electrolytic materials.

3.3.9 Zinc-Air Batteries

One of the future batteries which could be implemented in near future electric vehicles are the zinc-air batteries. It operates on air thereby reducing the emissions and also has high energy density (~9,780 Wh/L). This battery uses metallic zinc (Zn) as oxidizing electrode and the atmospheric oxygen as reducing electrode. In the presence of hydroxide ions as electrolytes, the anodic zinc undergoes oxidation where the Zn^+ ions are formed which are accompanied with the release of electrons. Since the cathode is known for reduction, the oxygen is reduced and forms hydroxide ions [46,47]. The cell chemistries of the zinc-air battery (Figure 3.9) are shown as follows:

- *Anode reaction*: $Zn + 4OH^- \leftrightarrow Zn(OH)_4^{2-} + 2e^-$
- *Fluid reaction*: $Zn(OH)_4^{2-} \leftrightarrow ZnO + H_2O + 2OH^-$
- *Cathode reaction*: $1/2\ O_2 + H_2O + 2e^- \leftrightarrow 2OH^-$
- *Overall reaction*: $2Zn + O_2 \rightarrow 2ZnO$

The zinc electrode can be modified to avoid certain issues by combining with alloys, providing surface coatings, adding additives, and so on. The performance of the zinc electrode could be enhanced by alloying it with elements such as lead, cadmium, tin, bismuth, indium, magnesium, aluminum, nickel, etc. These alloys will minimize the dendrite formation and enhance cycle reversibility. Moreover, by providing

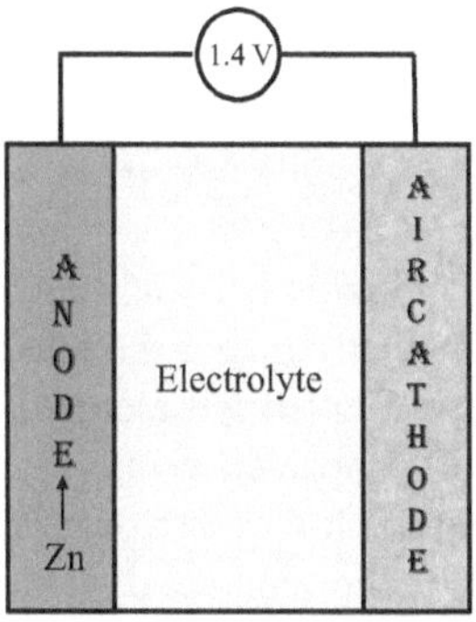

FIGURE 3.9 Schematic illustration of zinc-air battery.

aluminum oxide or lithium boron oxide as a surface coating for zinc electrodes, the hydrogen generation and the self-discharge rate could be minimized. By adding additives to the zinc electrode, the hydrogen evolution is further reduced followed by an increase in the performance and minimizes the corrosion. These additives can be inorganic (oxides of bismuth, lead, cadmium, gallium, indium, thallium and mercury, hydroxides of calcium and indium, and silicates), or polymer (polypyrrole, polycarbonate, ionomers, and polyaniline), or surfactant additives (CTAB, perfluorosurfactants, lignosulfonate and so on). Aqueous solutions such as potassium hydroxide and sodium hydroxide solutions are mostly preferred electrolytes [48,49]. Also, the separators used for the zinc-air batteries should have the following properties:

i. Low ionic resistance
ii. Greater electrical resistance
iii. High adsorption capacity
iv. Better corrosion resistance
v. Strong structural resistance to withstand the dendrite formation

3.3.10 Zinc-Bromine Flow Batteries

Researchers consider this battery type could serve as the effective replacement for Li^+ ion battery considering the energy storage capacity. The major features of the zinc-bromine flow battery include higher storage capacity, deep discharge without any damage to the battery pack, faster charging, no overheating, applicable for a wide range of service temperatures, discharge is not affected after deep cycles, long cycle life, low fire risk, no need for cooling systems, easy recycling, low cost, abundant and readily available raw materials. These advantages paved the way for these batteries to get implemented in the future electric vehicles [50,51].

In this battery, the metallic zinc is used as the anode material whereas the bromine is used as the cathode material. The anodic zinc is oxidized to Zn^+ ions and electrons whereas the cathodic bromine is reduced to bromide. The overall cell chemistries behind the zinc-bromine flow batteries are given below:

- *Anode reaction*: $Zn_{(s)} \leftrightarrow Zn^{2+}_{(aq)} + 2e^-$
- *Cathode reaction*: $Br_{2(aq)} + 2e^- \leftrightarrow 2Br^-$
- *Overall reaction*: $Zn_{(s)} + Br_{2(aq)} \leftrightarrow 2Br^-_{(aq)} + Zn^{2+}$

The electrode material can be of metallic electrode or any carbon-based electrode having one of these forms namely antisymmetric or asymmetric structure. Zinc bromide solution could be used the electrolyte. By adding chlorine-rich salts to this solution could increase the ionic conductivity and the ions are able to transfer in a rapid manner. The microporous and ion exchange membrane can be used as separators since they have better adsorption capacity, minimal ionic resistance, higher resistance toward electrons, greater oxidation (corrosion) resistance, and should sustain in the zinc dendrite formation. Since zinc dendrite growth is unavoidable in zinc-bromine flow batteries, the separators and the electrode materials should have sustainable resistance against the zinc dendrite formation. Moreover, the zinc dendrite

TABLE 3.1
Nominal Cell Voltage of Different Batteries

Battery Type	Nominal Cell Voltage (V)
NiMH	1.35
NiZn	1.76
Sodium air	2.75
Lithium sulfur	2.56
Lithium air	3.1
Sodium ion	2.5
Magnesium	3.09
Magnesium air	3.1
Lithium fluoridc	3.46
Zinc air	1.4
Zinc bromine	1.8

formation could be controlled by adding surfactants, inorganic and polymer as additives [52,53]. The nominal cell voltage for various battery types is listed in Table 3.1.

3.4 FUEL CELL ELECTRIC VEHICLES

A fuel cell is an electrochemical cell that transfers electron/charges in a fast manner. This technology is often referred as the future because of its emission-free transportation [54]. The fuel cell has the anode and cathode chamber instead of electrodes. The raw materials are hydrogen and oxygen which serves as the intake for anode and cathode chamber. In the anode chamber, the hydrogen is oxidized to hydrogen ions and electrons. The H^+ ions react with the O^{2-} ions to form H_2O as by-product. Hence, the fuel cell is going to rule the entire transportation industry in near future [55]. The cell chemistries behind the fuel cell are given below:

- *Anode reaction*: $H_2 \rightarrow 2H^+ + 2e^-$
- *Cathode reaction*: $1/2O_2 + 2H^+ \rightarrow H_2O$

As per the existing demand, different kinds of fuel cells can be implemented in electric vehicles. The major types which could be used to produce the fuel cell stacks are SOFC, PEMFC, AFC, DMFC, and so on [56]. Recently these fuel cell electric vehicles (FCEVs) can achieve high top speed and they can supply the desired amount of energy to travel long distance. This can be done by having some modifications and these modifications paved the way for new type of HEVs which are termed as fuel cell HEVs [57].

3.5 FUEL CELL HEVs

In order to achieve long distance transportation and to deliver the required energy, the fuel cell electric vehicle powertrain had undergone certain modifications that arise

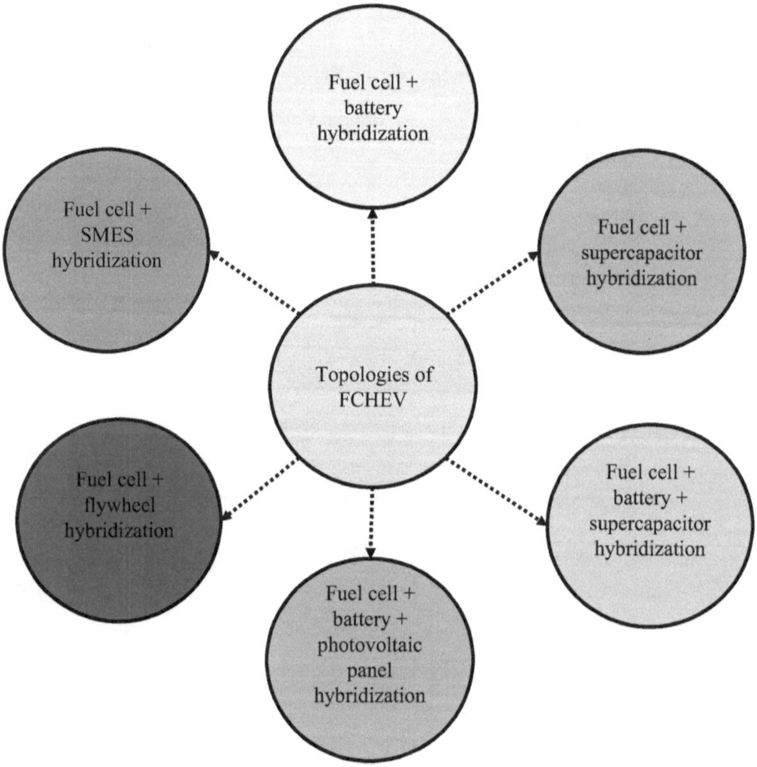

FIGURE 3.10 Topologies of fuel cell hybrid electric vehicle [71].

to a new class of electric vehicles called as fuel cell HEVs. This type has another energy storage system which acts as a support system in order to meet the demands such high top speed, efficient performance, and so on [58]. Based on the energy storage system, this electric vehicle is topologized into six types as shown in Figure 3.10. These six topologies are briefly explained as follows:

3.5.1 Fuel Cell + Battery Hybridization

This topology is often referred as the most common type which a battery pack is installed in the FCEV. The fuel cell and the battery are connected to the DC bus through unidirectional and bidirectional DC-DC converter [59]. To keep the fuel cell from running in the low-efficiency zone, the fuel cell and battery hybridization procedure include an start-up device attached to the battery. As a result, the motor is actuated and the fuel cell pack is reported to be stimulated in order to keep the motor in the working (running) stage. The battery is now being charged in accordance with the charge status criteria. The battery pack can be made from any one of the advanced batteries as discussed before [60].

3.5.2 Fuel Cell + Supercapacitor (Ultracapacitor) Hybridization

Ultracapacitors are a kind of capacitors that are derived from the conventional capacitor. The conventional capacitor has two conducting electrodes separated by a dielectric substance whereas the ultracapacitor (supercapacitor) has electrodes separated by a separator membrane and electrolyte. The use of electrolytes is intended to make it easier to store ions that carry electrostatic charges. Compared to traditional battery or capacitor electrodes, ultracapacitor electrodes have a significantly greater surface area. The electrodes in the ultracapacitors provide a substantially greater energy density when compared with the traditional capacitors. Also, ions move more slowly in ultracapacitors than electrons, they charge and discharge more slowly than the conventional capacitors. As a result, the capacity of the supercapacitor is improved in this fashion [61]. Ultracapacitors were initially employed in military applications for the activation of the motors of battle tanks and submarines. The cost of ultracapacitors has greatly dropped and capacitance has been grown due to the nano-scale technology. As a result, ultracapacitor applications are now more widespread. Currently, it is utilized for memory backup, pitch control, and the starting of diesel engines, locomotives, and wind turbines. Here, the supercapacitor is combined with the fuel cell in order to fulfill the required demand in emergency situations [62].

3.5.3 Fuel Cell + Battery + Supercapacitor Hybridization

This type of topology makes use of both battery and supercapacitor in order to experience the combined advantages of the aforementioned topologies. Fuel cell stack serves as the primary source whereas the battery and supercapacitors are termed as the secondary source. In this type, the unidirectional DC converter is used to link the fuel cell with the DC whereas the secondary sources are coupled with DC bus by using a bidirectional DC converter [63]. This combination has more advantages than the previous topologies where this type is able to deliver continuous energy followed by enhancing the dynamic performance of the vehicle in transient conditions [64].

3.5.4 Fuel Cell + Battery + Photovoltaic Hybridization

Photovoltaic panels convert light energy into electrical energy which can be combined with the FCEVs. Fuel cell stack serve as the primary source whereas the battery and photovoltaic panels serve as the secondary source [65]. In this architecture, unidirectional DC converter links the fuel cell and the photovoltaic cells to DC bus whereas the bidirectional DC converter is employed in coupling the battery to the DC bus. The output of the photovoltaic panel varies depending on the temperature, intensity and direction of sunlight. Hence the generated power from the photovoltaic panels is used to energize the motor or battery. Also by implementing supercapacitor, the power fluctuations can be minimized because of the high power density of the supercapacitor [66].

3.5.5 Fuel Cell + Flywheel Hybridization

Flywheel is reported as an energy storage device that has two modes of operation namely energy storage and release. The rotational energy is stored due to the influence of applied torque. The flywheel discharges the stored energy in the form of torque when it is coupled to a mechanical device. Here, the fuel cell is combined with the flywheel in order to provide the desired output [67]. This topology does not need any batteries or supercapacitors since the flywheel has excellent power ratings, high efficiency, fast charging capacity, and so on. Moreover, flywheels are eco-friendly, adapted to a broad range of service temperatures, extended lifetime, and greater energy storage capacity [68].

3.5.6 Fuel Cell + Superconducting Magnetic Energy Storage (SMES) Hybridization

This hybridization is reported to be implemented in near future electric vehicles. The SMES stores the energy by means of electromagnetic theory. SMES has greater charge and discharge cycles with better power conversion ratio (almost 95%) and is quite possibly implemented in future fuel cell HEVs [69–71].

3.6 SUPERCAPACITORS IN ELECTRIC VEHICLES

In recent years, supercapacitor-based energy systems are developed that could be an alternative energy storage systems for future electric vehicles. Supercapacitors are electrochemical storage devices well known for their higher storage capacity than the traditional capacitors. It also delivers greater power outputs compared to the batteries [72,73]. Some of the electrode materials are carbonaceous materials, metal-organic framework, bimetallic metal-organic framework, transition metal nitrides, redox polymers, and so on. The pros of supercapacitor-based energy storage systems are

i. Greater power density
ii. Better cycle life
iii. Rapid charging
iv. Greater current discharge
v. Holds good for low service temperature
vi. Compactness in terms of charging and discharging circuit

These advantages made the supercapacitors to get implemented in the future EVs as energy storage unit. Also, more research is going on on the battery-supercapacitor combination for the effective management of electric vehicles.

3.7 BIOFUELS BASED PROPULSION SYSTEMS FOR HEVs

Biofuels are fuels that are derived from the biological origin. These fuels are renewable and could be implemented for minimizing the pollution in internal combustion vehicles. Since hybrid, plug-in hybrid, and extended range electric vehicles use IC engines as one of the energy sources, these biofuels can be adapted for that IC engines to reduce the pollution effects caused by those vehicles.

3.8 CONCLUSION

Electric vehicles are unavoidable mode of transportation in the near future due to their numerous advantages. But the demands such as high top speeds and far distance transportation makes the researchers to work on the alternate propulsion system that could be implemented in future electric vehicles. HEV serves as the optimal solution in order to meet the above-mentioned demands. By incorporating advanced battery packs made from different materials, higher top speed with good energy density can be obtained. On the other hand, by combining fuel cell with battery, supercapacitors, photovoltaic cells, flywheel and SMES technology, far distance transportation along with high top speeds can be met. This chapter elaborates the HEV technology along with its different architecture which could work efficiently for nullifying the demands. The advanced battery technologies along with the cell chemistries were detailed. Moreover, the fuel cell-based various hybrid systems were presented which could be adopted as the alternative propulsion systems in future electric vehicles.

ACKNOWLEDGMENT

The authors like to acknowledge the editors of this book for providing an opportunity to write this section.

REFERENCES

[1] Sun X, Li Z, Wang X, Li C. Technology development of electric vehicles: A review. *Energies*. 2019;13(1):90.

[2] Sanguesa JA, Torres-Sanz V, Garrido P, Martinez FJ, Marquez-Barja JM. A review on electric vehicles: Technologies and challenges. *Smart Cities*. 2021;4(1):372–404.

[3] Patel N, Bhoi AK, Padmanaban S, Holm-Nielsen JB, editors. *Electric Vehicles: Modern Technologies and Trends*. Springer; 2021.

[4] Chan CC, Chau KT. *Modern Electric Vehicle Technology*. Oxford University Press on Demand; 2001.

[5] Mi C, Masrur MA. *Hybrid Electric Vehicles: Principles and Applications with Practical Perspectives*. John Wiley & Sons; 2017.

[6] König A, Nicoletti L, Schröder D, Wolff S, Waclaw A, Lienkamp M. An overview of parameter and cost for battery electric vehicles. *World Electric Vehicle Journal*. 2021;12(1):21.

[7] Khaligh A, Li Z. Battery, ultracapacitor, fuel cell, and hybrid energy storage systems for electric, hybrid electric, fuel cell, and plug-in hybrid electric vehicles: State of the art. *IEEE Transactions on Vehicular Technology*. 2010;59(6):2806–2814

[8] Hannan MA, Azidin FA, Mohamed A. Hybrid electric vehicles and their challenges: A review. *Renewable and Sustainable Energy Reviews*. 2014;29:135–150.

[9] Singh KV, Bansal HO, Singh D. A comprehensive review on hybrid electric vehicles: Architectures and components. *Journal of Modern Transportation*. 2019;27:77–107.

[10] Ehsani M, Gao Y, Miller JM. Hybrid electric vehicles: Architecture and motor drives. *Proceedings of the IEEE*. 2007;95(4):719–728.

[11] Lanzarotto D, Marchesoni M, Passalacqua M, Prato AP, Repetto M. Overview of different hybrid vehicle architectures. *IFAC-PapersOnLine*. 2018;51(9):218–222.

[12] Tran DD, Vafaeipour M, El Baghdadi M, Barrero R, Van Mierlo J, Hegazy O. Thorough state-of-the-art analysis of electric and hybrid vehicle powertrains: Topologies and integrated energy management strategies. *Renewable and Sustainable Energy Reviews.* 2020;119:109596.

[13] Chan CC. The state of the art of electric, hybrid, and fuel cell vehicles. *Proceedings of the IEEE.* 2007;95(4):704–718.

[14] Dhar SK, Ovshinsky SR, Gifford PR, Corrigan DA, Fetcenko MA, Venkatesan S. Nickel/metal hydride technology for consumer and electric vehicle batteries-a review and up-date. *Journal of Power Sources.* 1997;65(1–2):1–7.

[15] Ouyang L, Huang J, Wang H, Liu J, Zhu M. Progress of hydrogen storage alloys for Ni-MH rechargeable power batteries in electric vehicles: A review. *Materials Chemistry and Physics.* 2017;200:164–178.

[16] Sakai T, Uehara I, Ishikawa H. R&D on metal hydride materials and Ni-MH batteries in Japan. *Journal of Alloys and Compounds.* 1999;293:762–769.

[17] Fetcenko M, Koch J, Zelinsky M. Nickel-metal hydride and nickel-zinc batteries for hybrid electric vehicles and battery electric vehicles. In: Scrosati B, Garche J, Tillmetz W, editors, *Advances in Battery Technologies for Electric Vehicles.* Woodhead Publishing; 2015, pp. 103–126.

[18] Zhang H, Wang R, Lin D, Zeng Y, Lu X. Ni-based nanostructures as high-performance cathodes for rechargeable Ni–Zn battery. *ChemNanoMat.* 2018;4(6):525–536.

[19] Lai SB, Jamesh MI, Wu XC, Dong YL, Wang JH, Gao M, Liu JF, Sun XM. A promising energy storage system: Rechargeable Ni-Zn battery. *Rare Metals.* 2017;36(5):381–396.

[20] McBreen J. Nickel/zinc batteries. *Journal of Power Sources.* 1994;51(1–2):37–44.

[21] Scrosati B, Garche J, Tillmetz W, editors. *Advances in Battery Technologies for Electric Vehicles.* Woodhead Publishing; 2015.

[22] Fotouhi A, Auger DJ, Propp K, Longo S, Wild M. A review on electric vehicle battery modelling: From Lithium-ion toward Lithium-Sulphur. *Renewable and Sustainable Energy Reviews.* 2016;56:1008–1021.

[23] Manthiram A, Chung SH, Zu C. Lithium-sulfur batteries: Progress and prospects. *Advanced Materials.* 2015;27(12):1980–2006.

[24] Seh ZW, Sun Y, Zhang Q, Cui Y. Designing high-energy lithium-sulfur batteries. *Chemical Society Reviews.* 2016;45(20):5605–5634.

[25] Xu G, Ding B, Pan J, Nie P, Shen L, Zhang X. High performance lithium-sulfur batteries: Advances and challenges. *Journal of Materials Chemistry A.* 2014;2(32):12662–12676.

[26] Rahman M, Wang X, Wen C. A review of high energy density lithium-air battery technology. *Journal of Applied Electrochemistry.* 2014;44(1):5–22.

[27] Kraytsberg A, Ein-Eli Y. Review on Li-air batteries-opportunities, limitations and perspective. *Journal of Power Sources.* 2011;196(3):886–893.

[28] Balaish M, Kraytsberg A, Ein-Eli Y. A critical review on lithium-air battery electrolytes. *Physical Chemistry Chemical Physics.* 2014;16(7):2801–2822.

[29] Tan P, Jiang HR, Zhu XB, An L, Jung CY, Wu MC, Shi L, Shyy W, Zhao TS. Advances and challenges in lithium-air batteries. *Applied Energy.* 2017;204:780–806.

[30] Yabuuchi N, Kubota K, Dahbi M, Komaba S. Research development on sodium-ion batteries. *Chemical Reviews.* 2014;114(23):11636–11682.

[31] Vaalma C, Buchholz D, Weil M, Passerini S. A cost and resource analysis of sodium-ion batteries. *Nature Reviews Materials.* 2018;3(4):1.

[32] Nithya C, Gopukumar S. Sodium ion batteries: A newer electrochemical storage. *Wiley Interdisciplinary Reviews: Energy and Environment.* 2015;4(3):253–278.

[33] Su H, Jaffer S, Yu H. Transition metal oxides for sodium-ion batteries. *Energy Storage Materials.* 2016;5:116–131.

[34] Adelhelm P, Hartmann P, Bender CL, Busche M, Eufinger C, Janek J. From lithium to sodium: Cell chemistry of room temperature sodium-air and sodium-sulfur batteries. *Beilstein Journal of Nanotechnology.* 2015;6(1):1016–1055.
[35] Xu X, San Hui K, Dinh DA, Hui KN, Wang H. Recent advances in hybrid sodium-air batteries. *Materials Horizons.* 2019;6(7):1306–1035.
[36] Yadegari H, Sun Q, Sun X. Sodium-oxygen batteries: A comparative review from chemical and electrochemical fundamentals to future perspective. *Advanced Materials.* 2016;28(33):7065–7093.
[37] Chawla N. Recent advances in air-battery chemistries. *Materials Today Chemistry.* 2019;12:324–331.
[38] Saha P, Datta MK, Velikokhatnyi OI, Manivannan A, Alman D, Kumta PN. Rechargeable magnesium battery: Current status and key challenges for the future. *Progress in Materials Science.* 2014;66:1–86.
[39] Aurbach D, Gofer Y, Lu Z, Schechter A, Chusid O, Gizbar H, Cohen Y, Ashkenazi V, Moshkovich M, Turgeman R, Levi E. A short review on the comparison between Li battery systems and rechargeable magnesium battery technology. *Journal of Power Sources.* 2001 1;97:28–32.
[40] Li CS, Sun Y, Gebert F, Chou SL. Current progress on rechargeable magnesium-air battery. *Advanced Energy Materials.* 2017;7(24):1700869.
[41] Zhang T, Tao Z, Chen J. Magnesium-air batteries: From principle to application. *Materials Horizons.* 2014;1(2):196–206.
[42] Nowroozi MA, Mohammad I, Molaiyan P, Wissel K, Munnangi AR, Clemens O. Fluoride ion batteries-past, present, and future. *Journal of Materials Chemistry A.* 2021;9(10):5980–6012.
[43] Zhang N, Xiao X, Pang H. Transition metal (Fe, Co, Ni) fluoride-based materials for electrochemical energy storage. *Nanoscale Horizons.* 2019;4(1):99–116.
[44] Gopinadh SV, Phanendra PV, John B, Mercy TD. Fluoride-ion batteries: State-of-the-art and future perspectives. *Sustainable Materials and Technologies.* 2022:e00436.
[45] Gschwind F, Rodriguez-Garcia G, Sandbeck DJ, Gross A, Weil M, Fichtner M, Hörmann N. Fluoride ion batteries: Theoretical performance, safety, toxicity, and a combinatorial screening of new electrodes. *Journal of Fluorine Chemistry.* 2016;182:76–90.
[46] Li Y, Dai H. Recent advances in zinc-air batteries. *Chemical Society Reviews.* 2014;43(15):5257–5275.
[47] Caramia V, Bozzini B. Materials science aspects of zinc-air batteries: A review. *Materials for Renewable and Sustainable Energy.* 2014;3(2):1–2.
[48] Lee JS, Tai Kim S, Cao R, Choi NS, Liu M, Lee KT, Cho J. Metal-air batteries with high energy density: Li-air versus Zn-air. *Advanced Energy Materials.* 2011;1(1):34–50.
[49] Chen X, Zhou Z, Karahan HE, Shao Q, Wei L, Chen Y. Recent advances in materials and design of electrochemically rechargeable zinc-air batteries. *Small.* 2018;14(44):1801929.
[50] Lin H, Jiang T, Sun Q, Zhao G, Shi J. Research progress of zinc bromine flow battery. *Journal of New Materials for Electrochemical Systems.* 2018;21(2):63–70.
[51] Rajarathnam GP, Vassallo AM. *The Zinc/Bromine Flow Battery: Materials Challenges and Practical Solutions for Technology Advancement.* Springer; 2016.
[52] Jiang T, Lin H, Sun Q, Zhao G, Shi J. Recent progress of electrode materials for zinc bromide flow battery. *International Journal Electrochemical Science.* 2018;13:5603–5611.
[53] Xu Z, Fan Q, Li Y, Wang J, Lund PD. Review of zinc dendrite formation in zinc bromine redox flow battery. *Renewable and Sustainable Energy Reviews.* 2020;127:109838.
[54] Jayakumar A, Chalmers A, Lie TT. Review of prospects for adoption of fuel cell electric vehicles in New Zealand. *IET Electrical Systems in Transportation.* 2017;7(4):259–266.
[55] Singla MK, Nijhawan P, Oberoi AS. Hydrogen fuel and fuel cell technology for cleaner future: A review. *Environmental Science and Pollution Research.* 2021;28:15607–15626.

[56] Ogawa T, Takeuchi M, Kajikawa Y. Comprehensive analysis of trends and emerging technologies in all types of fuel cells based on a computational method. *Sustainability*. 10(2):458.
[57] Song Z, Pan Y, Chen H, Zhang T. Effects of temperature on the performance of fuel cell hybrid electric vehicles: A review. *Applied Energy*. 2021;302:117572.
[58] Lü X, Wu Y, Lian J, Zhang Y, Chen C, Wang P, Meng L. Energy management of hybrid electric vehicles: A review of energy optimization of fuel cell hybrid power system based on genetic algorithm. *Energy Conversion and Management*. 2020;205:112474.
[59] Zeng T, Zhang C, Hao D, Cao D, Chen J, Chen J, Li J. Data-driven approach for short-term power demand prediction of fuel cell hybrid vehicles. *Energy*. 2020;208:118319.
[60] Xu L, Li J, Hua J, Li X, Ouyang M. Adaptive supervisory control strategy of a fuel cell/battery-powered city bus. *Journal of Power Sources*. 2009;194(1):360–368.
[61] Uzunoglu M, Alam MS. Dynamic modeling, design and simulation of a PEM fuel cell/ultra-capacitor hybrid system for vehicular applications. *Energy Conversion and Management*. 2007;48(5):1544–1553.
[62] Bi H, Wang P, Che Y. A capacitor clamped H-type boost DC-DC converter with wide voltage-gain range for fuel cell vehicles. *IEEE Transactions on Vehicular Technology*. 2018;68(1):276–290.
[63] Li Q, Chen W, Li Y, Liu S, Huang J. Energy management strategy for fuel cell/battery/ultracapacitor hybrid vehicle based on fuzzy logic. *International Journal of Electrical Power & Energy Systems*. 2012;43(1):514–525.
[64] García P, Fernández LM, Torreglosa JP, Jurado F. Operation mode control of a hybrid power system based on fuel cell/battery/ultracapacitor for an electric tramway. *Computers & Electrical Engineering*. 2013;39(7):1993–2004.
[65] Mokrani Z, Rekioua D, Rekioua T. Modeling, control and power management of hybrid photovoltaic fuel cells with battery bank supplying electric vehicle. *International Journal of Hydrogen Energy*. 2014;39(27):15178–15187.
[66] Ezzat MF, Dincer I. Development, analysis and assessment of fuel cell and photovoltaic powered vehicles. *International Journal of Hydrogen Energy*. 2018;43(2):968–978.
[67] Huang CN, Chen YS. Design of magnetic flywheel control for performance improvement of fuel cells used in vehicles. *Energy*. 2017;118:840–852.
[68] Boukettaya G, Krichen L. A dynamic power management strategy of a grid connected hybrid generation system using wind, photovoltaic and Flywheel Energy Storage System in residential applications. *Energy*. 2014;71:148–159.
[69] Yang B, Zhu T, Zhang X, Wang J, Shu H, Li S, He T, Yang L, Yu T. Design and implementation of Battery/SMES hybrid energy storage systems used in electric vehicles: A nonlinear robust fractional-order control approach. *Energy*. 2020;191:116510.
[70] Ren G, Ma G, Cong N. Review of electrical energy storage system for vehicular applications. *Renewable and Sustainable Energy Reviews*. 2015;41:225–236.
[71] İnci M, Büyük M, Demir MH, İlbey G. A review and research on fuel cell electric vehicles: Topologies, power electronic converters, energy management methods, technical challenges, marketing and future aspects. *Renewable and Sustainable Energy Reviews*. 2021;137:110648.
[72] Kouchachvili L, Yaïci W, Entchev E. Hybrid battery/supercapacitor energy storage system for the electric vehicles. *Journal of Power Sources*. 2018;374:237–248.
[73] Xu H, Shen M. The control of lithium-ion batteries and supercapacitors in hybrid energy storage systems for electric vehicles: A review. *International Journal of Energy Research*. 2021;45(15):20524–20544.

4 Recent Advancements and Challenges of Powertrain Technologies in Electric Vehicles Applications

Ahmad Syed
Chaitanya Bharathi Institute of Technology (A), Hyderabad

Tara Kalyani Sandipamu
Jawaharlal Nehru Technological University, Hyderabad

G. Suresh Babu
Chaitanya Bharathi Institute of Technology (A), Hyderabad

Freddy Tan Kheng Suan
University of Nottingham, Malaysia

Xiaoqiang Guo
Yanshan University, China

Huai Wang
Aalborg University, Denmark

4.1 INTRODUCTION

In recent days drastic changes in the present atmosphere conditions due to increased population and industries, which results in dangerous CO_2 emissions and effects the next. In this scenario, low-emission transport is a good choice to designing electric vehicle technology (EVT) because in the future it will become zero emissions if concentrated more on E-MT. EV technology is a major and vital role in the present EV market. In other words, the electric powertrain technology has led the world due to a lot of restrictions on CO_2 emissions with revised government vehicle regulations and declined battery prices. As per the Bloomberg reports in 2018 the utility of oil has

DOI: 10.1201/9781003354901-4

been reduced by 3% since 2011 onwards due to increased electric power trains [1]. In addition, 75% of consumption is mainly from electric buses such as more energy transition is from electrified transport only. In 2025 it is expected that electric power trains will increase significantly such that three times as many buses will be electrified [2].

Nonetheless, recently the uses of gasoline power vehicles are very interesting solutions for low carbon EVT with electric buses and public transport vehicles. And the price list for development of the electric buses is varied and mostly reliant on nation and the city as well. As per the recent survey, China leads the highest possibilities on the development of electric buses such as 99% of the e-buses are adopted in China in 2018. In 2021 [3] six lacks electric buses are running in the Chinese cities, which will be expected that more count is reached i.e approximately 13 lacks 23 thousand four hundred ninety. Moreover, various Asian countries namely Delhi are running approximately 1,000 electric buses from 2019 onwards without fault. Nonetheless, nowadays peoples living in American cities and big universities are interested to run their places with a greater number of e-train buses, which is given by California's Innovative Clean Transit Rule (ICTR) and other cities are also flowed by the same such as in 2023 most of the states are run with the share of 25% will be zero emissions e-buses [4] and in 2050 it will increase further to reach 100% [2]. At present, in Santiago, 30% is electrified and aim to reach 100% by 2050 with 200 e-buses with six million capacity. First electric shuttle project was installed in Ecuador with 20 E-buses and transport daily to 10,500 citizens [5]. By the new regulation framed by the European Union, 25% of e-buses are deployed with cleanliness, which will be increased further to 75% in 2030 [2]. Figure 4.1 represents the stages of e-buses. However, a lot of restrictions and regulations by the government's notable achievements towards e-buses are made to suppress consumption emissions in transport. On the other hand, EV required the support of charging capability and its infrastructure. Nonetheless, the rate and time taken to charging are the main factors to limit the EV application.

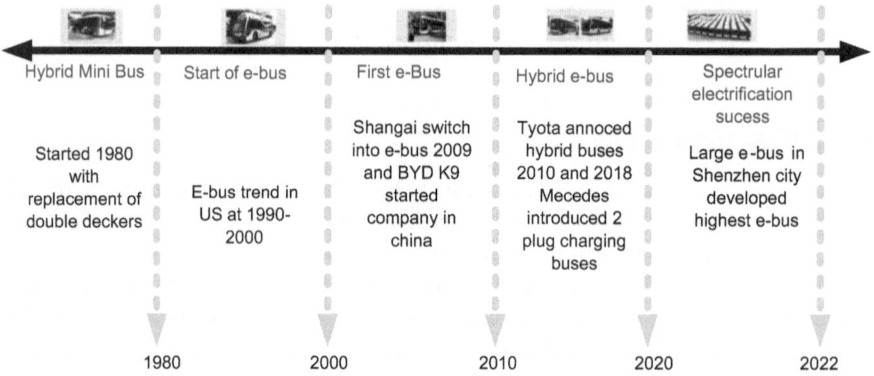

FIGURE 4.1 E-buses evaluation.

However, the issues related with the charging systems such as EV stations in general are adopted with the bulky transformer as compared with fuel systems, which leads to higher cost and longer time than normal stations. It is overcome through the off-board fast charger system (FCS) for better EV applications [1–6]. FCS is adopted with the line frequency transformer (LFT) between the grid and symmetrical filters of the converter, which results in the system being bulky and costly, schematic as shown in Figure 4.1. In addition to that power losses are higher and results efficiency is reduced significantly [3–5]. A typical structure is adopted between the grid to the load through higher frequency transformer (HFT), which reduces the cost and volume of the transformer. However, these types of solutions have no leakage current issues due to proper insulation and the parasitic direction in between ground of the Electric vehicle and the ground through excitation of the higher frequency common mode voltage (HF-CMV). If it is not maintained properly, the leakage current exceeds its limits, which will threaten human safety [4,7]. Figure 4.2 shows the higher frequency transformer charging system (HFT-CS), where the transformer is placed in between the converter and battery bank. It is accomplished by attenuating the leakage current stray paths with the help of an additional decoupling converter, which leads to reduced cost and volume, as compared with the LFT-CS [8,9]. However, the losses are more in HFT-CS and results less in efficiency [10,11]. Similarly, transformer converter configurations have been published on photovoltaic systems [11,12]. In [10] many structures have been realized with an additional switch on the direct current-DC and or alternating current-AC side with an equivalent pulse-width-modulation (PWM) scheme. In [7–12] several single-phase topologies.

Nonetheless, it is classified into two ways namely decoupling and non-decoupling systems using the location of the transformer during the power conversion mode. To attain more efficiency, smaller in shape and lower charge a universal solution is to eliminate the LFT in non-isolated inverters, which results in a direct path between the solar panel to the grid. Therefore, high common mode voltage oscillation is formed

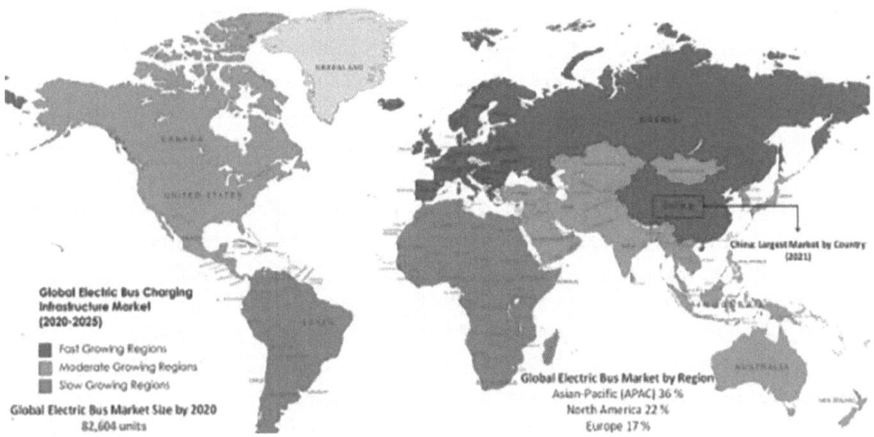

FIGURE 4.2 Global e-bus charging system.

by the converting of the solar inverter, which can stimulate through the resonant circuit and stray capacitance of the solar panels, filter, and impedance of the grid. As a result, common mode current or leakage current flows through parasitic capacitance of the solar panel to the ground. On the other hand, it creates issues related with waveform quality, electromagnetic compatibility (EMC), and electromagnetic compatibility interference (EMI), which leads to poor safety [11]. In addition, the higher frequency CMV fluctuations are generated through the conversion of the inverter and the resonant circuit is made a path in between the PV stray capacitor, filter, and grid resistor respectively. This results in generating higher dangerous leakage current through ground stray capacitor of solar panels. This causes serious affects on the overall system in terms of poor wave quality, safety, and increased electromagnetic compatibility (EMC) and hence it depreciates the overall system efficiency.

To overcome these issues many investigations have been done based on the conventional full-bridge (FB) configuration with the help of unipolar and bipolar modulation systems. But it fails to meet safety grid codes and standards due to higher leakage current and losses. To eliminate these issues various topologies were introduced by incorporating an additional circuitry to disconnect the PV panel and grid in the freewheeling period namely DC, AC decoupling methods. Here H5, HERIC, and H6 inverters are more popular and well-adapted inverters in the present market for transformerless PV applications [10–12].

In H5, the DC bypass is provided through one additional switch, which is added in series with the DC supply [10]. Similarly in AC bypass galvanic isolation is provided through two additional switches, which are added in parallel with the grid [10]. H6 topology is feasible with either DC or Ac bypass based on the locations of the devices (switch/diode) in series and /or parallel to input DC source and the grid [12], where all topologies are recorded more efficiently in practice. However, due to the lack of clamping ability, the leakage current issues are resolved properly because of the floating phase-neutral voltages during the freewheeling period. Nonetheless, higher leakage current is flow through resonant circuits because of the excitation of the PV stray capacitance and occasionally the changes in the weather and results the flow of leakage current exceeds the predefined benchmark levels because of recurrent behaviour of stray capacitance. However, it is still in the developing stage.

In this chapter in-depth analysis of transformerless inverters in e-bus charging systems is not covered, which is beyond the scope of this chapter. By the discussion in the above sections, it is revealed that e-buses are rising slowly due to increased cost in fossil fuels and CO_2 emissions. On the other hand, due to rising e-bus demand, it needs to charge the battery system through proper mechanisms as more charging stations are required to charge in a specific time. Figure 4.2 shows the e-bus charging infrastructure market worldwide. As per Figure 4.2. China is dominating Europe, North America in the e-bus market. One of the solutions is to charge the batteries through the medium voltage networks (MVN) such as number of vehicles charging and very dissolute charging using only one time constant and the same is not applicable for the lower charging systems, which accommodate the widespread use of EV technology. In addition, to realise these technologies it needs the support of power electronics systems such as converters are required to achieve effective charging. A conventional charger system is shown in Figure 4.3, which consists of two-stage

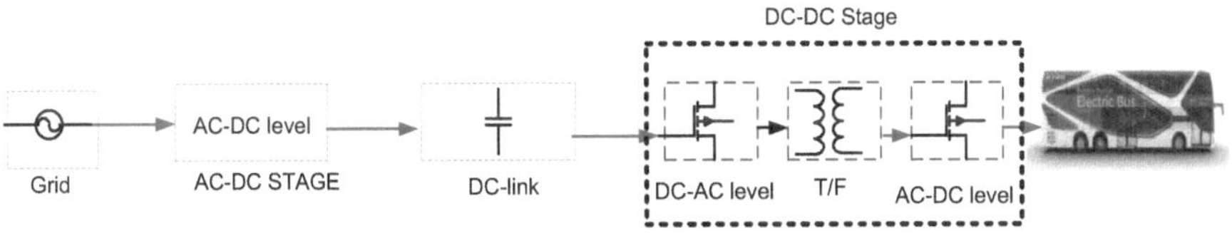

FIGURE 4.3 Conventional charger system.

converters such as AC-DC and a DC–DC conversion [13,14]. In Figure 4.3, the DC–DC conversion is the key role in MVN e-bus charging applications and it is an interesting technology for the adoption of EVT, where controlling the power through the voltage level is an important parameter in the grid-connected systems [15,16]. However, the DC–DC converter module is one best solution such as most popular in fast charging design stations due to simple structure and maintenance. Moreover, the selection of power devices plays a key role such as characterized with low voltage and low current than higher voltage current because it allows higher switching frequency, reduced size, weight of the converter. This chapter investigates in detail about various DC–DC converter topologies such as isolated and non-isolated and corresponding MVN applications. At the end the details route map and challenges for the requirements of medium voltage DC–DC converter networks in EV vehicle technology. This chapter gives a broad view to all young research scholars on MV technology and its related converter selection for EV application.

4.1.1 Components of Electric Power Train Technologies and its Classifications—A Broad Review

In general, an electric vehicle (EV) is a vehicle and it is powered by electricity. All EVs are incorporated with electric motor as an alternative of internal combustion engine. And EVs practices a huge traction battery pack to drive the motor, where it is plugged into the charging station. This makes EVs produces no consume from tailpipe and not contain the fuel pump, line or tank. For more clarity, here we first discuss the functions of various components of power train technologies such as Motor, DC–DC converter, motor controller, battery, on-board charger, and thermal system. On the other hand, EV structures are classified into five types based on the its modelling namely battery electric vehicles (BEVs), several hybrid electric vehicles (HEVs), and many types of plug-in hybrid electric vehicles (PHEVs). In this section, provides the basic variations in between the various power train technologies.

4.1.1.1 Motor

The major role of an electric motor is to convert electrical current into mechanical energy. In general, brushless DC motors are commonly incorporated into EVs by the manufacturers. The reasons behind the selection of such motors are that they are more effective in terms of price, size, power saving, good controllability, and higher efficiency than other conventional machines.

4.1.1.2 DC–DC Converter

It used a conversion mechanism such as change higher-voltage DC energy to the lower voltage DC energy to backup the battery pack.

4.1.1.3 Motor Controller

It is used to control motor properties such as voltage and current drop. And the motor controller does not require power during the vehicle's rest condition.

4.1.1.4 Battery

It is used to store electric energy. And in the market, lithium-ion batteries are used for hybrid and electric vehicles. It is required for both HEVs, PHEVs, and any other kinds of electric vehicles. Recently, rechargeable batteries have been used by automobile manufacturers due to more benefits in terms of charging time.

4.1.1.5 On-Board Charger

It takes the input power through the charge port, which converts into DC energy to charge the battery, and it controls battery parameters in terms of voltage, current, temperature, and state of charge, respectively.

4.1.1.6 Thermal System

It is used for cooling such as monitoring the temperature of the engine, motor, power electronics converters, and other components.

4.1.2 Conventional Electric Vehicles and its Working Operation

4.1.2.1 Battery Electric Vehicles Power Train (BHV-PT)

A BEV-PT is powered by electric supply, which consists of a big electric motor and a battery power pack. And the working mechanism through a clutch, gearbox, differential, and fixed gearing is run in sequential manner. And last, the variety of designs in BEV-PT with the help of more count of battery packs and motors are shown in Figure 4.4.

4.1.3 Mild Hybrid Electric Vehicles Power Train-(MHEV-PT)

MHEV-PT is designed into two types of energy sources such as an internal combustion engine (ICE) and an electric battery (EB) including either motor or generator. It is the smallest electrified model of EV. In this design, a large-size motor and hence used as a generator, which is also known as a belted alternator starter (BAS) and a large-size battery is used to recharge the motor, as shown in Figure 4.5. Here the engine is always in conditions during the vehicle moving and motor/generator is used for idle conditions during turn-off of the engine including high load conditions to increase the vehicle performance. And last for low loads conditions increases load and charge the EB.

4.1.4 Series Hybrid Electric Vehicles Power Train (EHEV-PT)

In this type of EV, only one path is provided to drive the wheels into two energy supplies, as shown in Figure 4.6. Here the fuel tank feeds are coupled to a machine for battery purposes and also used to provide the electrical energy to the generator

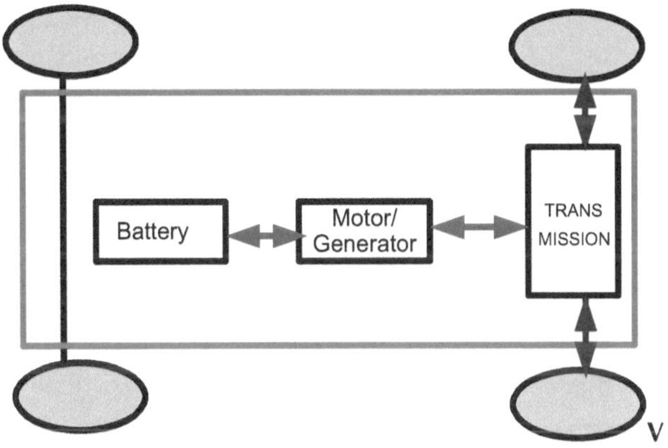

FIGURE 4.4 Structure of BEV power train.

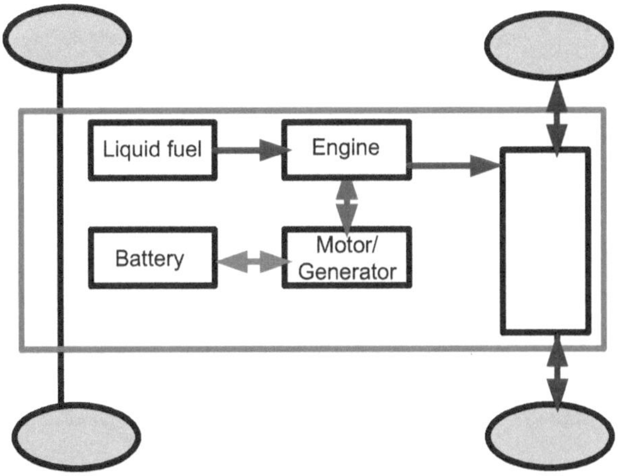

FIGURE 4.5 Structure of MH power train.

through a direct coupling transmission. And finally, the M-G set is utilized to charge the battery in the deceleration and braking conditions. And most of the SHEV-PT uses an ICE, which is powered by a hydrogen fuel cell such as a Fuel Cell Electric Vehicle powertrain (FCEV-PT).

4.1.5 Parallel Hybrid Electric Vehicles Power Train (PHEV-PT)

It consists of two ways of paths parallel and drive the wheels of the EV such as an engine route and electrical route, as shown in Figure 4.7. Here the engine path combines into the transmission through either motor or generator and the engine. However, controlling is very much complex compared to the HEV-PT due to coupling the motor/generator and engine maintain driveability and performance.

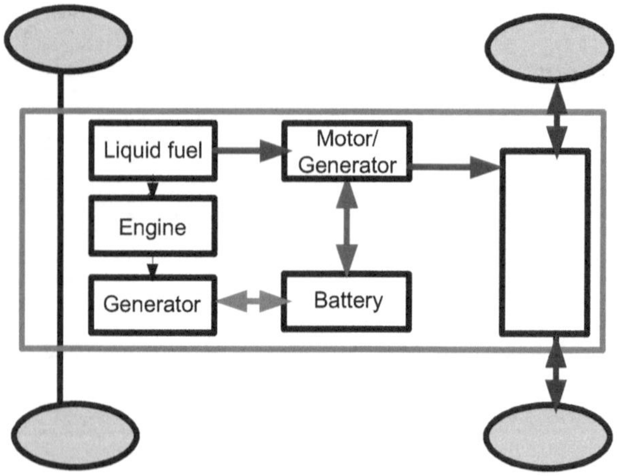

FIGURE 4.6 Structure of SH-power train.

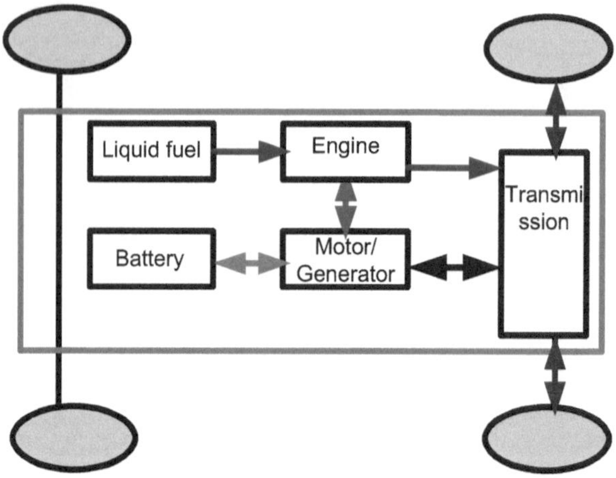

FIGURE 4.7 Structure of PH-power train.

4.1.6 Series-Parallel Hybrid Electric Vehicles Power Train (SPHEV-PT)

It consists of two energy paths namely series and parallel, as shown in Figure 4.8. Here motors-generators use a gearing or power split device sometimes to charge the battery. However, many variations are available in the market but few are simple and few are complex, based on the number of M-G and their performance in the EHV. And in the market at present two types of configurations are more popular which are complex hybrids (CH-EV), split-parallel hybrids (SP-HEV).

4.1.7 Plug-in Hybrid Electric Vehicles-Power Train (PIHEV-PT)

It is an HEV that is plugged-in or recharged through wall power. PHEVs are distinguished with more battery packs compared with other HEVs. It defines the shape of the

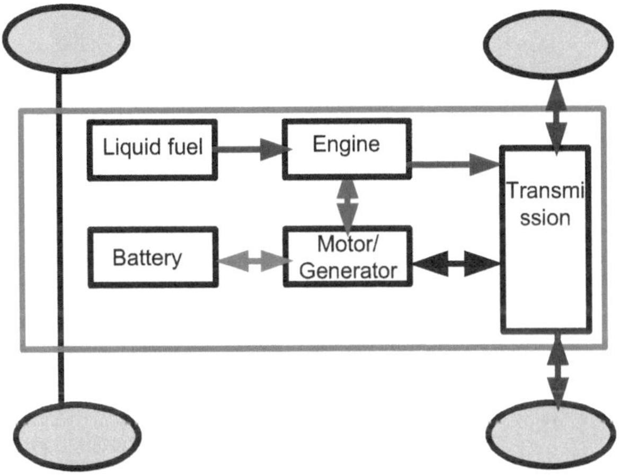

FIGURE 4.8 Structure of SPH-power train.

battery in all electric range (AER). It is commonly installed within the 30 to 50 miles range for electric hybrid vehicles. Plug-in H-electric vehicles can be of any hybrid configuration to utilize all components effectively in the engine and motor/generator sets. And there are no plug-in H-electric vehicles available in the market today because of the complexity and feasibility to design and model as per the present practical conditions with higher performance. So recently many companies in the market have been started and working on PHEVs to retail adaptation kits and facilities to convert a standard H-electric vehicle into a plug-in hybrid electric vehicle with the help of adding another battery capacity including the modified V-controller with E-management system

4.2 ISOLATED AND NON-ISOLATED CONVERTER TOPOLOGIES: AN OVERVIEW

In this section, the basic converter topologies are discussed in detail such as both decoupling and non-decoupling DC–DC converter configurations, as shown in Figure 4.9. Here, isolation is provided through the LFT/HFT systems. But this chapter is focused only on isolation converter topologies, and its performance parameters are clearly summarized in terms of output power range, cost, total switch count, and voltage stress, respectively. On the other hand, without isolation-based topologies namely boost, buck, and buck-boost cook converters [14,15] are not covered here due to their primary requirements are not listed in terms of isolation and low power. Again, the isolation converters are classified into two ways namely single ended and double ended, i.e., one is operated in first quadrant operation and the other is second quadrant only with BH-curve. Double-ended topologies are not required higher core than single-ended configuration.

Based on Figure 4.9, the converters are two ways with isolation and without isolation.

The conventional with-decoupling structures such as flyback-FB, forward-FW, push-pull-PP, half-bridge-HB, and full-bridge-FB with low power to high power are

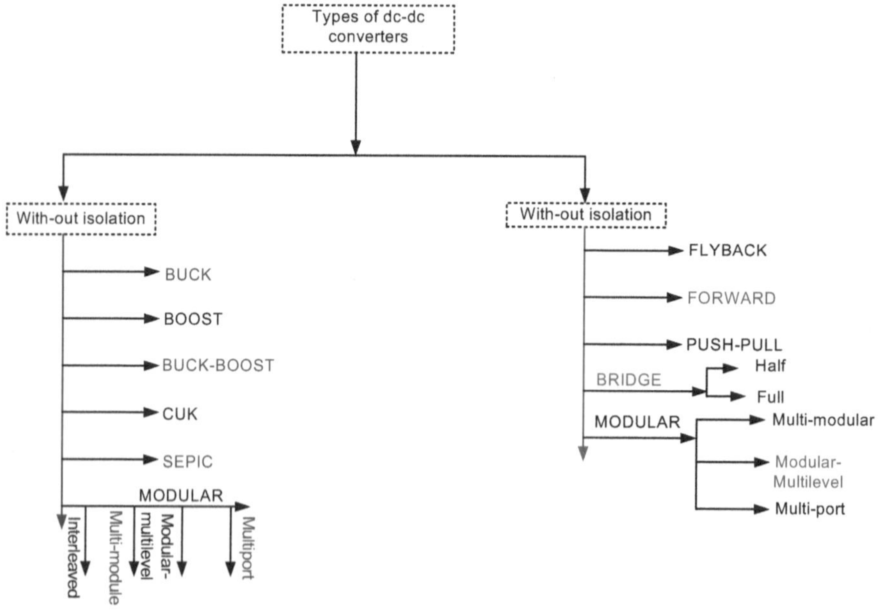

FIGURE 4.9 Types of conventional DC–DC converters [15].

demonstrated in Figure 4.10. As per the flyback converter, other overall cost is less due to low switching count during the active states, and it is designed with single end core, which means additional capacitors required to suppress the high ripples across the input and output capacitors and hence poor core utilization in the transformer. Similar characteristics in the both forward and active clamp except the duty cycle range limited in this configuration. On other hand, the push-pull (P-P), half-bridge-HB, and full-bridge-FB configurations are employed with high power application because it is designed with double end core and 100% core is utilized in the transformer [17].

However, the selection of the duty cycle (DC) is the main key parameter to optimize transformers further such as 50% duty cycle can be modelled with double-ended core and if DC creased to 100% then the filter size reduced effectively and hence transformer core. According to the voltage across the switches P-P configuration have higher switching stress, which is overcome through the H-B, F-B topologies such as main voltage stress is not outside the supply power and hence it has good transformer core utilization with single winding but design is more complex and higher cost than H-B topology due to more switching count in the active states. The detailed performance parameters are summarized in Table 4.1 [18].

4.3 CLASSIFICATIONS OF MEDIUM VOLTAGE DC–DC CONVERTERS [19]

MVDC converters are classified two ways based on the voltage and power ratings. In the first method 2 L converters with higher voltage and current ratings with series

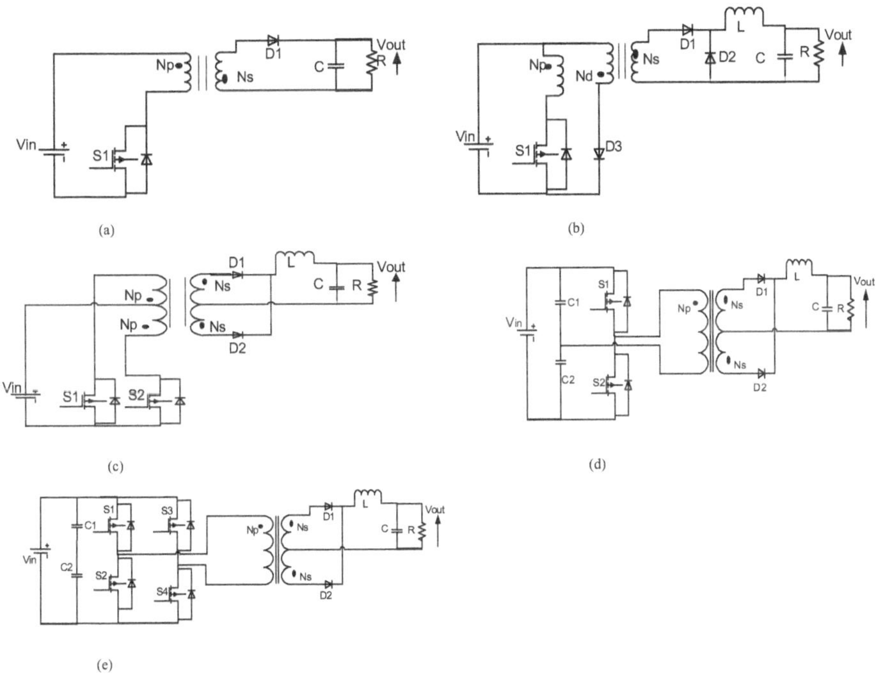

FIGURE 4.10 Conventional decoupling DC–DC converters: (a) flyback-FB, (b) forward-FW, (c) push-pull-P-P, (d) half-bridge-HB and (e) full-bridge-FB [16].

TABLE 4.1
Parameters Comparisons of Various Conventional DC–DC Converters

		Type								
		Non-isolated				Isolated				
S.NO	**Parameters**	**Buck**	**Boost**	**B-B**	**Ck-C**	**FB**	**FC**	**P-P**	**H-B**	**F-B**
1	Power Range	Low	High	Low	High	Low	Low	Low	Low	High
2	Winding	NA	NA	NA	NA	Single	Single	Double	Double	Double
3	Switching count	1	1	1	1	1	1	2	2	4
4	Voltage stress	Lower	Lower	Lower	Higher	Lower	Lower	Higher	Lower	Lower
5	Cost	Lower	Lower	Higher	Higher	Lower	Lower	Lower	Lower	Higher
6	Applications	LP	LP	MP	HP	MP	LW	HP	HP	HP

Note: NA, not applicable; LP, low power; MP, medium power; HP, high power.

and parallel connected power switches. Here based on the power exchange stages, the DC–DC converters are categorized in two ways namely single power stage and two power stage converters.

As per the single stage again it is classified into modular multilevel converter, medium voltage auto transformer and hybrid cascaded type converters. And the same way two-stage converters are two types such as with transformer and without transformer. Here without-transformer converters are more popular than with-transformer types converters due to reduced components in the whole structure. Nonetheless, in series type connection due to parameters mismatch unsymmetrical voltage is shared in the converter. In addition, to overcome these un-symmetrical issues, they need additional support such as voltage balance methods to avoid the converter filter, and hence, poor reliability. converters need proper voltage which leads to power switches. In the second method, modular converters are used in the medium voltage electric bus station technology.

It's a good alternative for the present trend EVT because if low power modules converters are provided to mitigate the converter failures and also for higher power rating more modules are connected to the system. However, to overcome converter failure due to the voltage, two popular converters are employed namely multi-modular (MM) and multi-modular-converters (MMC), which will be discussed in further sections, and more modules are connected to the system. However, to overcome converter failure due to the voltage, two popular converters are employed namely MM and MMC, which will be discussed in further sections, as shown in Figure 4.11.

4.4 SINGLE-STAGE MODULAR MULTILEVEL CONVERTERS

Figure 4.12 shows the MMC with two cascaded systems, such as power, that are exchanged with primary to the secondary systems at different levels of frequency. The modular-multilevel converter configuration is applicable to medium-voltage drives and also connects low-level voltage cells in series to produce the medium-level output voltage. Each cell consists of two legs and their corresponding phase leg inductors. It is simply made with either MOSFET or IGBT based on the level of voltage or application and one capacitor bank. However, it is suitable for MVEVT applications due to higher filter size and higher cost. Hence, losses are increased

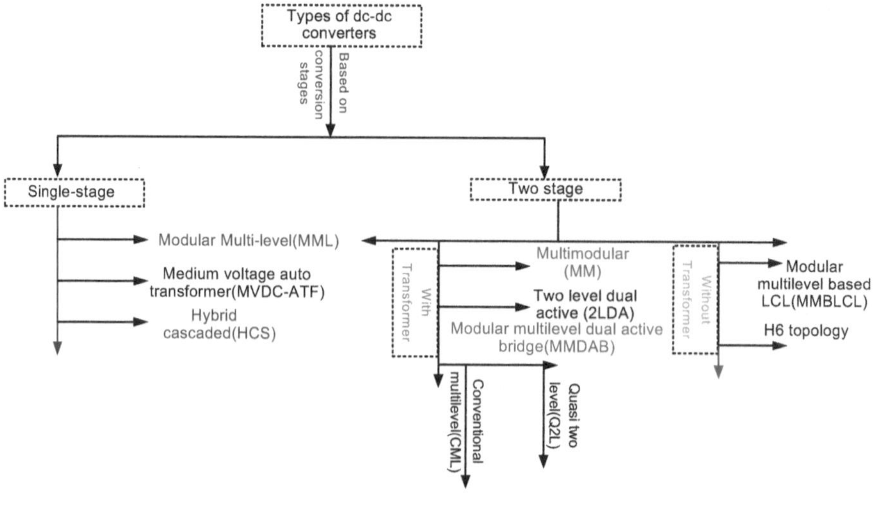

FIGURE 4.11 Types of conversion stages for conventional DC–DC converters [19].

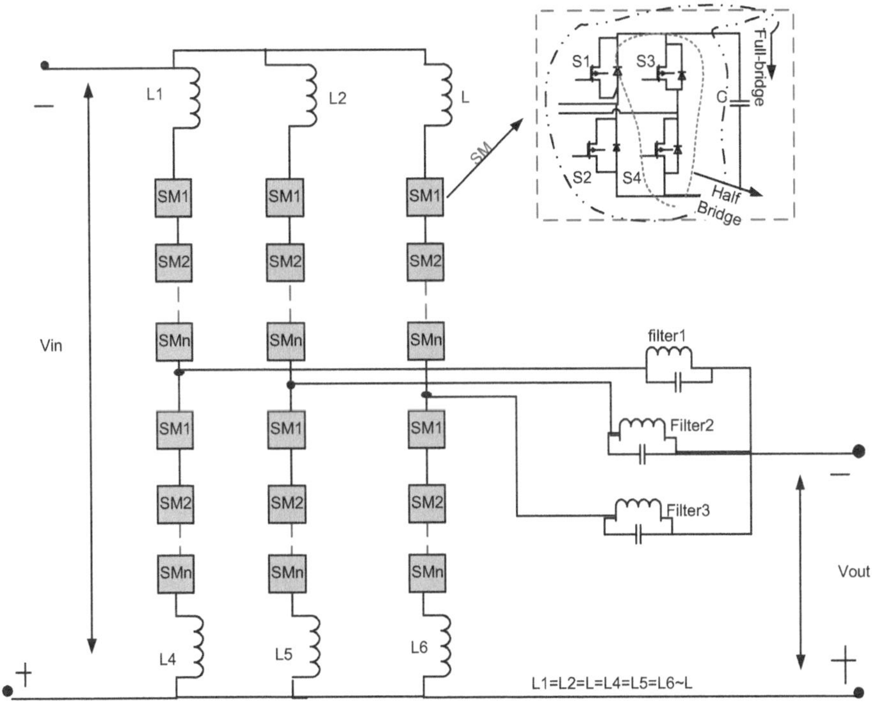

FIGURE 4.12 Single stage modular multilevel converter (MMC) topology [18].

significantly. On the other hand, compared to the dual active converters, MMC has huge switching devices in operating mode, which results in higher losses and poor system efficiency. And if the primary voltage and current ratio is high and without isolation, then only MMC is preferable because it is not a mandatory requirement [18].

4.5 MVDC-ATF

In [19], MVDC-ATF is proposed with two MMC converters in series with one to another through AC-link and transferring the power from lower to upper switch converters, as shown in Figure 4.13. It has a converter to the transformer core, and losses are reduced significantly due to only the DC portion of the converter power being transformed. Compared with MMC-based front-to-front topology [19], MVDC-ATF has higher DC power transfer capability due to the effort of low-rated power devices and hence higher voltage stress in the converter stage. The MVDC-auto transformer converter topology is modelled with a transformer, and this type of structure is good for reducing stress, but losses are more compared with other conventional topologies. However, MMC-ATF is a good choice for EVT application because low-size transformers are employed without galvanic isolation.

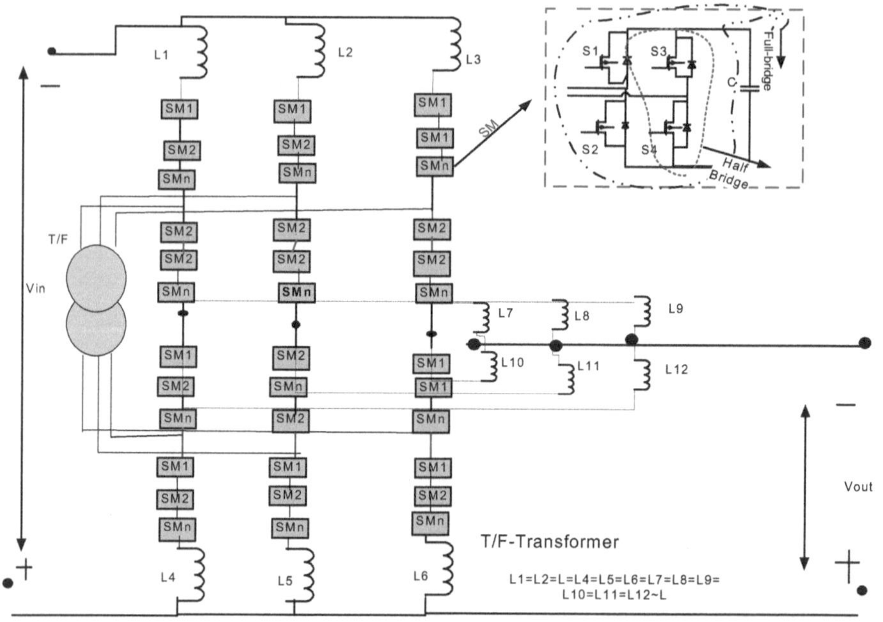

FIGURE 4.13 MVDC auto-transformer converter topology.

4.6 HYBRID CASCADED CONVERTER

In [20–22], HCC is employed with poly structure such as it consists of power devices with cascaded type converter group with energy storage unit, as shown in Figure 4.14.

In addition, it is in-built with the higher phase, which avoids the power flow interruption and ensures voltage is not balanced across the power device. The HCC topology is also modelled with a current source inverter, which is one of the emerging

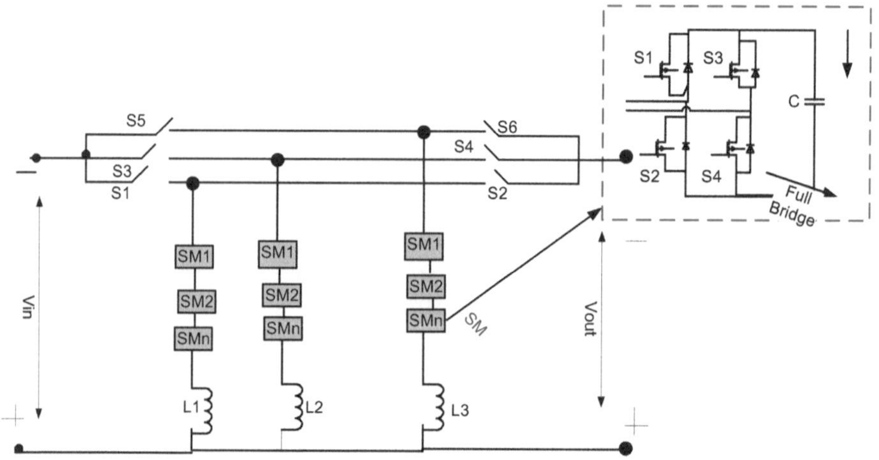

FIGURE 4.14 Hybrid cascaded converter topology.

topologies and is used in the CSI converters. It consists of a series of controlled voltage sources with a capacitive filter to provide higher voltage levels. This type of converter is very useful in the applications of non-conventional energy, voltage regulation-V-R, VAR regulation, compensation, and harmonic filtering in PV energy applications. A improved cascaded HB multilevel inverter (MLI) is realized for solar PV applications [2]. The main features of HCC are that it has to operate soft-switching techniques only and has higher switching losses due to more switching count in the active operations. Here storing is provided through the cascading HB-modules. And it is highly controllable due to soft switching devices in the bridge circuits. However, it demands high voltage and high power with the help of higher levels of MOSFET devices connected in series with topology phase legs [23–28].

4.7 MMC-BASED LCL TWO-STAGE CONVERTER

Another MMC-based LCL two-stage converter is connected with F2F through an intermediate AC-link and it made with help of DAB converter. It is classified based on the AC-link using with transformer and without transformerless converters and also filter L or LCL filter. However, transformerless converters are used for low transformation ratios and similarly high-power ratios are essential for with transformer applications but filter is removed from the structure, which is a great advantage for this type topologies. So here without transformer-based such as in other words it's MMC-based using symmetrical L-C-L-DC–DC converter and another is employed with AC transformers-based converters are discussed in the following sections.

Figure 4.15 shows the MMC-based LCL converter providing DC fault isolation, higher stepping ratios, and operating at a higher switching frequency. The main advantage of these converters is lower losses, ripple current, and higher reliability because they are employed in resonant configurations, which results in higher efficiency compared with the MMC-ATF.

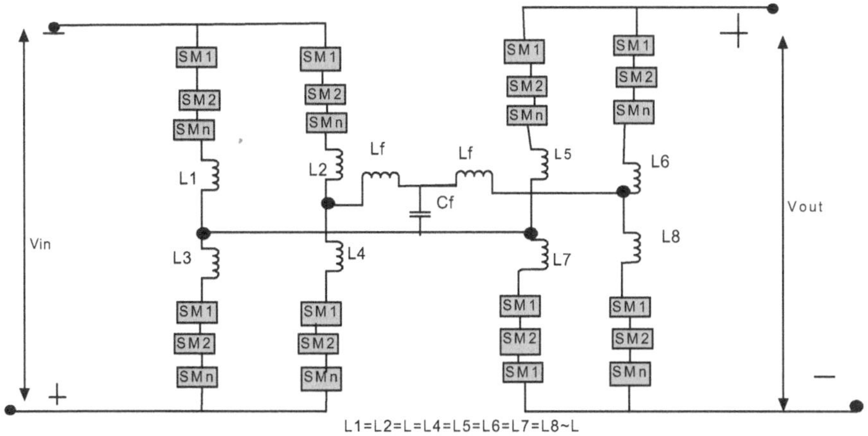

FIGURE 4.15 MMC-LCL converter topology [25].

4.8 TWO-LEVEL DUAL ACTIVE BRIDGE CONVERTER

Figure 4.16, shows the 2L-DAB, which is employed with two active bridges with one intermediate high-frequency transformer. And here the first half full bridge structure is realized for power conversion such as DC voltage-AC voltage with high frequency and the second F-bridge converter is back to AC-DC [19].

It is made with bidirectional power movement, which controlling is possible through the phase difference between primary and secondary bridges. And various control methods are presented in the literature to improve its device technology such as employed with soft switching device range and eliminate the transformer power. However, the initial cost was higher compared to other MMC due to more series and parallel units and hence more attention required in this issue. It is classified four ways, i.e., input-series (input-S) and output series (output-S), input-S and output-P (ISOP), input-P and output-P (IPOP), and input-P and output-S (IPOS) [20], which are shown in Figure 4.17 [29].

4.9 MULTI-MODULE DC–DC CONVERTERS (MCs)

It is adopted for medium-high voltage applications, which are depicted in Figure 4.18. And it is operating with switching frequency, and each module produces power while sacrificing system efficiency. One of the key features of this converter is that it operates in unidirectional and bidirectional power flow operations, and hence equal power distribution is required through proper control methods. In [30], several full bridge-based MCs are investigated in detail. However, the MC fails to operate if voltage exceeds the voltage levels such as higher voltage levels it is employed with multiple transformers. However, various research articles have investigated MMC, such as in [30], where modular DC–DC converters are discussed in terms of modelling and control techniques for ultra-fast charger EV systems. In [30], a low-speed EV charger system with ISOP Modular-based converter is discussed, and in [24], hybrid-based MMC is explored in detail. From the studies, it is confirmed that all MMC are adopted with modular-based units only but are not applicable for power electronics stage.

4.10 MODULAR MULTILEVEL CONVERTER (MMC-DAB/MMC-F2F) [31, 32]

MMC-DAB is introduced by the Lesnicar and Marquarft in 2003 through various low voltage ratings HB-SMs to produce AC voltage waveform [31], which makes

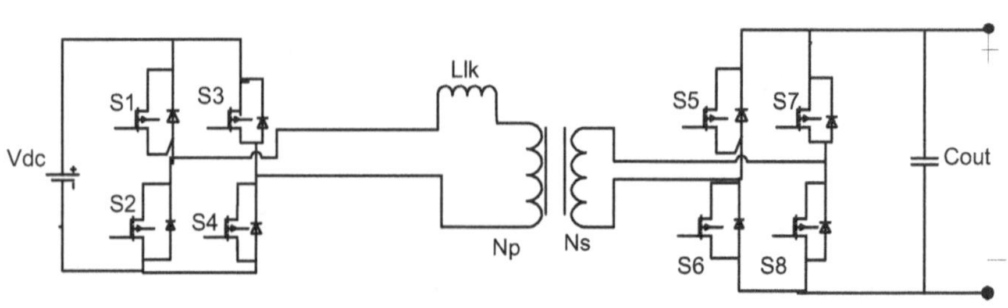

FIGURE 4.16 Dual active bridge (DAB) converter topology [26, 27].

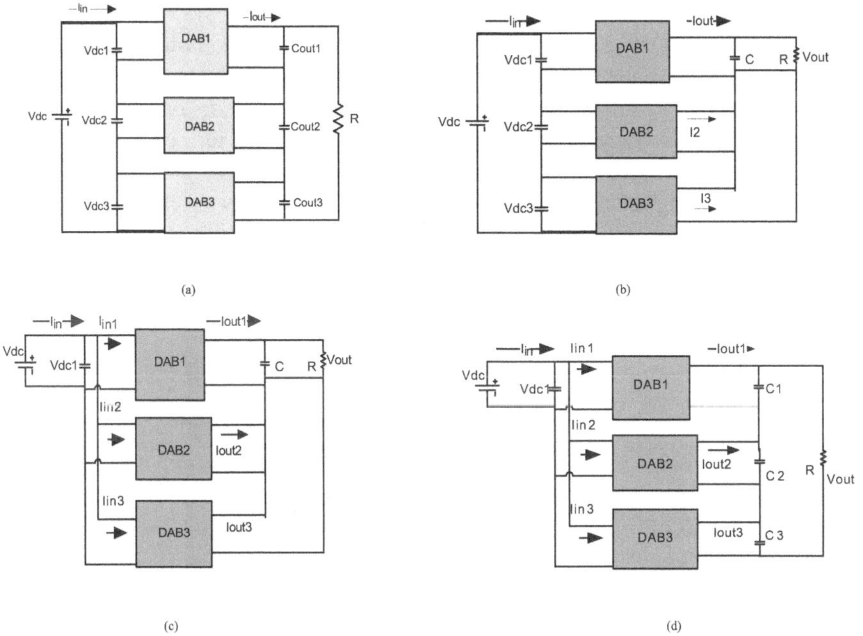

FIGURE 4.17 Modular DAB converter topologies: (a) input-S-output-S, (b) input-S-output-P, (c) input-P-output-P and (d) input-P-output-S.

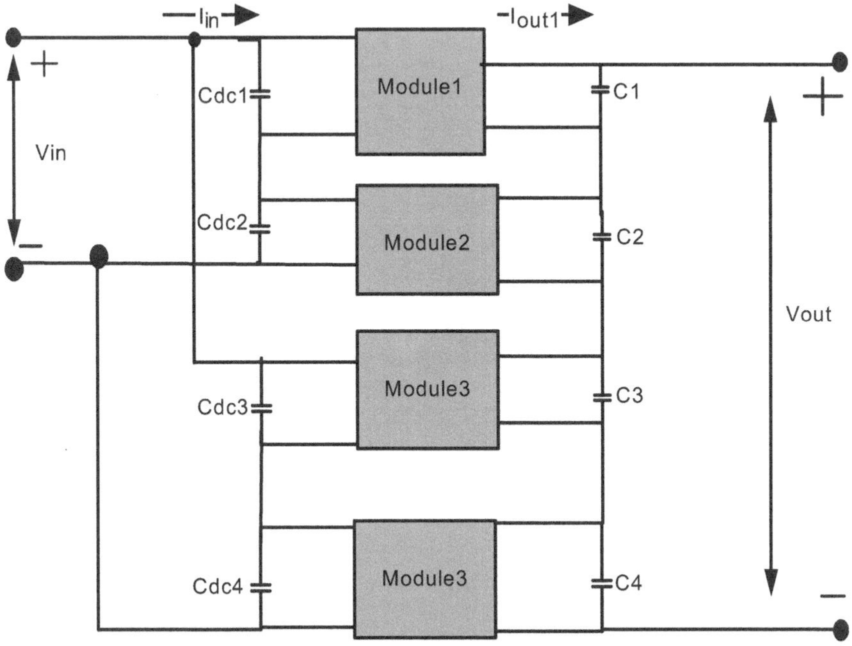

FIGURE 4.18 Multi-module DC–DC converter structure.

high quality and gained popularity with replacement of DAB systems, as shown in Figure 4.19. The MMC-F2F topology is one of most popular unit MV E-bus fast charging to eliminate the transformers from the converters, where multilevel or 2-L voltage waveform is generated across the transformer. It contains two cascaded full-scale(Full-S) converters with midway transformers. Here first unit converter DC-AC and second unit AC-DC voltage and voltage level can be changed through transformer. It is blocking fault capacity and series connected power devices also removed but poor utility of two converters due to more conversion stages and hence higher cost, weight, size and losses. And overall performance comparisons are depicted in Table 4.2 [33, 34]

As per the Table 4.2, it confirms that conventional DC–DC level topologies are good in low power applications, but in MMC converters, MC-DAB-based 2L converter structure is the finest selection for medium voltage DC applications because of its excellent performance characteristics in terms of size, magnetic components, switching count, cost, and efficiency, respectively.

The transformer design plays a critical role in the DC–DC level using medium voltage, such that high switching frequency-based units are more attractive than low frequency-based units and result in size, weight being reduced significantly. In other words, the design of MVDC with high-frequency transformers is not straightforward. And safety measurements are followed through the IEC standards only [39]. However, the main challenges and requirements factors for MVDC converters are power quality issues such as harmonic, inrush, fault currents, and ground level [35–38]. As per the standards, the harmonic content is limited as per their IEEE 519 standards. And

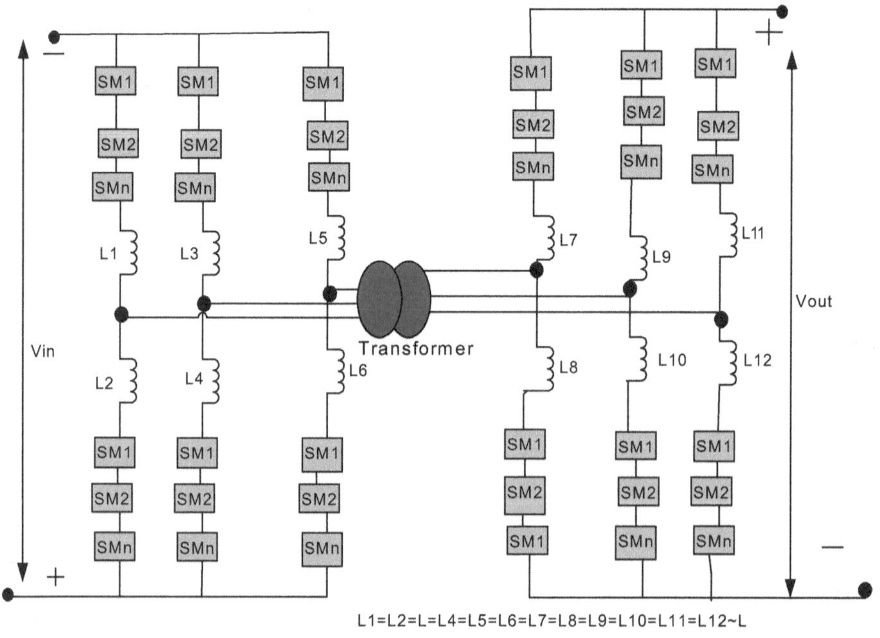

FIGURE 4.19 Modular-MC-based F-to-F converter topology.

TABLE 4.2
Performance Comparison of MMC Among MMC-2L/DAB Converter Topologies

S.no	Parameters	Topologies			
		MMC	MVDC-ATF	MMC-F2F/DAB	MC-DAB/2L
1	Stages	1	2	2	2
2	Isolation	NO	NO	Yes	Yes
3	Turn's ratio	Low-medium	Low-medium	High	High
4	Size	Large	Large	Large	Small
5	Magnetic components	Higher capacitors	Higher capacitors	Higher capacitors	Low capacitors
6	Switching count	Higher	Higher	Higher	Lower
7	Switching frequency	Low	Low	Low	High
8	Switching losses	Higher	Higher	Higher	Lower
9	Insulation	No	No	No	Yes
10	Cost	High	High	High	Low
11	Efficiency	Low	Low	Low	High

the inrush currents issues are resolved through the soft-start techniques, and faulty currents employed appear due to the absence of the period voltage in DC-networks, hence protection is required. Grounding is also required for the safety concern of power converters and in [39, 40] its prevention is discussed in detail. The detailed roadmap for selection of DC–DC converters and MVDC applications are explored clearly in this chapter. This review chapter is helpful to young research scholars for better understanding on DC–DC converters and its applications.

4.11 CONCLUSIONS

In this chapter, several conventional DC–DC configurations are investigated in detail made on the performance parameters such as power capability, transformer utility, total switching count, cost, respectively. On the other hand, single- and double-ended isolated DC–DC converters structures are explored clearly. The medium voltage e-bus station fast charging application using DC–DC converters such as DAB-based modular, multilevel converters and its features highlighted here. In addition, the detailed review has been given on DAB converters with various voltage level applications through both series-parallel connection MMC structure are explored clearly. And at the end, the detailed road map on best selection DC–DC converter for medium voltage bus station quick charging systems and its challenges are thoroughly discussed. In the future the following case studies are investigated further for better improvements in the EV technology such as (i) reduced switching count multilevel topologies with low losses, (ii) an advanced medium-level DC–DC converters, and (iii) an advanced battery management technologies are most key research areas for grid-connected applications.

REFERENCES

[1] Paddon, T.; Welch, D.; Silver, F. *California Transit Agencies Chart a Course to Zero Emissions: A Review of Proposed Zeb Pathways under the Innovative Clean Transit Regulation.* Calstart: Pasadena, CA, 2021; Report; pp. 1–18.

[2] Mathieu, L. *Electric-Buses Arrive on Time Marketplace, Economic, Technology, Environmental and Policy Perspectives for Fully Electric Buses in the EU.* Transport and Environment: Brussels, Belgium, 2018.

[3] Jiang, M.; Zhang, Y.; Zhang, Y. Multi-Depot Electric Bus Scheduling Considering Operational Constraint and Partial Charging: A Case Study in Shenzhen, China. *Sustainability* 2022, 14, 255.

[4] Al-Ogaili, A.S.; Al-Shetwi, A.Q.; Sudhakar Babu, T.; Hoon, Y.; Abdullah, M.A.; Alhasan, A.; Al-Sharaa, A. Electric Buses in Malaysia: Policies, Innovations, Technologies and Life Cycle Evaluations. *Sustainability* 2021, 13, 11577.

[5] Verbrugge, B.; Hasan, M.M.; Rasool, H.; Geury, T.; El Baghdadi, M.; Hegazy, O. Smart Integration of Electric Buses in Cities: A Technological Review. *Sustainability* 2021, 13, 12189.

[6] Arif, S.M.; Lie, T.T.; Seet, B.C.; Ahsan, S.M.; Khan, H.A. Plug-In Electric Bus Depot Charging with PV and ESS and Their Impact on LV Feeder. *Energies* 2020, 13, 2139.

[7] Vázquez-Guzmán Gerardo, Martínez-Rodríguez Pánfilo Raymundo, Sosa-Zúñiga José Miguel "High Efficiency Single-Phase Transformer-less Inverter for Photovoltaic Applications" Ingeniería Investigación y Tecnología, volumen XVI (número 2), abril junio 2015. DOI: 10.1016/j.riit.2015.03.002

[8] Tey K. S.; Mekhilef, S. A Reduced Leakage Current Transformer-Less Photovoltaic-Inverter. *Renewable Energy* 2016, 86, 1103–1112.

[9] Li, W.; & Gu, Y.; Luo, H.; Cui, W.; He, X.; Xia, C. Topology Review and Derivation Methodology of Single-Phase Transformerless Photovoltaic Inverters for ground leakage current Suppression. *IEEE Transactions on Industrial Electronics* 2016, 62, 4537–4551.

[10] Schmidt, D.; Siedle, D.; Ketterer, J. Inverter for Transforming a DC Voltage Into an AC Current or an AC Voltage. *EP Paten* 2003, 1, 369–985.

[11] Victor, M.; Greizer, K.; Bremicker, A. Method of Converting a Direct Current Voltage From a Source of Direct Current Voltage, More Specifically From a Photovoltaic Source of Direct Current Voltage, Into a Alternating Current Voltage. U.S. Patent 2005, 028-6281-A1, April 23, 1998.

[12] Khan, M.N.H.; Siwakoti, Y.P.; Li, L.; Blaabjerg, F. H-Bridge Zero-Voltage Switch Controlled Rectifier (HB-ZVSCR) Transformerless Mid-Point-Clamped Inverter for Photovoltaic Applications. *IEEE Journal of Emerging and selected Topics in Power Electronics* 2019, 99, 1

[13] Yang, H. Modular and Scalable DC-DC Converters for Medium-/High-Power Applications. Master's Thesis, Georgia Institute of Technology, Atlanta, GA, USA, 2017

[14] ElMenshawy, M.; Massoud, A. Hybrid multimodule DC-DC converters for Ultrafast Electric Vehicle Chargers. *Energies* 2020, 13, 4949.

[15] Beldjajev, V. Research and Development of the New Topologies for the Isolation Stage of the Power Electronic Transformer. Master's Thesis, Tallinn University of Technology, Tallinn, Estonia, 2013

[16] Engel, S.P.; Stieneker, M.; Soltau, N.; Rabiee, S.; Stagge, H.; De Doncker, R.W. Comparison of the Modular Multilevel DC Converter and the Dual-Active Bridge Converter for Power Conversion in HVDC and MVDC Grids. *IEEE Transaction Power Electronics* 2014, 30, 124–137.

[17] Papadakis, C. Protection of HVDC Grids Using DC Hub. Master's Thesis, Delft University of Technology, Delft, The Netherlands, 2017.

[18] Ferreira, J.A. The Multilevel Modular DC Converter. *IEEE Transaction Power Electronics* 2013, 28, 4460–4465.

[19] Andre Schon, Mark-M. Bakran. A New HVDC-DC Converter for the Efficient Connection of HVDC Networks. In: *PCIM Europe Conference Proceedings.* PCIM: Nürnberg, Germany, 2013.

[20] Schoen, A.; Bakran, M.M. Comparison of the Most Efficient DC-DC Converters for Power Conversion in HVDC Grids. In: *Proceedings of the PCIM Europe 2015; International Exhibition and Conference for Power Electronics, Intelligent Motion, Renewable Energy and Energy Management*, Nuremberg, Germany, 19–20 May 2015; pp. 1–9.

[21] Schön, A.; Bakran, M.M. Average Loss Calculation and Efficiency of the New HVDC Auto Transformer. In: Proceedings *of the 2014 16th European Conference on Power Electronics and Applications*, Lappeenranta, Finland, 26–28 August 2014; pp. 1–10.

[22] Yang, J.; He, Z.; Pang, H.; Tang, G. The Hybrid-Cascaded DC-DC Converters Suitable for HVdc Applications. *IEEE Transactions on Power Electronics* 2015, 30, 5358–5363

[23] Kish, G.J.; Lehn, P.W. A Modular Bidirectional DC Power Flow Controller With Fault Blocking Capability for DC Networks. In: *Proceedings of the 2013 IEEE 14th Workshop on Control and Modeling for Power Electronics (COMPEL)*, Salt Lake City, UT, USA, 23–26 June 2013; pp. 1–7.

[24] Soltau, N.; Stagge, H.; De Doncker, R.W.; Apeldoorn, O. Development and Demonstration of a Medium-Voltage High-Power DC-DC Converter for DC Distribution Systems. In: *Proceedings of the 2014 IEEE 5th International Symposium on Power Electronics for Distributed Generation Systems (PEDG)*, Galway, Ireland, 26–29 June 2014; pp. 1–8.

[25] Jovcic, D. Bidirectional, High-Power DC Transformer. *IEEE Transactions on Power Delivery* 2009, 24, 2276–2283.

[26] Kheraluwala, M.H.; Gasgoigne, R.W.; Divan, D.M.; Bauman, E. Performance Characterization of a High Power Dual Active Bridge DC/DC Converter. In: *Proceedings of the Conference Record of the 1990 IEEE Industry Applications Society Annual Meeting*, Seattle, WA, USA, 7–12 October 1990; vol. 2, pp. 1267–1273.

[27] Oggier, G.G.; García, G.O.; Oliva, A.R. Switching Control Strategy to Minimize Dual Active Bridge Converter Losses. *IEEE Transactions on Power Electronics* 2009, 24, 1826–1838.

[28] Qin, H. Dual Active Bridge Converters in Solid State Transformers. Master's Thesis, Missouri University of Science and Technology, Rolla, MO, USA, 2012.

[29] Adam, G.P.; Gowaid, I.A.; Finney, S.J.; Holliday, D.; Williams, B.W. Review of DC-DC converters for multi-terminal HVDC transmission networks. *IET Power Electronics* 2016, 9, 281–296.

[30] Wu, H.; Lu, Y.; Mu, T.; Xing, Y. A Family of Soft-Switching DC-DC Converters Based on a Phase-Shift-Controlled Active Boost Rectifier. *IEEE Transactions on Power Electronics* 2015, 30, 657–667.

[31] Xing, Z.; Ruan, X.; You, H.; Yang, X.; Yao, D.; Yuan, C. Soft-Switching Operation of Isolated Modular DC/DC Converters for Application in HVDC Grids. *IEEE Transactions on Power Electronics* 2015, 31, 2753–2766.

[32] Kish, G.J.; Ranjram, M.; Lehn, P.W. A Modular Multilevel DC/DC Converter With Fault Blocking Capability for HVDC Interconnects. *IEEE Transactions on Power Electronics* 2013, 30, 148–162.

[33] Kenzelmann, S.; Rufer, A.; Dujic, D.; Canales, F.; de Novaes, Y.R. Isolated DC/DC Structure Based on Modular Multilevel Converter. *IEEE Transactions on Power Electronics* 2014, 30, 89–98.

[34] Kusaka, K.; Orikawa, K.; Itoh, J.; Morita, K.; Hirao, K. Isolation System With Wireless Power Transfer for Multiple Gate Driver Supplies of a Medium Voltage Inverter. In: *Proceedings of the 2014 International Power Electronics Conference (IPEC-Hiroshima 2014-ECCE ASIA)*, Hiroshima, Japan, 18–21 May 2014; pp. 191–198.

[35] IEC 6i800-5-I; *Adjustable Speed Electrical Power Drive Systems-Part 5-1: Safety Requirements-Electrical, Thermal and Energy.* International Electrotechnical Commission (lEC): Genève, Switzerland, 2007

[36] Blooming, T.M.; Carnovale, D.J. Application of IEEE STD 519-1992 Harmonic Limits. In: *Proceedings of the Conference Record of Annual on Pulp and Paper Industry Technical Conference*, Appleton, WI, USA, 18–23 June 2006; pp. 1–9

[37] ETSI EN 301 605, *Environmental Engineering (EE); Earthing and Bonding of 400 VDC Data and Telecom (ICT) Equipment.* ETSI: Nice, France, 2011; vol. 1, pp. 1–85.

[38] Aoki, T.; Yamasaki, M.; Takeda, T.; Tanaka, T.; Harada, H.; Nakamura, K. Guidelines for Power-Supply Systems for Datacom Equipment in NTT. In: *Proceedings of the 24th Annual International Telecommunications Energy Conference*, Montréal, QC, Canada, 29 September–3 October 2002; pp. 134–139.

[39] Halgamuge, N.; Abeyrathne, C.; Mendis, P. Measurement and analysis of electromagnetic fields from trams, trains and hybrid cars. *Radiation Protection Dosimetry* 2010, 141, 255–268.

[40] Prabhala, V.A.; Baddipadiga, B.P.; Fajri, P.; Ferdowsi, M. An Overview of Direct Current Distribution System Architectures & Benefits. *Energies* 2018, 11, 2463.

5 Analysis of Recent Developments in Three-Phase Transformerless Inverter Topologies for Photovoltaic and Electric Vehicle Applications

Ahmad Syed
Chaitanya Bharathi Institute of Technology (A), Hyderabad

Tara Kalyani Sandipamu
Jawaharlal Nehru Technological University, Hyderabad

Freddy Tan Kheng Suan
University of Nottingham, Malaysia

Xiaoqiang Guo
Yanshan University, China

Huai Wang
Aalborg University, Denmark

B. Mouli Chandra
QIS College of Engineering and Technology, Ongole

5.1 INTRODUCTION

The applications of transformerless photovoltaic inverters in electrical vehicles are growing fast and attract many researchers to investigate various grid-connected single-phase and three-phase systems [1]. Recently, power generation through solar panels has been growing fast due to increased population and industries to meet the

DOI: 10.1201/9781003354901-5

world's huge demands in several applications. In that case, most of the PV systems are connected with the grid because of their impact in terms of compactness, lower price, and higher efficiency [2], and they supply the surplus power to the grid if solar power is higher than the local power demand. Moreover, this process's major goal is to attract policymakers and members to pay more attention to the detailed examination and growth on grid-type connected solar inverters [3]. Nonetheless, it is classified into two ways, namely disconnected and non-disconnected systems with the location of the transformer during the power conversion stage. To attain more efficiency, be smaller in shape, and lower price, a universal solution is to eliminate the line frequency transformer in non-isolated inverters, which results in a direct connection to the solar panel to the grid. Hence, high common mode voltage oscillation is formed by the operating of the solar inverter, which can stimulate the resonant branch elements between the stray capacitance of the solar panels, filter, and impedance of the grid. As a result, CM-leakage current circulates through stray capacitance from the solar panel to the ground. On the other hand, it creates issues related with waveform quality, electromagnetic compatibility (EMC), and electromagnetic compatibility interference (EMI), which leads to poor safety. However, the circulating of these CM-leakage currents depend on only one parameter such as high-frequency phase voltages, and its floating due to parasitic effects can excite the resonant branch elements, which are like stray capacitor of the PV panel, inductor, and ground resistor, respectively. And it causes and creates high and dangerous CM-leakage current to the resonant elements and the ground, which means it affects the overall system performance in the form of poor wave quality and results in higher harmonics such as EMC, and hence safety issues are higher than the normal inverter.

In this way to overcome the CM-leakage current issues various investigations were done based on the conventional three-phase configuration with several modulation schemes. So that three -phase transformerless PV topologies (TP-TLPVIT) are attracted towards the eminent scientists. Converter reliability in a three-phase photovoltaic system is mostly dependent on the size of DC-link capacitors [16]. However, in-phase balanced system the power is always constant only.

This process helps the inverter in the form of reduced size DC-link capacitors and results in reduced working stress, and the overall system lifespan is increased well [17,18]. In contrast with single-phase PV systems, three-phase systems are designed for higher power applications such as equal to 20kW inverter output. The standard three-phase six switch called H6 inverter is normally used in PV applications because it is built with simple, easy design and process of software implementation. The standard modulation techniques SVPWM, DPWM are designed in three-phase full-bridge TPV inverter [9]. But the utilization of these standard H6 inverters is not applicable for PV application because of the floating phase voltages and results in higher CM-leakage current. So, to mitigate these issues in literature many configuration and suitable modulation methods were introduced to reduce the CM-leakage current.

In these new schemes here discussed, all types one by one such as firstly to reduce the CM-leakage current a new method in terms of pulses namely reduced CM-voltage PWM (RCM-voltage-PWM) are introduced in several main system performance factors such as low CM-voltage and CM-leakage current [9–14]. This modulation is designed without zero states and it is one of the major factors to design this PWM scheme and results in reduced CM-voltage and CM-leakage current. On the other hand, another modulation scheme known as active zero voltage state PW-modulation (AZVS-PWM) [10] is introduced with active voltage state vectors only such as to set the desired output voltage. In these techniques, the selection of active states is the same as the standard SVPWM method. In other ways, zero voltage states are replaced with the help of a pair of two opposite states and same time duration. This method is good as compared with SP-PWM, RCM-V-PWM but the overall converter performance factor is reduced due to bipolar voltage at the filter parameters [13,14] and it increases the harmonics in the form of current, losses, price and shape of the system. And the behaviour of the CM performance also degraded such as higher floating than standard discussed methods. Further another method is investigated on their vectors such as three adjacent active voltage states are used to maintain the desired output voltage known as near voltage state pulse-width modulation (NVS-PW-M) [11]. In other ways, it is called the near-state method. Here to maintain the less device count it needs to handle the vectors carefully in NSPWM. However, it has a limited modulation index than DPWM because of switch devices during the operating modes and also the modulation is operating in between $0.62 < m < 0.92$ [14].

As a result, in this operating zone, the CM-voltage and CM-leakage current is within the limit after this range it fails to operate as per the standards of normal modulation schemes. After these methods, many researchers are investigated with improved performance parameters such as a new method namely remote state PW-modulation (RSPWM) [12] is designed with a group of odd and even active state vectors to maintain the desired output voltage in all the sectors. And it results in the CM-voltage performance being improved as compared with the other standard and discussed methods. In other way, the CM-voltage is reduced up to 33% V_{dc} and hence low CM-leakage current. However, in this method, the only highlights the CM-leakage current and ignores the complete response of the system inverter. And it gives 33% V_{dc} but the selection of the modulation index is limited such as $m < 0.53$ [13] and operates at higher voltage with the magnitude of 1,100 V_{dc} bus voltage for the standard 230 V grid supply. It causes major losses in the whole system and may be not possible to implement for practical approach PV applications. The main issues in RSPWM are bipolar voltage at the three phases [13] and hence higher losses, ripples and harmonic distortion and hence overall efficiency of the system with RSPWM is poor as compared with other schemes. And it is observed that PWM signals are similar in SVPWM and AZ-PWM and DPWM, NSPWM are also similar pulse schemes [17]. From the aforementioned discussions, it is revealed that the modulation pulses play a very key role in PV three-phase inverters with a suitable converter structure; otherwise, accurate results, such as the desired output voltage, are not programmed well, resulting in higher CM-leakage current. This process makes the system have good performance in any system. In addition, it is revealed that as contrast with three-phase topologies, single-phase topologies are modelled with galvanic isolation only.

In three-phase PV application, normal standard SVPWM and D-PWM schemes are used in many ways based on the power design and their performance parameters such as high-level voltage segment, reduced switching loss and harmonic distortion.

Nonetheless, these modulation schemes are not appropriate for PV applications due to higher fluctuations in CMV and hence higher CM-leakage current. It creates CM-leakage current that injects high ripples, and it increases the losses and ripples-harmonic distortion. In a further way, the whole PV inverter performance is reduced. In this way to eliminate the CMV to mitigate the CM-leakage current, many modulation schemes, which are named as RCMV-PWM, were introduced. The modulation schemes can be termed as two ways: *X* and *Y*, and here this way space vectors are divided in [11]. It comes to type X voltage state with SVPWM introduced by Hava et al. in 2011 [21], AZ-PWM and RSPWM. DPWM were introduced by Hava et al., in 2011 and NSPWM was introduced by Un and Hava in 2009 [22] using type B sectors state and the pulse states for many modulation methods are in [12].

AZ-PWM is introduced with one active state pulse to develop the output voltage. The selection of the active states is similar to the standard SVPWM. The zero states (*M*0 and *M*7) are used with a double complementary active state at the same time, i.e., *M*1 and *M*4, *M*2 and *M*5, or *V*3 and *V*6. The pulse scheme of which is given and demonstrated in [12] by the Hava and Un, in 2009. A couple of complementary active states that are together with the active modes is selected. For illustration, the *M*3 and *M*6 pair is selected to compensate for *M*0 and *M*7 respectively in region *X*1 as given in [12]. The *M*3 and *M*6 pair is chosen because it is adjacent to the active vectors *M*1 and *M*2. Similar sectors are used in all types of regions (*X*1–*X*6). Here the CM-voltage is reduced with a level of 66% without zero sectors such as floating between 33% and 66% of the input voltage. And hence CM-leakage current is reduced but the whole performance is reduced. In [12] shows the device level or states in *L-L* output voltage at the region of *X*1 in AZ-PWM. In this scheme similar device schemes are used such as two ways appeared, but at practical approach, this approach is not possible because device dead-time affects and along with hard power electronics devices schemes. This approach has increased the whole CM performance than standard SC-PWM and DPWM. And another issue with changing the zero sectors with opposite pair sectors causes bipolar output voltage in [12]. The 2-level voltage such as $-V_{dc}$ and $+V_{dc}$ also doubles the whole system losses and switching states at the filter requirements. It leads to higher losses, ripples and hence filter size, cost is increased more than standard topologies.

NSPWM was introduced by Un and Hava in 2009 and it is implemented with active sectors to obtain desired output voltage such as *M*1, *M*2, *M*3 or *M*2, *M*3, *M*4 and so on. And hence called it near-state PWM (NSPWM). The whole pulse pattern in every sector operation is illustrated in [12]. Here to decrease the device count in the operation and then need to tackle very carefully in the selection of active sectors during the operation. Similar to the standard DPWM technique, this one is implemented with a clamped DC bus in state. In this case study, the pulse pattern in selected *Y*1 is *M*6, *M*2 *M*1, *M*2, and *M*6 are given in [12].

And such that phase A is ON and clamped in positive power DC bus state like *Y*1. And hence losses in devices are quite low because all phases are not modulated at a time in each leg. In addition to that, the issues like device losses of AZ-PWM are eliminated in NSPWM. And it is possible to select the active states to reduce the loss and with minimum switching count. And the same way in phase A the clamped positive DC bus region $B1$. Hence, losses are less due to less active phase legs. As compared with the AZ-PWM and NSPWM without using the zero sectors, the CM-voltage is decreased by the factor of 66% and hence floating in between 33% and 66% of the input DC voltage. In [12], the output voltage pattern and CM performance using AZ-PWM in the $X1$ state are clearly demonstrated. Nonetheless, the reduction of zero states is obtained by the bipolar results in the form of L-L voltage given in [12]. The two-level bipolar voltage is caused to raise the losses, voltage pressure across the devices, ripple in current, and voltage. It is made and designed with higher filter ratings and hence the whole system cost and size, weight is increased significantly. The overall voltage range in the form of modulation index is limited in the range of $0.62 < m < 0.91$.

In RSPWM, introduced by Cacciato et al., in 1999 [20]; Cavalcanti et al. [2], here pair of odd active sectors such as $M1$, $M3$ and $M5$ or $M2$, $M4$ and $M6$ type even vectors are used to obtain the desired voltage across the output side in the whole sectors. This makes RSPWM have superior performance factors in the form of CM-voltage compared to other standard methods, such as CM performance index up to 33% of the input DC voltage, as given in [12]. And hence CM-leakage current is reached to approximately zero magnitude, and above these specific regions, it fails to obtain the standard output performance factors in the form of CM-voltage and CM-leakage current, such as being higher than normal compared with standard PWM. So that RSPWM is one of the best among all methods in the case of CM performance and CM-leakage current such as constant CMV at 33% [12]. However, this method ignores the whole system's performance than other key factors. It gives L-L output voltage pattern with the pulse as per the switches and also suffers with voltage linearity issues. The possible modulation index margin is $m < 0.52$ and maximum voltage is feasible upto 1,100 V for grid-connected 230 V applications. And it leads to higher losses and hence feasibility with a practical approach is very difficult due to bipolar L-L voltage. Here all phases two-level voltage appears in all systems. However, this PWM limitation increases the losses, ripples and hence poor system efficiency.

As per the discussion above, it confirms that most of the three-phase system and its modulation methods are employed with galvanic isolation schemes only, and its practicality is also a very difficult approach. So, all structures are introduced with their corresponding pulse pattern [17–19]. A few such topological configurations are discussed in the below section.

In [14], three-phase PV inverter is introduced with a split capacitor method and similar pulse pattern scheme by Kerekes et al. And this approach has poor CM performance in the form of higher CM-leakage current due to the inductor line in the neutral line [18,19]. The output ground current is beyond the limits as per the standards. In [15], another approach is with freewheeling current path circulates type typology is proposed namely full-bridge three-phase zero voltage rectifier topology (FB-TP-ZVR) and hence less floating in phase voltages and low CM-leakage current. It is modelled with the idea of a conventional three-phase inverter with the help of two extra freewheeling lines to obtain the isolation between the PV and the grid, specifically in the freewheeling stage. The main proposed freewheeling branch is employed with two power devices and 14 diodes and results in higher losses, weight, size and price as well. But, the CM-leakage current is dangerously higher than normal as per the limited standards. Similarly, in [19], a modified three-phase inverter with seven switches with galvanic isolation is introduced and it is known as the H7 structure. Here a simple extra switch is adopted on DC-side to isolate the power circuit to disconnect during abnormal conditions. It is built with one switch and four diodes but the cost, losses, size and difficulty to operate than normal structures. And the whole system CM-leakage current is much higher than normal and limits as per the German VDE-AR-N 4105 standard.

But it is not reached to meet safety grid codes and its standards due to higher leakage current and losses. Unfortunately, in the literature, there is not much attention on the structure of the three-phase topologies (TPT) such as a wide gap between the configuration and its suitable control strategy for leakage current suppression. To enhance the above issues, new space vector modulation schemes (SVMS) are investigated to suppress the leakage current for TPT in [1,2]. Similarly, in [4] another T-type TPT topology is proposed with new SVMS. However, the conventional SVMS and discontinuous modulation scheme (DMS) is failed for three-phase transformerless PV (TPT-PV) applications. Nonetheless, in [7], a reduced common mode voltage modulation schemes (RCMVMS) namely active zero-state MS (AZMS) [7], near-state MS (NSMS) [7] and remote-state MS (RSMS) [7], have been proposed without zero vectors and results reduced CMV and the leakage current. But the performance parameters in the form of voltage linearity, DC-link current ripples, L-L voltage, losses and harmonic distortion (%HD) of the output current are degraded. Recently, one popular single-phase extended H5 topology was proposed in [4] but it has a higher switching count as well as more independent PV sources and results in higher losses [5].

In [6–9], a modified H7 topology is proposed by adding another switch at DC-side with boost converter, namely H8 inverter. But due to more switching count and two-stage operation the overall system efficiency is degraded. Similarly, another interesting topology is proposed in [10–14] using DC-bypass methods namely H7 three-phase transformerless inverter with proper modulation scheme. The authors have designed a very well topology such as each phase is clamped for 1,200 to achieve a common zero state. Once it's reached the PV and load are isolated through DC-bypass switch *S*5. However, it is limited to two-level output voltages only. Nonetheless, higher CM-leakage current will circulate through resonant circuits due to the active PV stray capacitance, and a few times due to climate changes, the flow of CM-leakage current will exceed the expectations of the German VDE 0126-1-1 standards due to the intermittent behaviour of stray capacitance. From the aforementioned discussions, it is clearly mentioned that single-phase with clamping topologies are as good as the unclamping topologies in terms of leakage current reduction. But now this statement is valid for TPT? So, in this chapter a better or proper conclusion will be given to the readers such as a simple modified three-phase topology and proper modulation scheme with clamping is introduced and investigated in detail to cover the gap between single-phase to three-phase concepts. Finally, the proposed structure is validated through the simulation. The performance characteristics in terms of CMV, CM-leakage current, total switching count and %THD are summarized in detail.

5.1.1 Single-Phase With-Transformer Inverter Working Principle

In general, converters DC voltage to single-phase AC voltage. For this purpose, two power devices are employed to change one source of power to another source of power, i.e., DC-AC. The other elements such as diodes and capacitors are used to supply or operate power supply to smooth manner such as filter the harmonics. As per the inverter function in the name of the half-bridge converter, the output voltage is between +0.5 and 0.5 V_{dc}. In the power module, all devices are employed with phase supply namely phase legs to the ground. Here switching devices are many ways depending on the application and power such as SCR, MOSFET and IGBT. In the PV applications, MOSFET and IGBT are used with high switching frequency to mitigate the harmonics. In most of cases, MOSFET is used in recent solar applications due to its excellent features in the form of device and conduction losses such as low switching and high conduction losses. In this method, a normal inverter is operating, similar to the transformerless PV inverter with load or using grid loop application [7]. Similar to without a transformer inverter, in positive power mode, two devices are ON and remain OFF to obtain the specified output voltage to balance grid power during abnormal power conditions. For one cycle of operation, two devices are ON. Similarly, in the negative power mode, remaining devices (*S*2, *S*3) are ON and remaining are OFF. As per the modulation scheme, it must follow the minimum time delay requirements; otherwise there is no proper output at the load side, and higher harmonics are generated, decreasing the efficiency of the PV system. However, here the output voltage is altered between +0.5 and 0.5 V_{dc} with the given time period. It is noted that the output voltage looks like stepped square wave because in inverters it also tries to alternate into two levels to convert into sine wave and the same principle in three-phase inverters [8].

5.2 COMMON MODE ANALYSIS OF THREE-PHASE TRANSFORMERLESS INVERTER TOPOLOGIES

In this module, a detailed investigation is carried out on common mode behaviour of various conventional and recently proposed three-phase transformerless inverters without clamping and clamping topologies for the minimization of the CM-leakage current in three-phase PV systems. In a three-phase PV system, the reliability of the system is mostly dependent on the size of the DC-link capacitor. The standard three-phase full-bridge inverter, shown in Figure 5.1, is commonly used in PV applications because of its simple design and easy implementation. It gives clarity on which topology is suitable for reduced CM-voltage and CM-leakage current in various grid-type applications

5.3 COMMON-MODE BEHAVIOUR OF H6 TOPOLOGY

As per the 1-phase systems, the design of the three-phase systems is very much difficult to achieve the desired output voltage with less CM-leakage current. In other ways, the modulation pulse pattern is designed properly with corresponding topology structure and gives better performance of whole PV system. As per the literature

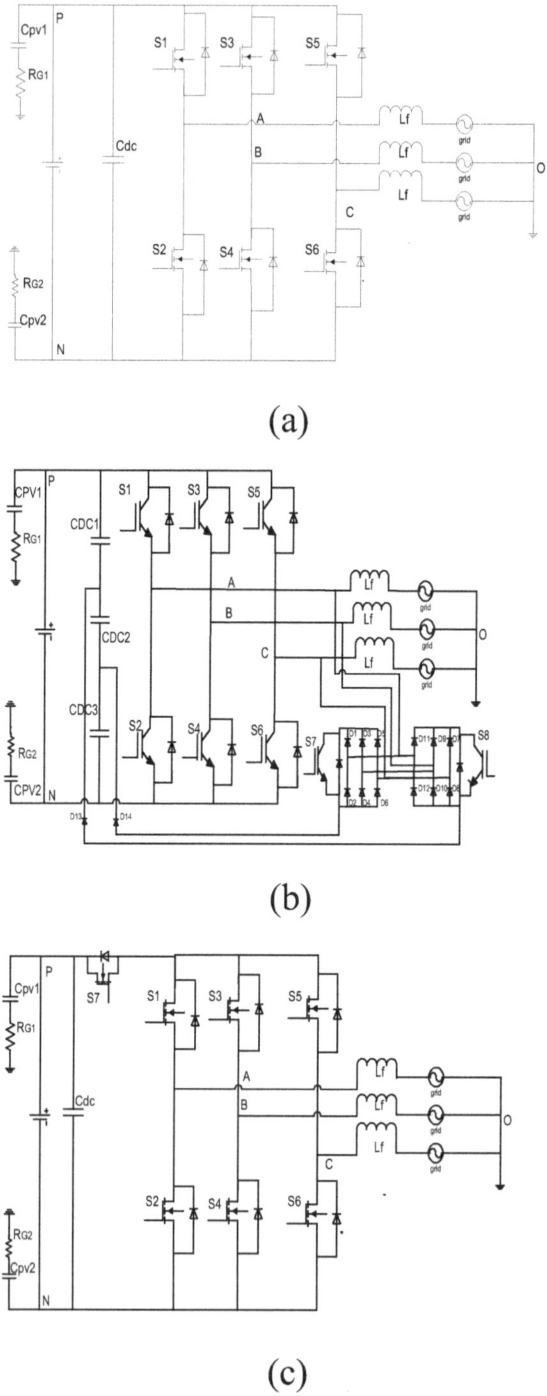

FIGURE 5.1 Existing three-phase TL-PV topologies: (a) H6, (b) modified-H6, and (c) H7 [7].

RCMV-schemes are only discussed and designed for CM-leakage current elimination and not validate whole performance parameters. In this chapter demonstrated the high-frequency CM model of the all topologies and proposed configurations with valid modulation pulse patterns for three-phase PV applications. And three-phase system is also designed without transformer such as resonant elements are employed at DC-side to flow the abnormal CM-leakage currents through PV to the ground. Here the CM-leakage current is purely depended on the power module elements and hence stray elements. To find the high-frequency circuit model of inverter structure, CM-voltage, DM-voltage are calculated for all given three-phase legs such as A and B, phase B and C, phase C and A only. Here the case study the identical filter inductors ($L1 = L2 = L3$) are employed and reached the desired CM-voltage, which is same to the 1-phase PV system. And all equations in the forms of V_{CM}, V_{dm}, V_{an} and V_{bn} is given in equation (5.1)–(5.3). And all the derivations are same in all pairs of phases. The modified HF-CM of the three-phase PV system is given in detailed. The CM-voltage is calculated the average of the two-phase legs voltage in [16] and given equation (5.1). And from the aforementioned discussions, it is confirmed that the limitation of the CM-leakage current is highly depends on the CM-voltage performance. Hence, proper and focused modulation pulse pattern for specific structure is required to CM-leakage current reduction.

And the switching pulse can be obtained through the universal pulse generator, which is made with single carrier-based modulation and corresponding logic functions. Here sinusoidal modulating signals can be represented by the *ma*, *mb*, *mc* and zero sequence signal is represented by the $V0$, as shown in Figure 5.2a. And a simplified CM model of the three-phase H6 TL-PVI system is given in Figure 5.2b. It is noticed that the CM-leakage current is increased only with rise of high-frequency CM-voltage and defined as average of three-phase leg voltages (V_{an}, V_{bn}, V_{cn}) is known as CMV such as [15]

$$V_{CM} = \frac{V_{an} + V_{bn} + V_{cn}}{3} \tag{5.1}$$

For understanding to the readers mode 4 switching operation along with equivalent circuit is represented in Figure 5.2. As per the mode 4 such as in active mode phase B and phase C switches $S3$, $S5$ are on and top switch of leg A is off. And the evaluation of CMV can find-out through the equation (5.1) [15].

where $V_{an} = 0$, $V_{bn=}$ V_{dc}, $V_{cn} = V_{dc}$

$$V_{CM} = \frac{0 + V_{dc} + V_{dc}}{3} = \frac{2V_{dc}}{3} = 0.66\,\mathrm{V_{dc}} \tag{5.2}$$

However, from the observation it is confirmed that CMV is varying between 0 and V_{dc}, which is equal to the DC bus voltage throughout the operating modes, as shown Table 5.1. Hence, the conventional three-phase H6 topology is not suitable for both EV and PV applications.

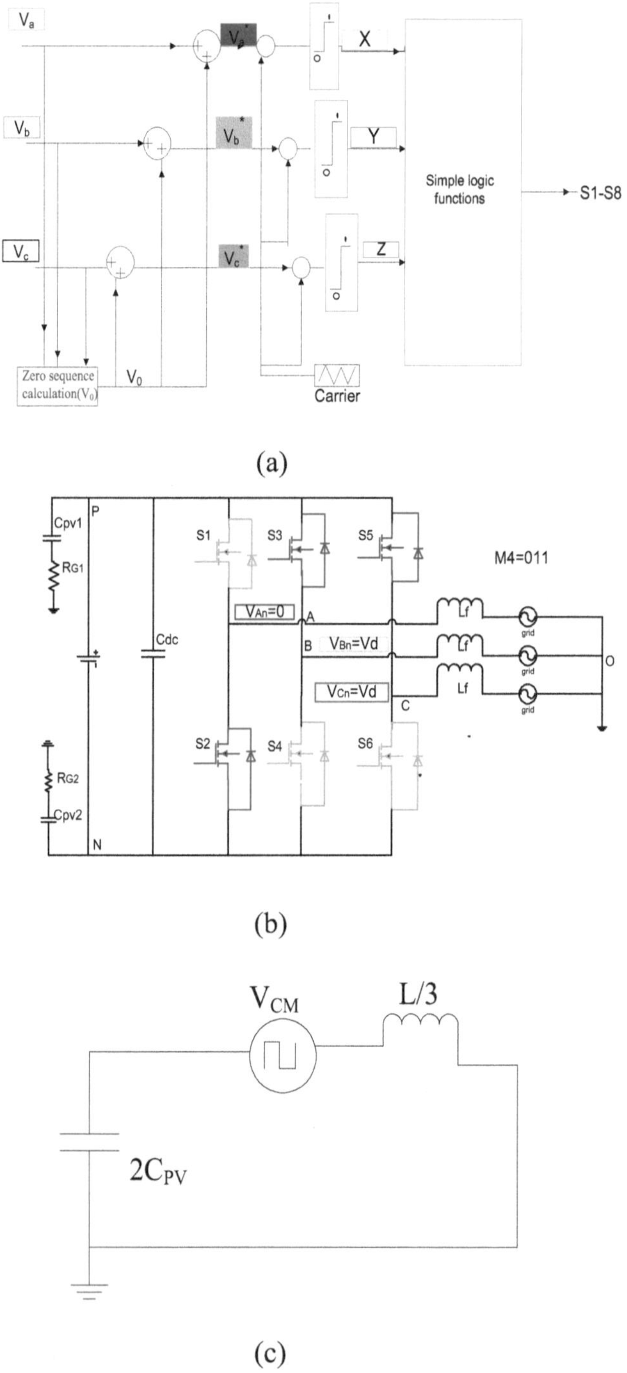

FIGURE 5.2 Three-phase H6 transformerless inverter: (a) pulse generator, (b) active switching state, and (c) CM model.

TABLE 5.1
CMV Evaluation

Mode	V_{An}	V_{Bn}	V_{Cn}	V_{cm}	V_{max}
M1(100)	V_{dc}	0	0	$V_{dc}/3$	V_{dc}
M3(010)	0	V_{dc}	0	$V_{dc}/3$	
M5(001)	0	0	V_{dc}	$V_{dc}/3$	
M2(110)	V_{dc}	V_{dc}	0	$2V_{dc}/3$	
M4(011)	0	V_{dc}	V_{dc}	$2V_{dc}/3$	V_{cmv_max}
M6(101)	V_{dc}	0	V_{dc}	$2V_{dc}/3$	
M7(111)	V_{dc}	V_{dc}	V_{dc}	V_{dc}	
M8(000)	0	0	0	0	0

5.4 COMMON-MODE EVALUATION OF H7 TOPOLOGY

$$\begin{cases} S2 = X + \left(\overline{X+Y+Z}\right), S1 = \overline{S2} \\ S4 = Y + \left(\overline{X+Y+Z}\right), S3 = \overline{S4} \\ S6 = Z + \left(\overline{X+Y+Z}\right), S5 = \overline{S6} \end{cases} \tag{5.3}$$

$$S4 = Y + \left(\overline{X+Y+Z}\right), S3 = \overline{S4}$$

$$S6 = Z + \left(\overline{X+Y+Z}\right), S5 = \overline{S6}$$

$$S7 = \overline{XYZ}$$

Based on the conventional three-phase H6 topology, an additional switch has been added on the DC-side namely 3-H7-TL-PVI topology, as given in Figure 5.1c. Here an interesting modulation is implemented with a few simple logic functions to generate the pulse to the switches ($S1$–$S7$), as shown in equation (5.3) [7,11].

The detailed operating modes and its CMV variations are shown in Table 5.2. For understanding to the readers here mode 8 such as during the freewheeling period now the isolation switch ($S7$) is operating and its equivalent circuit along with the CMV model is represented in Figure 5.3. As per the mode 8 phase A, B and C switches

TABLE 5.2
CMV Evaluation

Mode	V_{An}	V_{Bn}	V_{Cn}	V_{cm}	V_{max}
M1(1001)	V_{dc}	0	0	$V_{dc}/3$	V_{dc}
M3(0101)	0	V_{dc}	0	$V_{dc}/3$	
M5(0011)	0	0	V_{dc}	$V_{dc}/3$	
M2(1101)	V_{dc}	V_{dc}	0	$2V_{dc}/3$	
M4(0111)	0	V_{dc}	V_{dc}	$2V_{dc}/3$	V_{cmv_max}
M6(1011)	V_{dc}	0	V_{dc}	$2V_{dc}/3$	
M7(1111)	V_{dc}	V_{dc}	V_{dc}	V_{dc}	
M8(1110)	$3V_{dc}/4$	$3V_{dc}/4$	$3V_{dc}/4$	$3V_{dc}/4$	$3V_{dc}/4$

*S*2, *S*4, *S*6 are on and lower devices of phase leg *A*, *B*, *C* and *S*7 are off. However, there is no such line to circulate the CM-leakage current in this mode because V_{an}, V_{cn} decrease and V_{bn} is increase till to reach the values are same and it no practical possibility. And the CMV can be obtained as follows through equation (5.1).
where $V_{an} = 3V_{dc}/4$, $V_{bn} = 3V_{dc}/4$, $V_{cn} = 3V_{dc}/4$

$$V_{CM} = \frac{3V_{dc}/4 + 3V_{dc}/4 + 3V_{dc}/4}{3} = \frac{3V_{dc}}{4} = 0.75\,\mathrm{V_{dc}} \tag{5.4}$$

Figure 5.3a, shows the capacitances of devices *C*1–*C*7. And equivalent circuit of H7 and corresponding CM model are given in Figure 5.3b and c, where *R* is the ground resistance [15] and its initial voltages are represented clearly. It is clearly shows that the capacitor *C*3 is charged by *C*1, *C*5, and *C*7 in parallel.

Hence, the V_{an}, V_{bn}, V_{cn} *and* V_{cm} are as follows.

$$V_{CM} = V_{an} = V_{bn} = V_{cn} = \frac{(C_1 + C_5 + C_7)}{C_1 + C_3 + C_5 + C_7} V_{dc} \tag{5.5}$$

From Table 5.2, it is clearly revealed that H7 CMV amplitude is lower than to H6 topology. Hence, the leakage current performance in H7 is reduced significantly, which will be verified through the simulation results.

5.5 COMMON-MODE EVALUATION OF H8 TOPOLOGY

$$S1 = X + \overline{YZ}, S2 = \overline{X} + YZ, S3 = Y + \overline{XZ}, S4 = \overline{Y} + XZ$$

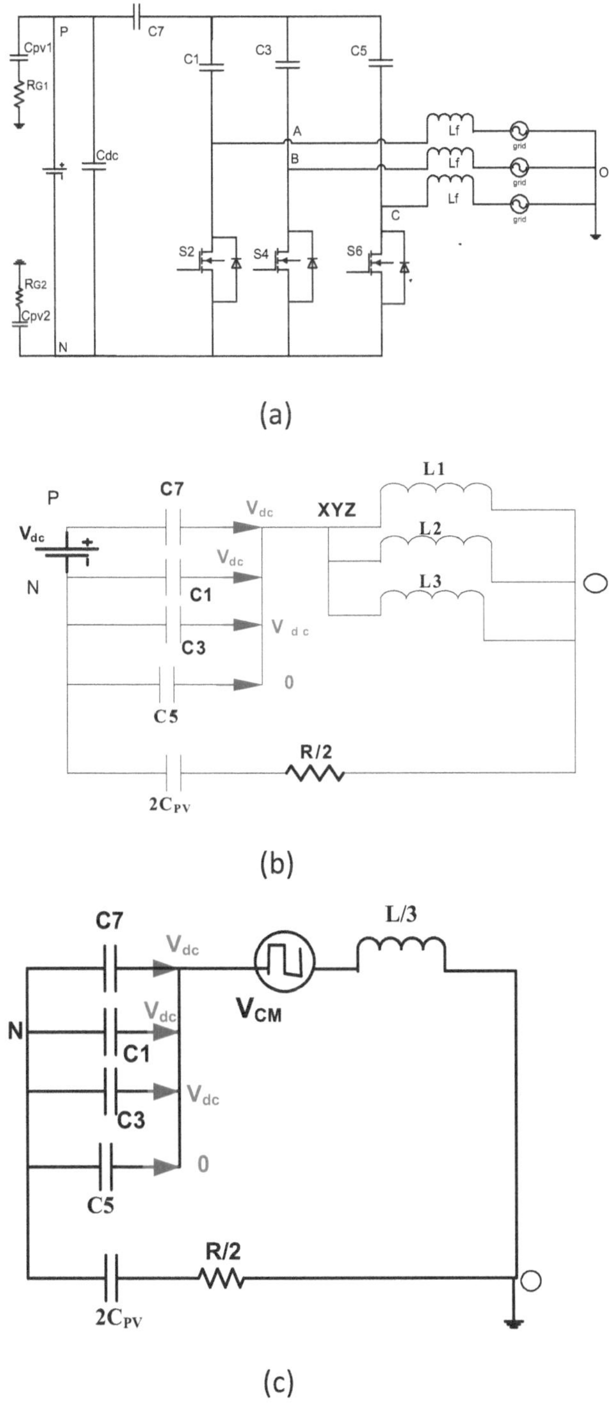

FIGURE 5.3 H7 topology: (a) freewheeling mode, (b) equivalent circuit, and (c) CM model.

$$S3 = Y + \overline{XZ}, S4 = \overline{Y} + XZ, S5 = Z + \overline{XY}, S7 = \overline{Z} + XY \quad (5.6)$$

$$S5 = Z + \overline{XY}, S7 = \overline{Z} + XY, S7 = S8 = \overline{X}Z + X\overline{Y} + Y\overline{Z}$$

Based on the three-phase H7 topology, an additional switch has been added on the DC-side PV-array namely DC-decoupling three-phase H8-TL-PVI topology, as shown in Figure 5.4a [15]. Here an interesting modulation is implemented with a few simple logic functions to generate the pulse to the switches ($S1$–$S7$), as shown in equation (5.4).

The detailed operating modes and its CMV variations are shown in Table 5.3. For understanding to the readers here mode 8 such as during the freewheeling period the isolation switches ($S7$, $S8$) is operating and its equivalent circuit along with the CMV model is represented in Figure 5.4. As per the mode 8 phase DC-decoupling switches are tuned off and remaining phase leg A, B, and C, top switches $S1$, $S3$, $S5$ and bottom switches $S2$, $S4$, and $S6$ are turned on. However, no line to circulate the CM-eakage current in this mode because V_{an}, V_{cn} decrease and V_{bn} is increase till to achieve their specified values and not applicable in practice. And the CMV can be obtained through equation (5.1).

where $V_{an} = V_{dc}/2$, $V_{bn} = V_{dc}/2$, $V_{cn} = V_{dc}/2$

$$V_{CM} = \frac{\frac{V_{dc}}{2} + \frac{V_{dc}}{2} + \frac{V_{dc}}{2}}{3} = \frac{V_{dc}}{2} = 0.5\,V_{dc} \quad (5.7)$$

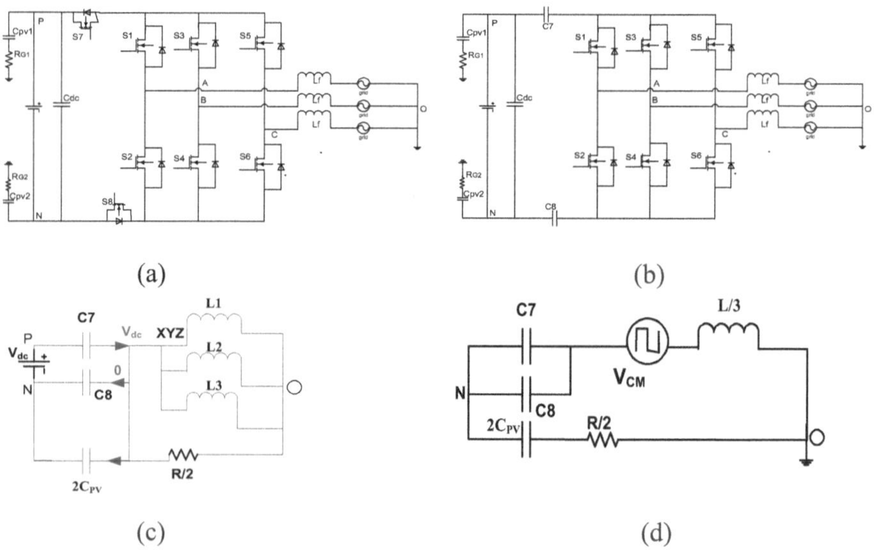

FIGURE 5.4 H8 topology: (a) structure, (b) freewheeling mode, (c) equivalent circuit, and (d) CM model.

TABLE 5.3
CMV Evaluation

Mode	V_{An}	V_{Bn}	V_{Cn}	V_{cm}	V_{max}
M1(1001)	V_{dc}	0	0	$0.33V_{dc}$	V_{dc}
M3(0101)	0	V_{dc}	0	$0.33V_{dc}$	
M5(0011)	0	0	V_{dc}	$0.33V_{dc}$	↑
M2(1101)	V_{dc}	V_{dc}	0	$0.66V_{dc}$	
M4(0111)	0	V_{dc}	V_{dc}	$0.66V_{dc}$	
M6(1011)	V_{dc}	0	V_{dc}	$0.66V_{dc}$	↓ V_{cmv_max}
M7(***0)	$0.5V_{dc}$	$0.5V_{dc}$	$0.5V_{dc}$	$0.5V_{dc}$	$0.5V_{dc}$

Figure 5.4a, shows the capacitances of devices *C*1–*C*7. And the equivalent circuit of H8 and corresponding CM model are given in Figure 5.4b and c, where *R* is the ground resistance [15] and its initial voltages are represented clearly. It clearly shows that the capacitor *C*3 is charged by *C*7 and *C*8 in parallel.

Hence the V_{an}, V_{bn}, V_{cn} *and* V_{cm} are as follows.

$$V_{CM} = V_{an} = V_{bn} = V_{cn} = \frac{C_7}{C_8} V_{dc} \tag{5.8}$$

From Table 5.3, it is clearly revealed that the H8 CMV amplitude is lower than that of the H6, H7 topologies if junction capacitances are identical; otherwise, it is oscillating during the whole grid period, which will be verified through the simulation results. Hence, it will be expected that the leakage current in H8 will be reduced significantly.

5.6 PROPOSED THREE-PHASE CLAMPING TOPOLOGIES: AN INVESTIGATION

In this section, single-phase extended three-phase clamping topologies and corresponding modulation strategies are investigated in detail. At the end, a proper conclusion has been given to readers, which strengthens the theoretical findings of both DC-decoupling and the feasibility of clamping-based topologies in the future. For investigations here, single-phase H5 and H6 topologies are considered in terms of three-phase transformation, as an additional switch has been added on the DC-side

PV-array namely clamping H7-Switch (H7-S) and H8-switch (H8-D), as shown in Figure 5.5a. Here similar modulation techniques are adopted in clamping topologies to check the effectiveness among DC-decoupling topologies such as logic functions are shown in equation to generate the pulse to the switches (*S*1–*S*7), as shown in equation (5.3) and Figure 5.6.

As discussed in Sections 5.3 and 5.4, the clamping branch switches are operated in freewheeling modes to prevent the CM-voltage oscillations. And the key role of clamping switches *S*7 and *S*8 are operated complementary to one to another to keep half of the DC voltage (0.5 V_{dc}) at phase legs A, B, and C points. Above, the CM-voltage oscillates in freewheeling mode such as the current circulates through the DC-side capacitor and the clamping device. From the observation it is noted that operating modes of H7, H7-S are almost same in active states except during freewheeling periods and CM model, as given in Figures 5.3c and 5.5d.

Similarly, in H8-D the clamping section employed with two diodes (*D*1, *D*2) at centre of the DC-link capacitor, which is given in Figure 5.5.

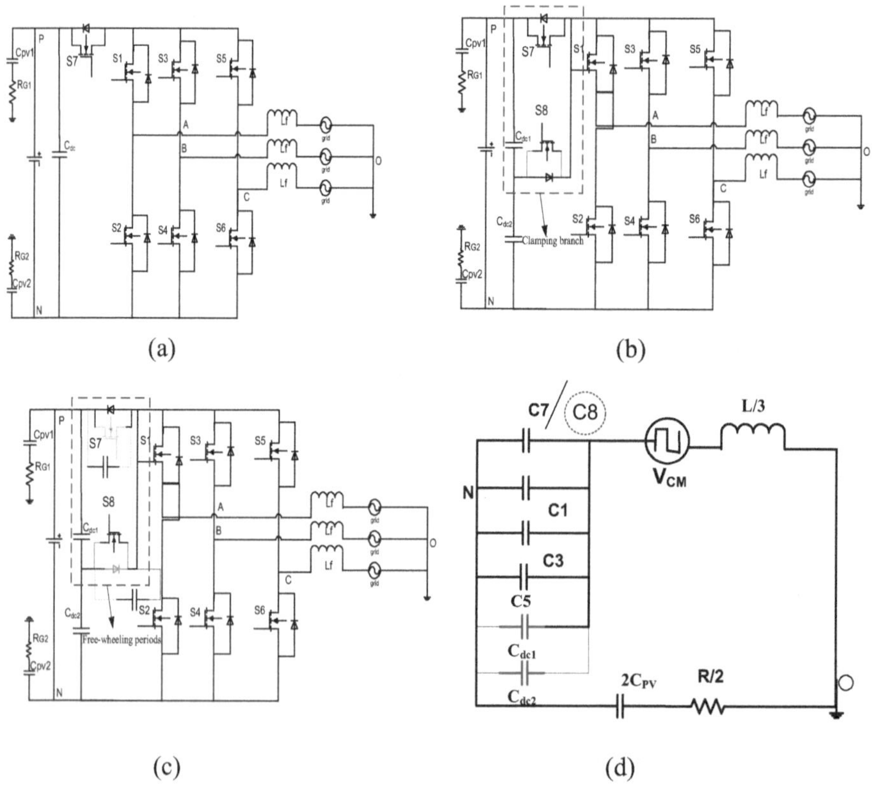

FIGURE 5.5 H7-S topology: (a) structure, (b) freewheeling mode, and (c) CM model.

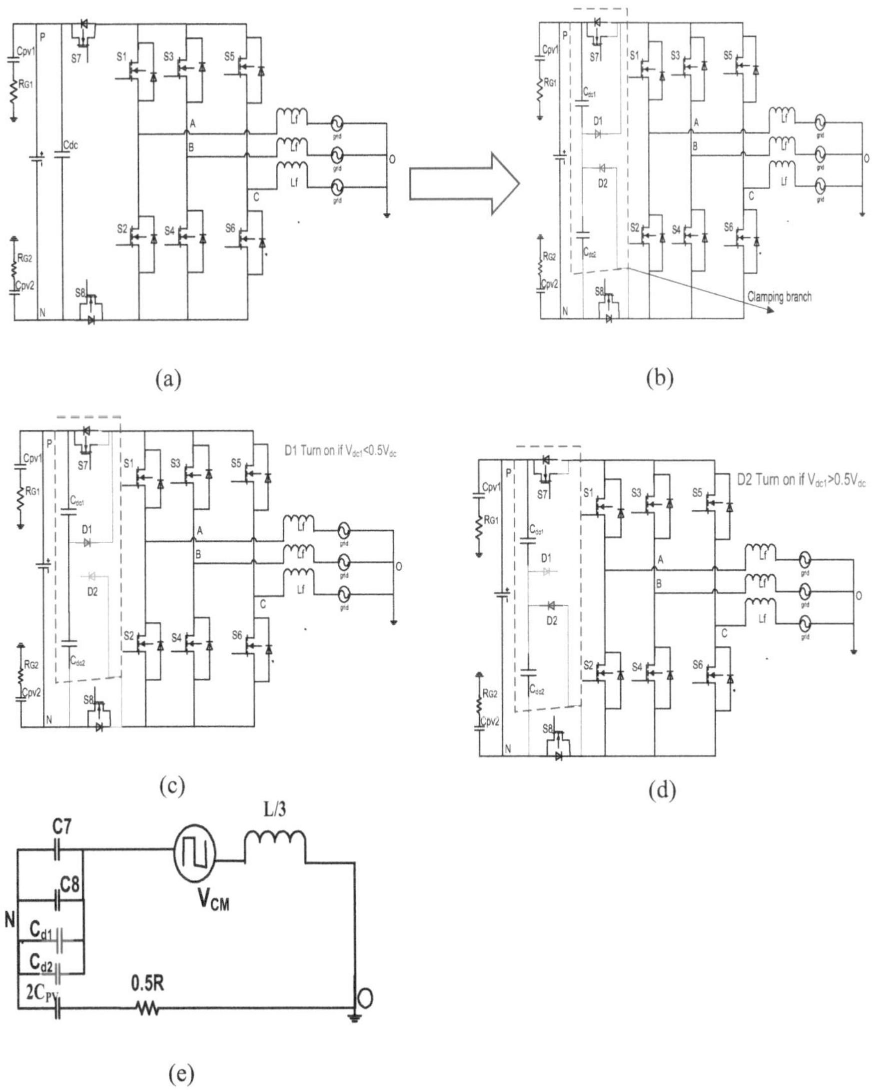

FIGURE 5.6 H8-D topology: (a) existing H8, (b) configuration, (c) freewheeling mode $V_{dc1} < 0.5\ V_{dc}$, (d) freewheeling mode $V_{dc1} > 0.5\ V_{dc}$, and (e) CM model.

In the freewheeling mode, the *D*1 is turned on if unclamped branch voltage is less than 0.5 V_{dc} else *D*2 is active when DC-link capacitor voltage is more than 0.5 V_{dc}. And the CM model is as follows in Figure 5.6c and d.

Hence in Tables 5.2 and 5.3 gives the above modulation schemes, it clearly shows that clamping-based topologies are better common mode performance among DC-decoupling topologies, which will be verified in simulation results in the next section (Table 5.4).

TABLE 5.4
Performance Analysis of Conventional and Clamping-Based Three-Phase Topologies

	Conventional and DC-Decoupling			Clamping	
Topology	**H6**	**H7**	**H8**	**H7-S**	**H8-D**
Output voltage	Unipolar	Unipolar	Unipolar	Unipolar	Unipolar
No. of switching count	6	7	8	8	9
No. of diodes	0	0	0	0	2
CMV	0–V_{dc}	0–V_{dc}	0–V_{dc}	$3V_{dc}/10$–$5V_{dc}/6$	$V_{dc}/3$–$2V_{dc}/3$
CMC (mA_{rms})	>0.3	>0.3	>0.3	<0.3	<0.3
THD_i (%)	5.23	5.24	4.17	5.36	4.38

5.7 SIMULATION RESULTS

To validate the theoretical findings of the conventional TPT with DC-decoupling and clamping inverters are performance through the simulation results. The parameters are used in the simulation as follows; input DC voltage of 300 V, stray capacitors are considered (C_{PV1}, C_{PV2}) with 220 nF, ground resistance (R_{G1}, R_{G2}) is 11 Ω, three symmetrical filter inductors ($L1$, $L2$, $L3$) with value of 5 mH, switching frequency is 10 kHz respectively [7].

5.8 OUTPUT VOLTAGE AND LEAKAGE CURRENT PERFORMANCE

The output voltage and CM-leakage current V_{out}, $I_{out,}$ V_{cm}, i_{leak} of DC-decoupling and clamping topologies are presented in Figures 5.7 and 5.8. It reveals that all topologies (DC-decoupling and clamping) are three-level ($+V_{dc}$, 0, $-V_{dc}$) output voltage with sinusoidal output current, as given in Figure 5.7.

In addition to that, conventional H6, H7, and H8 topologies fail to attenuate the CMV such as oscillate between 0 and V_{dc} and higher the leakage current, as demonstrated shown in Figure 5.8a–c. It is reduced through the proposed clamping-based H7-S and H8-D such CMV is reduced in H7-S is $3V_{dc}/10$–$5V_{dc}/3$ and H8-D is $V_{dc}/3$–$2V_{dc}/3$ and henceforth low CM-leakage current, follows as per the VDE0126-1-1 standards i.e. below 300 mA, as shown in Figure 5.8d and e and which is in good agreement with the theoretical findings [16]. The total harmonic distortion (%THD) also measured through the simulation platform, which are with magnitudes of 5.26%, 5.24%, 4.17%, 5.36%, and 4.38%, respectively, as shown in Figure 5.9. However, the %THD is higher in H7-S and H8-D, as compared with the H7, H8 topologies due to parasitic capacitor effect and more count topological structure but followed as per IEEE1547 standards [17].

5.9 CONCLUSION

This chapter has demonstrated in detail the analysis of various three-phase transformerless inverter topologies, such as conventional-H6, DC-decoupling-H7, H8, and

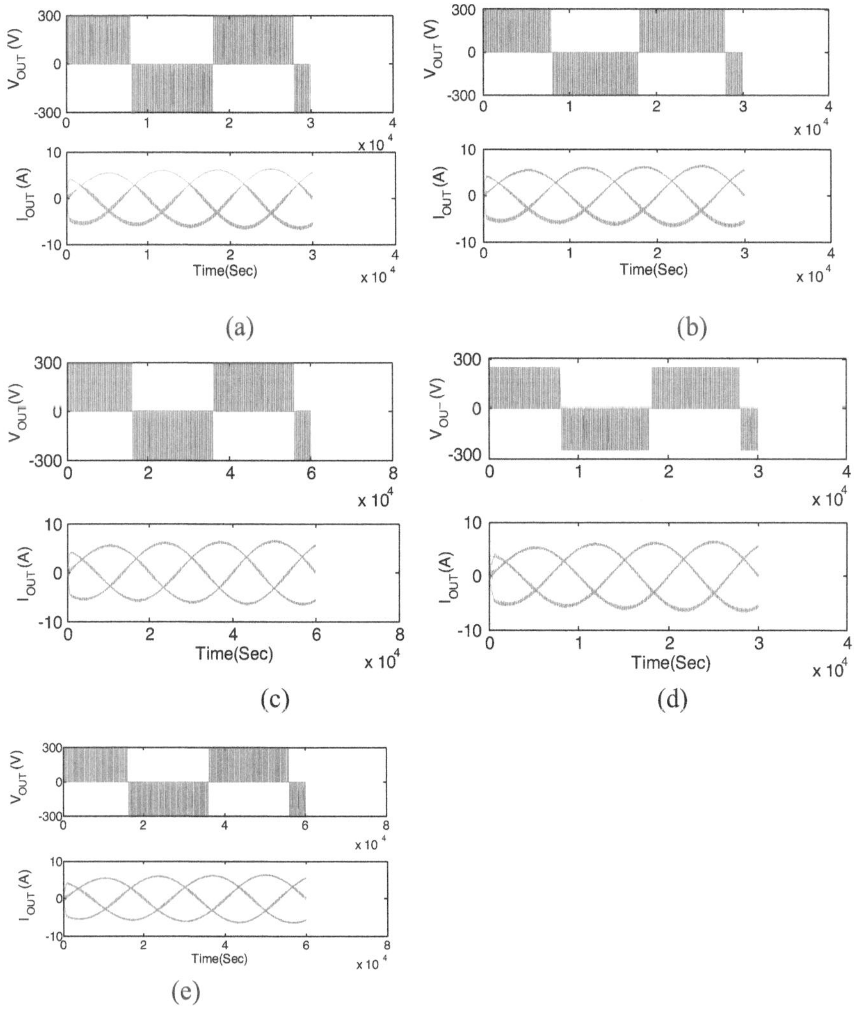

FIGURE 5.7 Output test parameters of V_{out} and I_{out}, for (a) H6, (b) H7, (c) H8, and (d) proposed H7-S, and (e) H8-D topologies.

clamping H7-S and H8-D, respectively. For readers understanding, a detailed step-by-step theoretical procedure for mode of operation along with the common mode model are presented and discussed thoroughly. All discussed topologies theoretical findings are validated through simulation results. And a new insight conclusion has arrived, such as that the DC-decoupling topologies are excellent than clamping topologies in the forms of CM-leakage current reduction. On the other hand, H8 topology is better than the H7 topology in the form of reduced CMV, and leakage current is reduced significantly. Finally, the DC-decoupling H8 and clamping H8-D topologies are good choices among all DC-decoupling topologies for the CM-leakage current attenuation. Hence it is feasible to both PV and EV applications [18,19].

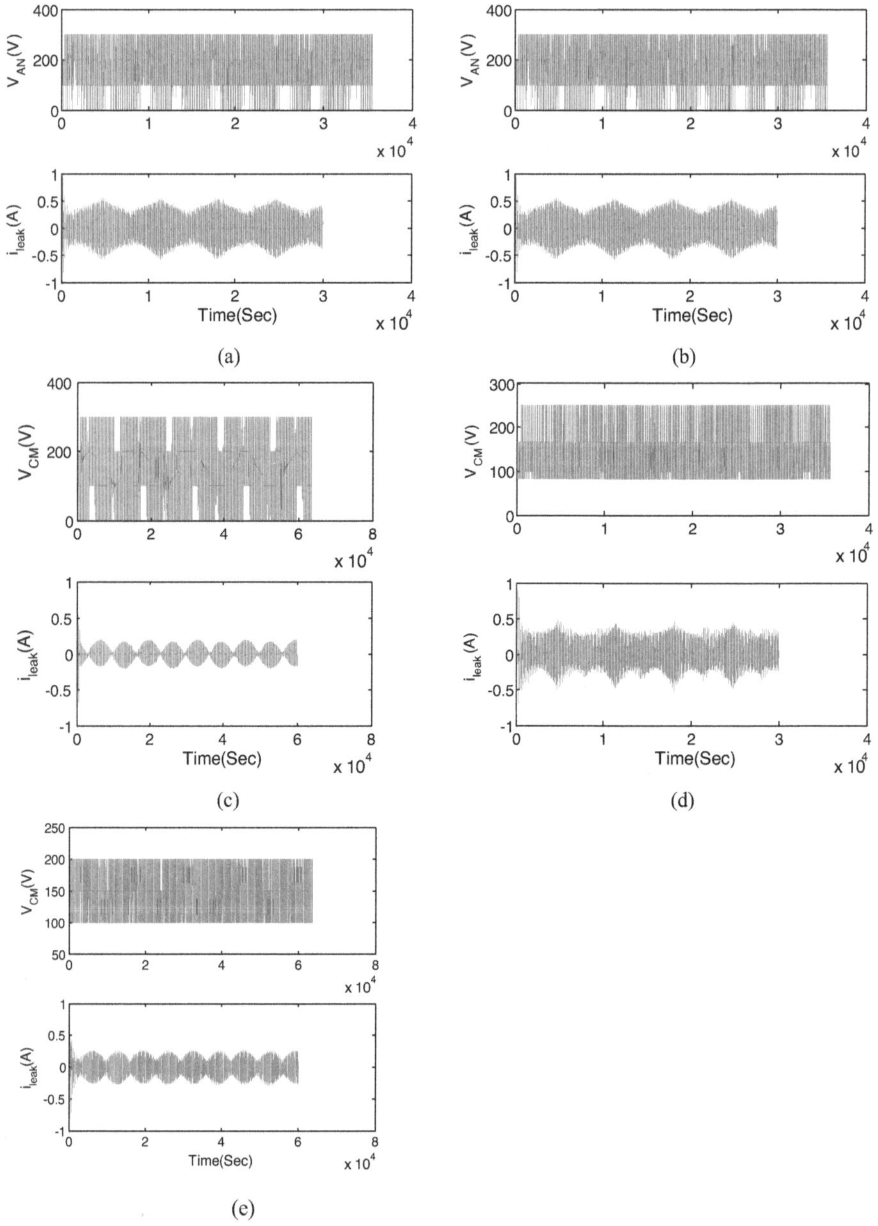

FIGURE 5.8 Common mode test parameters of V_{CM}, i_{leak} for (a) H6, (b) H7, (c) H8, (d) proposed H7-S, and (e) H8-D topologies.

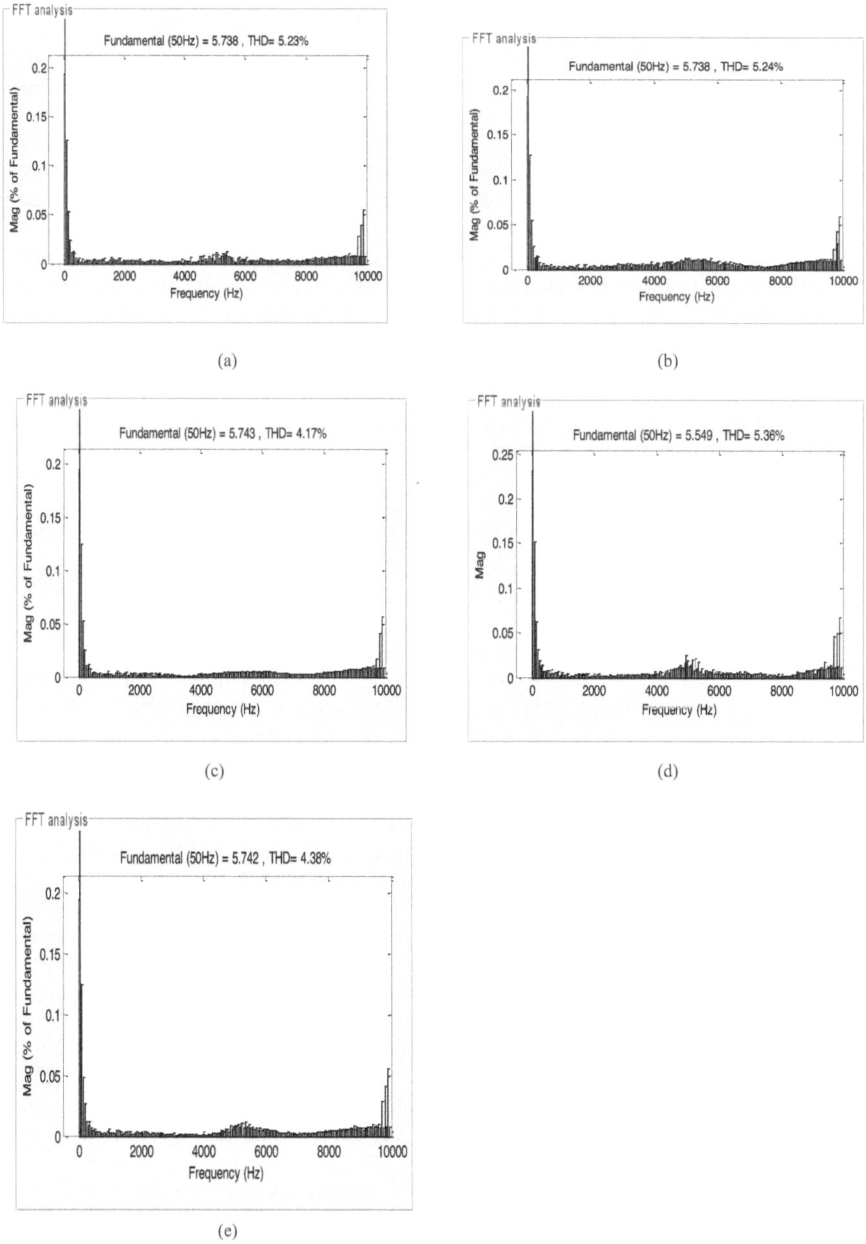

FIGURE 5.9 Total harmonic distraction (a) H6, (b) H7, (c) H8, (d) proposed H7-S, and (e) H8-D topologies.

REFERENCES

[1] G. Carli and S. S. Williamson, "Technical considerations on power conversion for electric and plug-in hybrid electric vehicle battery charging in photovoltaic installations," *IEEE Transactions on Power Electronics*, vol. 28, pp. 5784–5792, 2013

[2] M. C. Cavalcanti, K. C. de Oliveira, A. M. de Farias, F. A. S. Neves, G. M. S. Azevedo and F. C. Camboim, "Modulation techniques to eliminate leakage currents in transformerless three-phase photovoltaic systems," *IEEE Transactions on Industrial Electronics*, vol. 57, no. 4, pp. 1360–1368, 2010.

[3] X. Li, X. Xing, C. Zhang, A. Chen, C. Qin and G. Zhang, "Simultaneous common-mode resonance circulating current and leakage current suppression for transformerless three-level T-type PV inverter system," *IEEE Transactions on Industrial Electronics*, vol. 66, no. 6, pp. 4457–4467, 2019.

[4] F. Faraji, S. M. Mousavi G. A. Hajirayat, A. A. Motie Birjandi and K. Al-Haddad, "Single stage-single phase three-level neutral-point-clamped transformerless grid-connected photovoltaic inverters: Topology review," *Renewable and Sustainable Energy Review*, vol. 80, pp. 197–214, 2017.

[5] H. Xiao, L. Zhang and Y. Li, "An improved zero-current-switching single-phase transformerless PV H6 inverter with switching loss-free," *IEEE Transactions on Industrial Electronics*, vol. 64, no. 10, pp. 7896–7905, 2017.

[6] X. Guo, "A novel CH5 inverter for single-phase transformerless photovoltaic system applications," *IEEE Transaction on Circuits and Systems II: Express Briefs*, vol. 64, no. 10, pp. 1197–1201, 2017.

[7] T. K. S. Freddy, N. A. Rahim,W. P. Hew and H. S. Che, "Modulation techniques to reduce leakage current in three-phase transformerless H7 photovoltaic inverter," *IEEE Transactions on Industrial Electronics*, vol. 62, no. 1, pp. 322–331, 2015.

[8] L. Zhou, F. Gao and T. Xu, "A family of neutral-point-clamped circuits of single-phase PV inverters: Generalized principle and implementation," *IEEE Transactions on Power Electronics*, vol. 32, no. 6, pp. 4307–4319, 2017.

[9] X. Guo, J. Zhou, R. He, X. Jia and C. A. Rojas, "Leakage current attenuation of a three-phase cascaded inverter for transformerless grid-connected PV systems," *IEEE Transactions on Industrial Electronics*, vol. 65, no. 1, pp. 676–686, 2018.

[10] A. Hota, S. Jain and V. Agarwal, "An improved three-phase five-level inverter topology with reduced number of switching power devices," *IEEE Transactions on Industrial Electronics*, vol. 65, no. 4, pp. 3296–3305, 2018.

[11] X. Guo, D. Xu and B. Wu, "Three-phase seven-switch inverter with common mode voltage reduction for transformerless photovoltaic system," In: *Proceedings of 40nd Annual Conference, IEEE Transactions on Industrial Electronics*, 2014, pp. 2279–2284.

[12] F. Faraji, S. M. Mousavi, G. A. Hajirayat, A. A. M. Birjandi and K. Al-Haddad, "Single stage-single phase three-level neutral-point-clamped transformerless grid-connected photovoltaic inverters: Topology review," *Renewable and Sustainable Energy Review*, vol. 80, pp. 197–214, 2017.

[13] G. S. Chandini, G. Shiny and M. R. Baiju, "Common-mode voltage eliminated 2-level PWM inverter based on a cascaded 3-level inverter," In: *Proceedings of International Conference on Electrical, Electronical and Optimized Technologies*, pp. 561–566, 2016.

[14] V. Sonti, S. Jain and S. Bhattacharya, "Analysis of the modulation strategy for the minimization of the leakage current in the PV grid-connected cascaded multilevel inverter," *IEEE Transactions on Power Electronics*, vol. 32, no. 2, pp. 1156–1169, 2017.

[15] X. Guo, N. Wang, B. Wang, Z. Lu and F. Blaabjerg, "Evaluation of three-phase transformerless DC-bypass PV inverters for leakage current reduction," *IEEE Transactions on Power Electronics*, vol. 35, no. 6, pp. 5918–5927, 2020.

[16] DIN VDE 0126-1-1, *Automatic Disconnection Device Between A Generator and the Public Low-Voltage Grid.* DIN Electro Technical Standard DIN VDE 0126-1-1, 2005.

[17] IEEE Std 1547.1-2005, *IEEE Standard Conformance Test Procedures for Equipment Interconnecting Distributed Resources with Electric Power Systems.* IEEE, 2005, pp.442–449.

[18] V. Sonti and S. Jain, "DC decoupling-based three-phase three-level transformerless PV inverter topology for minimization of leakage current," *IEEE Transactions on Industrial Electronicsm,* vol. 66, no. 10, pp. 8273–8278, 2019.

[19] X. Guo, R. He, J. Jian, Z. Lu, X. Sun and JM Guerrero, "Leakage current elimination of four-leg inverter for transformerless three-phase PV systems," *IEEE Transactions on Power Electronics*, vol. 31, no. 3, 1841–1846, 2016.

[20] M. Cacciato, A. Consoli, G. Scarccella, and A. Testa, "Reduction of common-mode currents in PWM inverter motor drives," IEEE Trans. Ind. Appl., vol. 35, no. 2, pp. 469–476, Mar./Apr. 1999.

[21] A. M. Ilava and E. Ün, "A high-pcrformancc PWM algorithm for common-modc voltage reduction in three-phase voltage source inverters," IEEE Trans. Power Electron., vol. 26, no. 7, pp. 1998–2008, Jul. 2011.

[22] A. M. Hava and E. Ün, "Performance analysis of reduced common-mode voltage PWM methods and comparison with standard PWM methods for three-phase voltage-source inverters," IEEE Trans. Power Electron., vol. 24, no. 1, pp. 241–252, Jan. 2009.

6 Evaluation of Single-Phase Rectifier Bridge Clamping Circuit Solar Power Transformerless Inverter Topologies for Electric Vehicle Applications

Ahmad Syed
Chaitanya Bharathi Institute of Technology (A), Hyderabad

Tara Kalyani Sandipamu
Jawaharlal Nehru Technological University, Hyderabad

Freddy Tan Kheng Suan
University of Nottingham, Malaysia

Xiaoqiang Guo
Yanshan University

Huai Wang
Aalborg University

B. Mouli Chandra
QIS College of Engineering & Technology, Ongole

6.1 INTRODUCTION

In the present century, the development of human civilization is facing two major challenges: energy and environmental conditions. However, the world demands more energy generation and consumption as well because, day by day, the increased

DOI: 10.1201/9781003354901-6

population and industries lead to peak energy demands and result in environmental failure in terms of pollution. It is required to protect the environment as per the relevant laws and regulations and corresponding modifications based on the present century. In other words, this energy advancement technology has been modified by government policies and regulations to support "energy saving" and "zero emission". Therefore, one of the most popular solutions in the 21st century is electric vehicle technology (EV-T), which is a leading technology and also supports the two major concerns of reducing energy consumption and achieving zero emissions. In future, it will lead to an innovative load of power grid if more EV charging stations are increased, but this will require attention for grid integration, such as equilibrium load, capacity, and power quality, which are impacted on the grid hence need more in-depth study and analysis on the EV charging. On the other hand, EV-T contributes to the following categories: first, urban environment pollution; second, getting rid of petroleum resources; and last, industry-related development. It has been selected in three ways: pure EV, fuel cell vehicle, and hybrid EV. And a lot of developments have been made in the EV-T to promote and establish their own EV in the present market for future demands and policies. And a lot of developments in home and abroad areas such as an American EV Research and Development Programs (ARDP) has partnership for a New Generation of Vehicles Plan (PNGV) and Freedom Cooperative Automotive Research Partnership (FCARP). Japan has implemented the plans for the reduced pollution vehicles and Hydrogen and Fuel Cell Demonstration Project. Similarly, EU Plans are Research Development and Design Plan(R&DD) and Fuel Cell Bus Demonstration Program (FC-BDP). And lastly China has an EV special sequencer of China 973 plan Development of EVs.

In addition, battery charging has been categorized into three ways: constant current (CC) charging (CC-charging), constant voltage (CV) charging (CV-charging), and pulse charging (PC). In the CCC, it requires a small amount of current for a long time of charging, and this is the main drawback of the method as well. Similarly, in CVC, the voltage is constant and charging current is decreasing, and the key benefit of this scheme is that charging time is short, it is simple to control, it is not easy to fully charge, and the initial current is high, which results in reduced battery life. Another excellent method is pulse current charging, which is fast charging the battery and hence reduces the charging time. It is not fully charged and has an effect on battery life and hence the battery management system plays a main role in the EV-T.In addition, the plug-in hybrid EV (P-I-HEV) applications are popular in grid-connected applications due to their eminent features in terms of zero emissions and novel functionalities [1]. In addition to that, the growth of EV technology is increasing because of declined manufacturing costs and advancements in power electronics technology such as converters with two stages and battery chargers [2], which leads to factors of vehicle-to-grid technology (V-G). Nonetheless, all EVs are designed with onboard chargers with different levels (1, 2). Compared to level 2, level 1 chargers are slower but more available. And other end bidirectional onboard power conditioners are ac-dc converters [3,4] are highly efficient in grid-connected applications. Another popular technology is voltage source inverters (VSI) uses dc-dc converters, which are cost-effective and efficient. However, one universal solution can be provided through transformerless inverter topologies, such as high-frequency isolation, which has been employed in V-G inverters and leads to common mode and leakage current issues between the DC-link and the grid. And hence DC-link losses increase significantly during the freewheeling periods. Based on that, to improve the efficiency of transformerless inverters, special arrangements are required compared to the conventional VSI topologies. And few authors have investigated different converter topologies with a variety of structures in [1–4] but its derivation methodology and working principle are not clearly revealed.

In addition to those photovoltaic applications, it is also one of the green resources for power generation to meet the world's demands. In that case, most PV systems are connected with the grid because of their impact in terms of compactness, lower price, and higher efficiency [5,6], and they supply the surplus power to the grid if solar power is higher than local claim. Moreover, this process's major goal is to attract policymakers and members to pay more attention to the detailed examination and growth of grid-embedded solar inverters [7,8]. Nonetheless, it is classified into two ways, namely disconnected/isolated and non-disconnected/isolated systems with the help of location of the transformer during the power conversion step. To accomplish more efficiency, be smaller in shape, and be cheaper, a universal solution is to eliminate the line frequency transformer in non-isolated inverters, which results in a direct path between the solar plates and the grid. Therefore, high CM-voltage oscillation is formed by the shifting of the solar inverter, which can stimulate the resonant elements among the stray capacitance of the solar panels, filter, and impedance of the grid. Hence, CM-current or leakage current circulates through stray capacitive element of the solar panel to the ground. On the other hand, it creates issues related with waveform quality, electromagnetic compatibility (EMC), and electromagnetic compatibility interference (EMI), which leads to poor safety [9,10]. In this way, the resonant circuit elements are excited such as PV high level switching frequency oscillations are formed through inverter and hence currents oscillate through the stray capacitive element, filter components and lastly through the grid impedances to the ground. It creates the dangerous issues in terms of earth currents known as CM-current through parasitic elements of PV panels through the ground. And hence poor output wave quality such as increased distortions, electromagnetic compatibility (EMC) and also poor safety on grid side coupling. To overcome these issues, many investigations have been done with the help of normal full-bridge configurations with unipolar and bipolar modulation schemes. But it has not met safety grid-codes and standards due to higher leakage current and losses. In order to eliminate these issues various topologies were introduced by incorporating an additional circuitry to isolate the solar panel from the grid during the freewheeling period, namely DC and AC bypass methods. Here H5, HERIC, and H6 configurations are more famous and well-adapted inverters in the present market for transformerless PV applications. In H5, the DC-bypass is provided through one additional switch, where it is added in series to the dc supply [18], which is discussed clearly here.

Single-phase de-decoupling topology, namely an H5 circuit, is proposed using an H4-bridge inverter, and it was invented by SMA technology to limit the common-mode leakage current issues in transformerless PV inverter systems [5]. This configuration, as presented in Figure 6.1a, and from the observations, it clearly looks that H5 structure is similar to the H4 except the one additional branch is incorporated

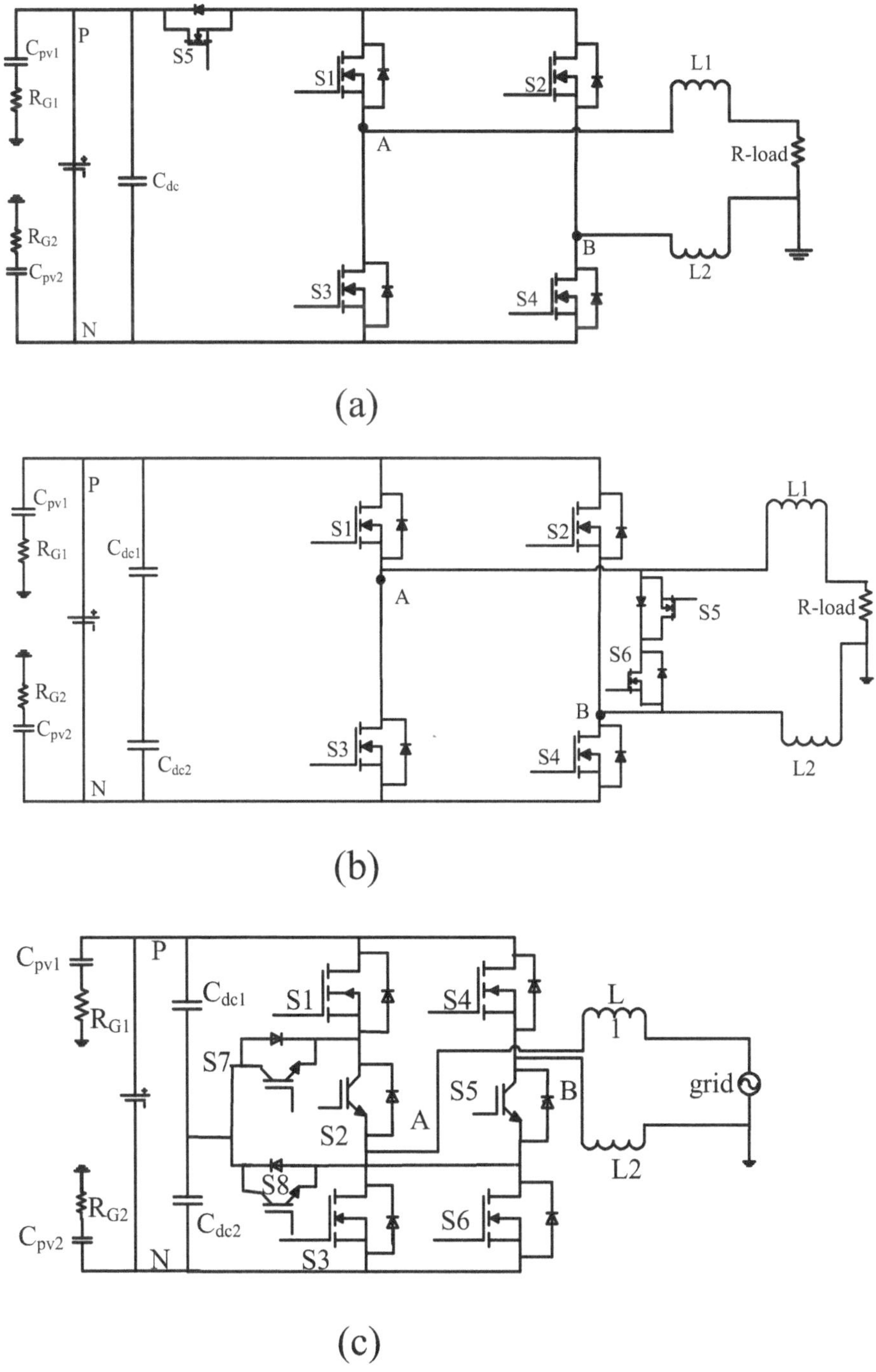

FIGURE 6.1 Existing single-phase SP-TLI topologies: (a) H5 [18], (b) HERIC [19] and (c) H6-family (PN-NPC) [23].

at the DC side of the normal topology like H4 such as an extra power device (*S*5) is added between input DC side to the DC-link capacitive element and the phase-leg bridge. This extra switch *S*5 is working as an isolator such as to disconnect the PV from the grid under abnormal conditions. In other words, an extra switch disconnects the dc side PV panel if zero states are achieved through the grid side, it is an operative approach to diminish the common mode-LC in transformerless PV inverter systems [7,8]. And the H5 structure is capable of bidirectional power flow based on suitable PWM methods, as it also follows the conventional H4 topology in the bidirectional case study. In the positive grid cycle the switches *S*1, *S*4, *S*5 are turned ON and remain turned OFF. In this mode, the CMV is matched to the input dc voltage (V_{dc}). And negative grid cycle *S*2, *S*3, *S*5 are ON and remains are OFF. In the negative mode also, the CMV is matched to the $-V_{dc}$. Through the freewheeling grid periods the direction of current is circulating through the diode of switch *S*2 and switch *S*1 and vice versa. But in the freewheeling periods the output voltage in terms of CMV is matched to half of the input dc voltage ($V_{dc}/2$). Similarly in ac bypass galvanic isolation is provided through two additional switches, which are added in parallel with the grid [19]. However, the topologies with decoupling methods fail to retain constant CM-voltage but still the common mode ground current (CM-ground current).

The HERIC structure is given in Figure 6.1b. From the observations, two back-to-back *S*5 and *S*6 devices are employed in conventional H4-bridge to disconnect the input dc side from the grid side during zero voltage states. During active grid power modes, HERIC works similarly to H4-bridge, which is capable of transferring bidirectional power flow in these modes of operation. In most of the literature it is revealed that all modulation methods are limited as per the reactive power capability such as proposed PWM for the HERIC topology, which is not capable of generating reactive power. It is one of the major thorns in the future or present inverter structure, which is similar issue in the H5 topology and needs to be modified accordingly. The switching technique is described in [5]. As per the PWM scheme, the operating *S*1–*S*4 are at fired with high frequency and others *S*5, *S*6 at grid power frequency. The operation sequence is same as conventional H4 bridge such as in the positive power grid mode *S*1, *S*4 is ON and others are OFF and negative power grid mode *S*2, *S*3 is ON and others are OFF but the difference is easily found in terms of its freewheeling actions. In positive power grid freewheeling modes, the current circulates through grid frequency either *S*5 and/or diode of *S*6 and vice versa. And many novel PWM methods have been proposed and illustrated recently to permit the supply of the reactive power to the grid. In [6], during the intervals grid path voltage is injected but the polarity is not the same, such as when switches *S*6 and *S*5 are switched at high-level frequency. *S*5 is operated quite in reverse to *S*2, *S*3, while *S*6 is conducted in complementarity to *S*1, *S*4 devices. The devices *S*6, *S*5 are only conducted at high power frequency through non-unity power factors, and hence the direction of voltage and current is employed with reverse polarity; And switch improvements are similar to the conventional PWM method using unity power factor (UPF).

H6 topology is feasible with either dc or ac bypass based on the locations of the devices (switch/diode) are employed either in series and/or parallel to supply and the grid [15], where all configurations are recorded more efficiently in practice approach.

Similarly, the FB-DCB-based structure here is a modified conventional H-bridge, which was announced by Ingeteam [16]. It is a bidirectional AC-DC converter that, with the help of a conventional H4-bridge converter, can be utilized to energy storage applications. It is implemented with two additional switches and diodes, and it is used to isolate the input DC end from the grid end during zero-state voltage. In general, these power devices are employed in the H4-bridge converter to alleviate the CMLC i.e. current circulation path is provided in freewheeling path. Hence key methods are useful to mitigate the CM-leakage current of transformerless PV inverter and hence boost the total system efficiency by anticipation of reactive power conversation between the symmetrical line filter $L1$, $L2$ and $C1$, $C2$ dc-link capacitors. According to the PWM methods here many modulation schemes are proposed for these configurations. As per the first conventional PWM schemes, switches $S1$, $S4$ are ON in positive power grid mode and $S2$, $S3$ are ON in negative power grid mode. Therefore, $S1$, $S2$, $S3$, and $S4$ are conducted at grid power frequency. However, zero voltage states are provided in both positive power and negative power modes, $S5$, $S6$ are conducted at the switching frequency [8]. The switching method for this modulation scheme is in [8]. On the other hand, another modulation method, $S1$, $S4$ are ON in positive power grid mode and $S2$, $S3$ are conducted together at high-frequency complementarity to $S5$, $S6$. In dissimilarity, during negative half cycles, $S2$ and $S3$ are turned ON and $S1$, $S4$ are switched together at switching frequency complementarity to $S5$ and $S6$ [8]. The switching method of the modulation technique is depicted in [8]. However, due to the absence of clamping ability, the leakage current issues are resolved properly because of the floating terminal phase voltages in the freewheeling grid period. Nonetheless, higher CM-leakage current will circulate through resonant circuits due to the excitement of the PV stray capacitance and occasionally due to climate changes the flow of CM-leakage current beyond the German VDE 0126-1-1 levels due to intermittent behaviour of stray capacitance. [10].

Similarly, a modified FB-DCB-based structure is proposed with a modified H-bridge with clamping ability namely positive-negative neutral-point clamped (PN-NPC) [16]. It is also a high-power converter with bidirectional capability using conventional H4-bridge topology utilized for energy storage applications. It is implemented with two additional switches, and it is used to isolate the input dc voltage on the PV panel from the grid power side during zero voltage level states. In general, these power devices $S7$, $S8$ are employed to H4-bridge topology in order to overcome the CMLC issue through a circulating current such as freewheeling path. These key measurement techniques are employed to mitigate the common mode leakage current of transformerless PV topology systems and hence the improved system total efficiency with no reactive power exchange between the symmetrical line filter $L1$, $L2$ and dc-link capacitors $C1$ and $C2$. According to the PWM methods here three key modulation techniques are proposed for these configurations. As per the first conventional PWM schemes, devices $S1$, $S2$, $S5$, $S6$ are ON in positive power half cycles and $S3$, $S4$ are ON in negative power half cycles. Therefore, $S2$ and $S5$ are conducted at grid frequency. To provide zero voltage states in both positive power and negative power half cycles, $S7$, $S8$ are switched at the switching frequency [8]. The switching method for this modulation scheme is in [8]. On the other hand, the second modulation scheme, $S1$, $S2$, $S5$, and $S6$ are ON in positive power half cycles and $S2$, $S5$ are conducted with the help of both at high-frequency complementarity to $S7$, $S8$. In dissimilarity, during negative power half cycles, $S3$ and $S4$ are ON and $S1$, $S3$, $S4$ are switched both at switching frequency complementarity to $S7$ and $S8$ [8]. The switching scheme of another modulation technique is employed in [8]. Further to improve the quality of the waveform and eliminate leakage current issues in a relative manner such as a clamping circuit branch and split-level dc-link capacitors are employed to prevent the oscillation during freewheeling periods, where CMV is operating with the same value in four modes of operations. In the literature many clamping-based topologies namely improved H5 and HERIC are proposed with corresponding modulation schemes and UPF to achieve the constant CMV during all modes of operations by incorporating a diode or switch at the centre of the dc-link capacitor [7–17].

It makes, the leakage current is disregarded significantly. Correspondingly in literature many clamping ability topologies are proposed and investigated based on the device selection such as if clamping branch is made with switches, then called active clamping topology and if clamping branch is made with diode, then it is called passive clamping topology. In the literature, passive clamping topologies are given clear information about clamping ability during the freewheeling modes but fail to the dead time operation and ignore its considerations. In H-bridge, zero voltage rectifier-diode (HB-ZVR-D) is proposed in [14], where the clamping branch is made with one simple diode and results in improved common mode voltage and reduced leakage current and it is close to zero in all modes of operations. But it fails to keep the constant common mode voltage during the dead time periods. In [18], the same type of topology is proposed with active clamping switches, namely the HERIC-S topology. Here the clamping ability is possible through a single switch at the centre of the DC-link capacitor. And the generations of common-mode voltages during freewheeling mode are not clearly explained, but low CM voltage and leakage current are reduced as per the standards, such as close zero. But in today's generations, most of the inverters are followed as per the VDE-AR-N4105 standard, where reactive power support is the key factor for TL-PVI design [7]. It can be achieved through improved modulation schemes, which are reported in detail in [9] but not concluded properly on the switch junction capacitance effect and its leakage current because they require additional components and filters, which increase system complexity and increase the cost and size of the system.

To highlight this problem in this chapter an improved T-type topology is invented with power modulator like rectifier branch is adapted at centre of the dc-link capacitor to get constant CMV during the freewheeling period and a modified modulation technique is introduced to reach and support reactive power capacity. Finally, the proposed topology is validated through the simulation environment with results. The performance characteristics in terms of CMV, CM-leakage current, losses, total switching count and %THD summarized in detail.

6.2 ANALYSIS OF COMMON-MODE-CURRENT IN TL-PVITs

In general, the magnitude of CMC in transformerless PV inverter topologies (TLIs) is more dependent on charging and discharging of the stray capacitors. Further input dc Voltage is referred as V_{dc}, PV module parasitic capacitance is referred as C_{PV} and P, N and a, b are the positive-phase and negative-phase terminals of the dc source, phase and neutral respectively. However, the relation between CMV and CMC is explained through the following equations [8–13].

The average of two-phase leg voltages (V_{an}, V_{bn}) is known as CMV such as

$$V_{CM} = \frac{V_{an} - V_{bn}}{2} \tag{6.1}$$

$$V_{DM} = V_{an} - V_{bn} \tag{6.2}$$

Similarly, the equations for $V_{CM\text{-}DM}$ and total CMV are shown below.

$$V_{CM-DM} = V_{DM}\frac{L_2 - L_1}{2(L_2 + L_1)} \tag{6.3}$$

$$V_{tCM} = V_{CM} + V_{CM-DM} \tag{6.4}$$

$$V_{CMC} = V_{PV}\frac{dV_{tCMV}}{dt} \tag{6.5}$$

From equation (6.3), if $L1 = L2$ (i.e., a well-designed magnetic-filed branch), the voltage $Vtcmv = Vcm$ [10]. From equation (6.5), it is clearly conveyed that the magnitude of CPV and floating in $Vtcmv$ are maintained within the level of the CMC. Finally, the magnitude of CPV is not controllable because it completely depends on the environmental conditions. Hence, to eliminate the CMC, one must reduce the oscillations in $Vtcmv$, which depend on the structure and PWM scheme of an inverter.

6.3 PROPOSED MOSFET RECTIFIER BRIDGE CLAMPING TOPOLOGY

This section proposes a simple modified inverter topology based on the existing methodology, namely MOSFET rectifier bridge clamping topology (MRBCT). Corresponding modulation techniques are examined thoroughly, as shown in Figure 6.2. The MRBCT is made with a rectifier bridge, which consists of four diodes and one switch at the centre of the dc-link to maintain half the input dc voltage during freewheeling modes. Nonetheless, the MRBCT control scheme is able to support the reactive power during the system's operations.

6.4 OPERATING MODES OF PROPOSED MOSFET RECTIFIER BRIDGE CLAMPING TOPOLOGY

The proposed topology operation modes are designed with unipolar output voltages such as $+V_{dc}$, 0 and $-V_{dc}$ and it is shown in Figure 6.2. The detailed current path and its CMV, DM are as follows (Figure 6.3);

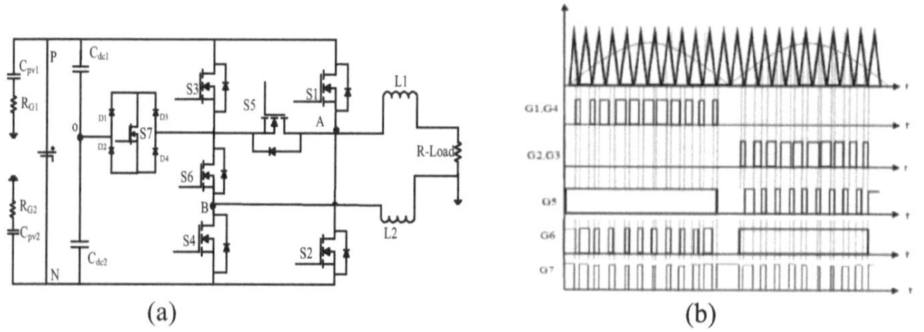

FIGURE 6.2 Proposed MRBCT: (a) converter structure and (b) modulation signals ($G1$–$G7$).

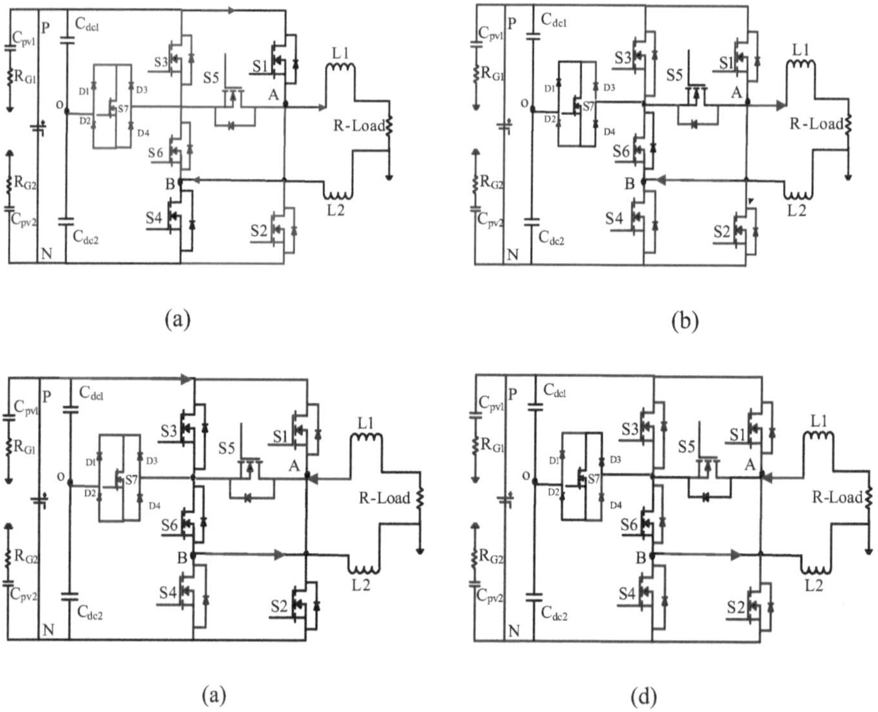

FIGURE 6.3 Operation modes of MRBCT: (a) positive half cycle, (b) positive freewheeling, (c) negative half cycle, and (d) negative freewheeling periods.

Mode 1: During Positive power Half Cycle mode (PHC) $S1$, $S4$ are turn on and others are turned off. Then current path circulates through $S1$, filter, load and $S4$, which is presented in Figure 6.4a. Hence, CMV and differential mode voltage become

$$V_{CM} = \frac{V_{an} + V_{bn}}{2} = \frac{1}{2}(V_{dc} + 0) = \frac{V_{dc}}{2} = 0.5\,\mathrm{V_{dc}} \tag{6.6}$$

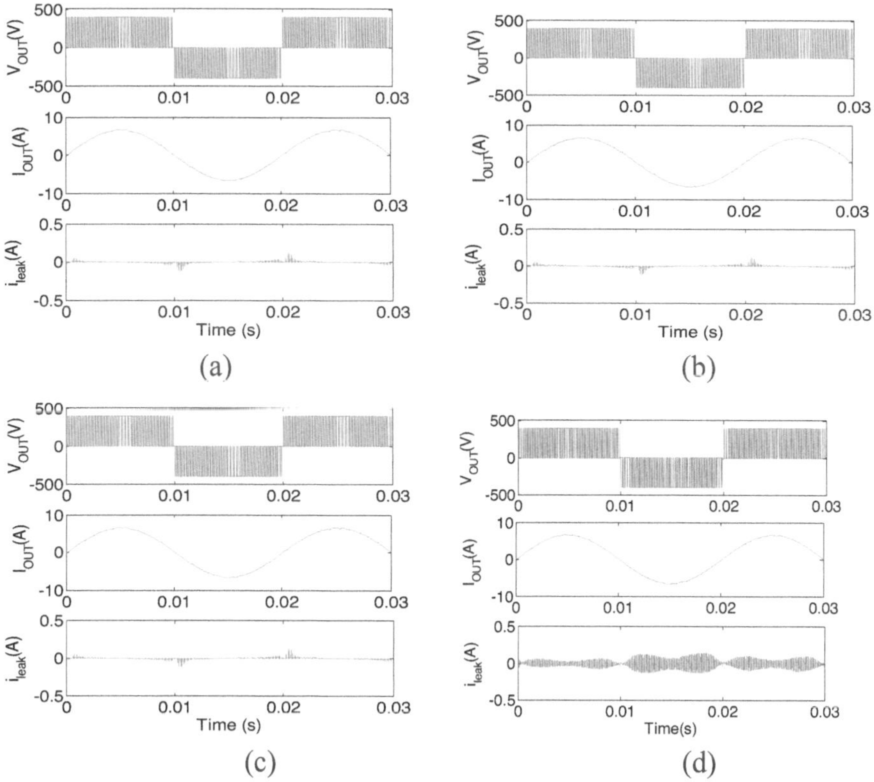

FIGURE 6.4 Output parameters of V_{out}, I_{out}, i_{leak} for (a) H5, (b) HERIC, (c) H6-family, and (d) proposed topologies.

$$V_{ab} = V_{an} - V_{bn} = V_{dc} - 0 = V_{dc} \tag{6.7}$$

$$V_{CM} = \frac{V_{an} + V_{bn}}{2} = \frac{1}{2}(0 + V_{dc}) = \frac{V_{dc}}{2} = 0.5\,V_{dc} \tag{6.8}$$

$$V_{DM} = 0 - V_{dc} = -V_{dc} \tag{6.9}$$

Mode 2: During the positive power freewheeling mode, the dc supply is completely disconnected from the grid. And *S*5, body diode of *S*6 and *S*7 is ON and remains are OFF. And the current circulates and freewheels through the inductors and the load, as shown in Figure 6.3b. The decreased V_{AN} voltage and increased V_{BN} voltage to reach the desired values of common point, $V_{dc}/2$ and thus V_{cm} and V_{dm} become

$$V_{CM} = \frac{V_{an} + V_{bn}}{2} = \frac{1}{2}\left(\frac{V_{dc}}{2} + \frac{V_{dc}}{2}\right) = \frac{V_{dc}}{2} = 0.5\,V_{dc} \tag{6.10}$$

$$V_{DM} = \frac{V_{dc}}{2} - \frac{V_{dc}}{2} = 0 \tag{6.11}$$

Mode 3: During negative power Half Cycle mode (NHC) $S2$, $S3$, $S6$ is ON and remains are OFF. And the current circulates through source to load or load source is given in Figure 6.3c. Thus, CMV and differential mode voltage become

Mode 4: During the negative freewheeling period, the dc source is completely disconnected from the grid. And the $S6$, body diode of $S5$ and $S7$ is ON and remains are OFF. And the current circulates through the inductors and the load, as shown in Figure 6.3d. The voltage (Table 6.1)

Increased V_{AN} voltage and decreased V_{BN} voltage until its values reach the common point, $V_{dc}/2$ and hence V_{cm} and V_{dm} write as follows

$$V_{CM} = \frac{V_{an} + V_{bn}}{2} = \frac{1}{2}\left(\frac{V_{dc}}{2} + \frac{V_{dc}}{2}\right) = \frac{V_{dc}}{2} = 0.5\,\mathrm{V_{dc}} \tag{6.12}$$

$$V_{DM} = \frac{V_{dc}}{2} - \frac{V_{dc}}{2} = 0 \tag{6.13}$$

From the equations (6.6)–(6.13), it is concluded that CMV has the same magnitude in four modes (mode 1–mode 4) because of the clamping branch with rectifier bridge during freewheeling periods, and hence the magnitude of CMC is completely eliminated. Nonetheless, proposed MRBCT-based TL-PVI configurations with four MOSFETs operating at lower voltage ratings and simple design clamping branches with low count devices in the freewheeling periods and then proposed MRBCT based clamping topologies are well suitable for high-efficiency PV applications.

6.5 SIMULATION RESULTS

To validate the parameters-level of the proposed clamping branch with conventional un-clamping structures are run in the MATLAB/SIMULINK platform. The magnitudes are used in the simulation as follows; input dc voltage of 400 V, stary capacitors are considered (C_{PV1}, C_{PV2}) with 100 nF, ground resistance (R_{G1}, R_{G2}) is 11 Ω, two symmetrical filter inductors ($L1$, $L2$) with value of 3 mH, switching frequency is 10 kHz and 230 V, 50 Hz grid voltage (*rms*) system [22] respectively.

6.6 OUTPUT PERFORMANCE AND COMMON-MODE BEHAVIOUR

The output and common-mode parameters in terms of V_{out}, I_{out}, i_{leak} and V_{an}, V_{CM}, V_{bn} of non-NPC, NPC circuits are presented in Figures 6.4 and 6.5. It reveals that all topologies (un-clamping and clamping) are three-level ($+V_{dc}$, 0, $-V_{dc}$) output voltage with quality sinusoidal output current. In addition to that, conventional H5, HERIC topologies are fails to get the constant voltage during the freewheeling periods, which results in the oscillatory phase leg voltages (V_{AN}, V_{BN}) at ~218 V, which is clearly given in Figure 6.4a and b. Hence the leakage current is not eliminated completely

TABLE 6.1
Switching Modes and CMV Evaluations on Proposed MRBCT

	Proposed MRB-TLIT	*S1*	*DS1*	*S2*	*DS2*	*S3*	*DS3*	*S4*	*DS4*	*S5*	*DS5*	*S6*	*DS6*	*S7*	*D1*	*D2*	*D3*	*D4*	V_{AN}	V_{BN}	V_{cm}
H6	Mode1	1	0	0	0	0	0	1	0	1	0	0	0	0	0	0	0	0	*Vdc*	0	0.5*Vdc*
	Mode2	0	0	0	0	0	0	0	0	1	0	0	1	1	1	0	0	1	0.5*Vdc*	0.5*Vdc*	0.5*Vdc*
	Mode3	0	0	1	0	1	0	0	0	0	0	1	0	0	0	0	0	0	0	*Vdc*	0.5*Vdc*
	Mode4	0	0	0	0	0	0	0	0	0	1	0	1	1	0	1	1	0	0.5*Vdc*	0.5*Vdc*	0.5*Vdc*

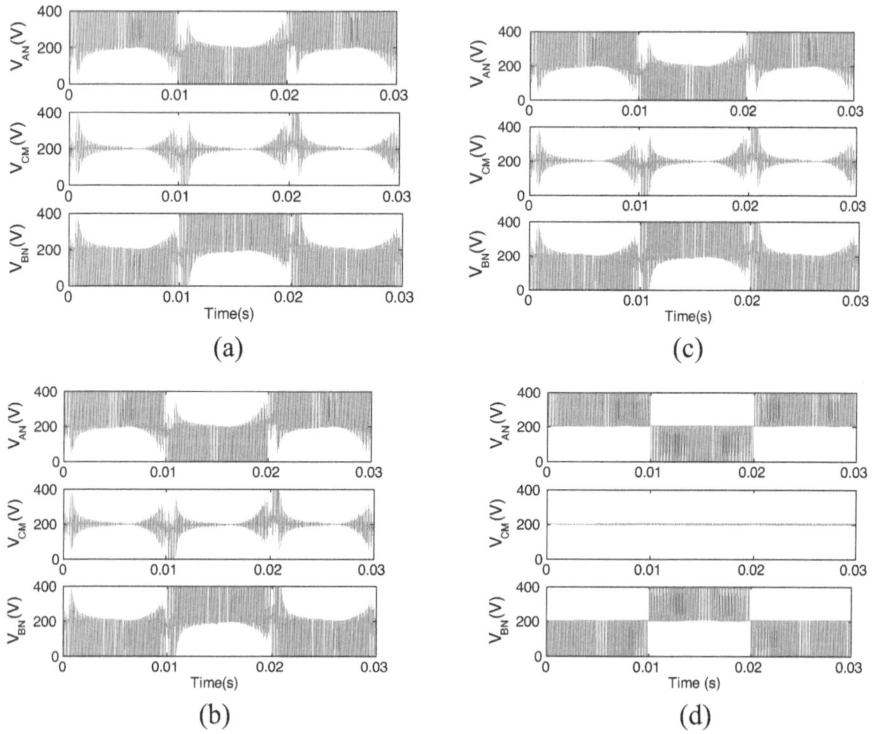

FIGURE 6.5 Common mode parameters of V_{an}, V_{CM}, V_{bn} for (a) H5, (b) HERIC, (c) H6-family, and (d) proposed topologies.

in the whole grid cycle, which illustrates in Figure 6.5a and b. It can be overcome through the proposed MRBCT-based clamping branch, which clamps at 0.5 V_{dc} during whole operating modes and hence CMC is completely eliminated, as shown in Figure 6.5d. To know the quality of the output-level voltage, here total harmonic distortion (%THD) is also measured through the simulation platform, which is with the magnitude of 1.01%, 1.41%, 0.92%, and 0.88, respectively, as shown in Figure 6.6. And all topologies are followed as per IEEE1547 standards [24].

6.7 LOSS ANALYSIS AND COMPARISONS

The analysis of conduction and freewheeling losses of both IGBT and MOSFET devices for both non-NPC and NPC topologies rated with 1 kW is explored in this section. All losses are measured through the PSIM software using thermal modules [27]. Table 6.2 shows the device parameters [18], which are used during simulation. For clear understanding to readers losses are simply presented in histogram, which is shown in Figure 6.7 and Table 6.3. The detailed calculation procedure and theoretical studies have been discussed in [14–20]. And few equations are computed to analyze the losses of both IGBT and MOSFET, which may be useful to the readers.

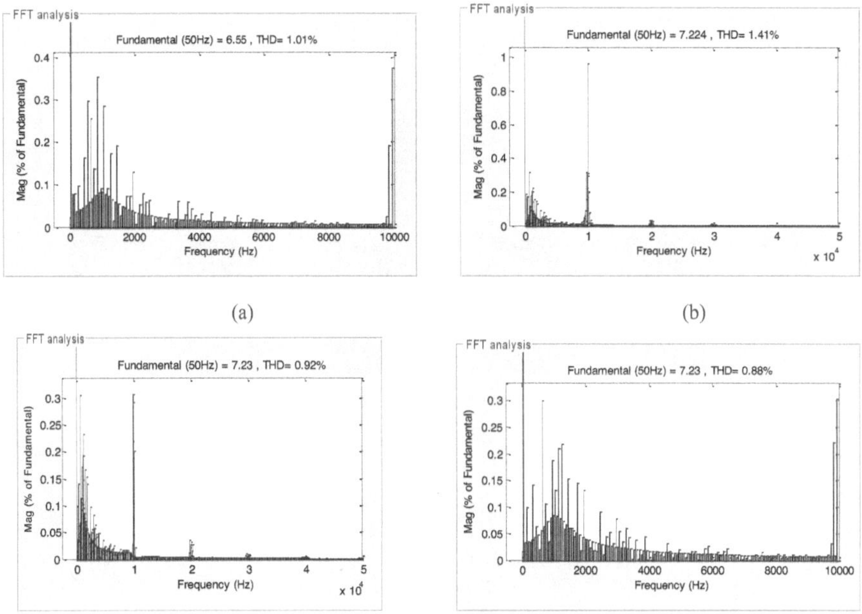

(b) (d)

FIGURE 6.6 Total harmonic distraction: (a) H5, (b) HERIC, (c) H6-family, and (d) proposed topologies.

TABLE 6.2
Power Devices Loss Distributions Parameters

Parameters	**Value**
IGBT (600 V)	HGTG20N60A4D
MOSFET (600 V)	SPN04N60C2
SiC diode	IDD08SG60C
Frequency	50 Hz
Junction temperature, $T_{j(max)}$	150°C
Pcond_Q calibration factor	1
Psw_Q calibration factor	1
Pcond_D calibration factor	1
Psw_D calibration factor	1

For IGBTs conductions, diode and switching losses are calculated by the following equations (6.13)–(6.20) [17].

$$P_{conduction} = V_{CE(SAT)}\ I_C \tag{6.14}$$

$$P_{conduction} = V_F\ I_F \tag{6.15}$$

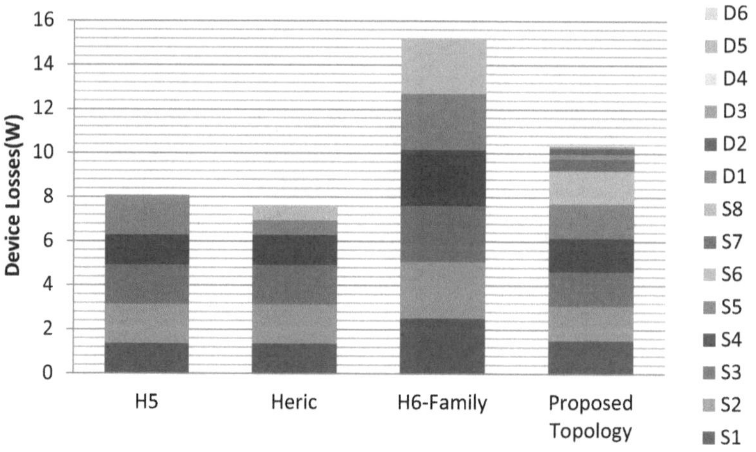

FIGURE 6.7 Loss analysis.

$$P_{SW} = P_{SW\_ON} + P_{SW\_OFF} \tag{6.16}$$

$$P_{SW\text{-}ON} = \frac{E_{on}\ f\ V_{dc}}{V_{dc\text{-}datasheet}} \tag{6.17}$$

$$P_{SW\text{-}OFF} = \frac{E_{off}\ f\ V_{dc}}{V_{dc\text{-}datasheet}} \tag{6.18}$$

where I_C defines the on-state current, V_F is forward voltage drop, I_F defines free-wheeling current and $V_{dc\text{-}datasheet}$ is actual dc-bus voltage respectively.

And MOSFET losses are calculated by equations (6.19 and 6.20)

$$P_{Conduction} = \frac{I_m^2 R_{ds}\ 2m}{3\pi} \tag{6.19}$$

$$P_{switching} = f_{sw} E_{oss} \tag{6.20}$$

where R_{ds} defines the on-state drain-source resistance, I_m is peak output current, M for modulation index, f_{sw} is switching frequency and E_{oss} is energy loss from the device data sheet respectively.

However, the power loss calculations depend on the accuracy of the device data sheet, which is provided by the manufacturer. For the readers understanding, a few sample loss calculations for H5, HERIC, and H6-family power circuits are given below.

The operating conditions of the HERIC topology with MOSFET-SPN04N60C2 is V_{in} = 600 V, V_{out} = 540 V, T_J = 25°C, 100°C, 150°C, R_{ds} = 0.8 Ω, f_{s1} = 10 kHz,

TABLE 6.3
Measured Converter Devices Losses of Non-NPC and NPC SP-TLITs

Topology		Losses (W)														Total losses(W)
		Conduction Losses								External Diode Losses						
		S1	*S2*	*S3*	*S4*	*S5*	*S6*	*S7*	*S8*	*D1*	*D2*	*D3*	*D4*	*D5*	*D6*	
Non-NPC	H5	1.37	1.78	1.78	1.37	1.78	0	0	0	0	0	0	0	0	0	8.08
	HERIC	1.37	1.78	1.78	1.37	0.66	0.64	0	0	0	0	0	0	0	0	7.6
	H6-family	2.52	2.56	2.56	2.52	2.56	2.51	0	0	0	0	0	0	0	0	12.67
NPC	MRCT-H6	1.52	1.56	1.56	1.52	1.56	1.51	0.53	N.C	0.21	0.28	0.08	0.06	0	0	10.39

$M = 0.95$, COS $= 0.994$, $I_m = 0.8–4$ A and the drain source on state resistance [$V_{GS} = 10$ V, $I_D = 0.65$,$T_J = 25°$C, $R_{\mathrm{DS(ON)}} = 0.95$] and E_{oss} from cool MOSFET at 400 V is 28 μJ [25].

For H5

From the equation (6.19)

$$P_{Conduction\text{-}H5} = \frac{2^2 * 0.95 * 2 * 0.95}{3\pi} = 7.55\,\text{W}$$

From equation (6.20)

$$P_{switching} = 10000 * 28\,\mu\text{J} = 10000 * 0.000028 = 0.28\,\text{W}\ \text{[Note: 1 μJ = 0.000001 W]}$$

$$P_{Total=P_{Condu}+P_{switch}} = 7.55 + 0 + 28 = 7.83\,\text{W}$$

From the equation (6.19)

$$P_{Conduction\text{-}Heric} = \frac{2^2 * 0.8 * 2 * 0.95}{3\pi} = 6.36\,\text{W}$$

From equation (6.20)

$$P_{switching} = 10{,}000 * 28\mu\text{J} = 10{,}000 * 0.000028 = 0.28\,\text{W}\ \text{[Note: 1 μJ = 0.000001 W]}$$

$$P_{Total=P_{Condu}+P_{switch}=6.36+0+28=}\ 6.64\,\text{W}$$

$$P_{Conduction\text{-}Heric} = \frac{2^2 * 0.8 * 2 * 0.95}{3\pi} = 6.36\,\text{W}$$

For H6-family

From the equation (6.19)

$$P_{Conduction\text{-}H5} = \frac{2.5^2 * 0.95 * 2 * 0.95}{3\pi} = 11.80\,\text{W}$$

From equation (6.20)

$$P_{switching} = 10000 * 46\,\mu\text{j} = 10000 * 0.000054 = 0.54\,\text{W}\ \text{[Note: 1 μJ = 0.000001 W]}$$

$$P_{Total=P_{Condu}+P_{switch}} = 7.55 + 0.54 = 12.26\,\text{W}$$

However, the theoretical losses are lower than the simulation results because all values are running at real time environment and forward voltage of the device diode. But theoretical values are closer to the simulation results. Here losses are included with the conduction loss, switching loss and freewheeling loss respectively. Table 6.3 gives the device operation of the discussed non-NPC and NPC topologies to justify the performance among the MRCB-based TL-PVITs. As expected, non-NPC H5, HERIC TL-PVITs has low losses due to reduced switching count (only two MOSFETs) in the active modes (mode 1, mode 3) and only one switch and one diode is active (mode 2, mode 4) during the freewheeling periods. On the other hand, proposed MRCB-based TL-PVIT is higher device losses as compared with non-MRBCT-based TL-PVITs due to excessive switching count during freewheeling periods. However, the device losses of proposed MRBCT topologies are higher as compared with non-NPC topologies (H5, HERIC, H6) but poor performance interms of CMV and leakage current which is discussed in the earlier section. However, loss analysis is one of the critical issues in power electronics devices because to predict the exact losses in both theoretical and simulation is very difficult but must be relative equal to the practical measurement of the device losses [18–21].

6.8 EXPERIMENTAL RESULTS

To verify the competency of proposed MRBCT based TL-PVITs, a universal experimental setup has been designed in the laboratory with the same simulation device parameters used in the simulation [8].

The control algorithms are incorporated through FPGA SPARTAN-6. As per the theoretical validation, the NPC and non-NPC TL-PVITs give the sinusoidal output current (I_{out}), which is presented in Figure 6.8. And the common mode performance of both non-NPC and NPC circuits is shown in Figure 6.8. As expected, the non-NPC HERIC TL-PVITs have oscillating CMV such as floating at 267 V and hence higher leakage current with magnitudes of 29.10, 11.31 and 11.30 mA, as given in Figure 6.9. From the observation of the leakage current magnitudes, it is clearly shown that decoupling topologies have double leakage current compared with the

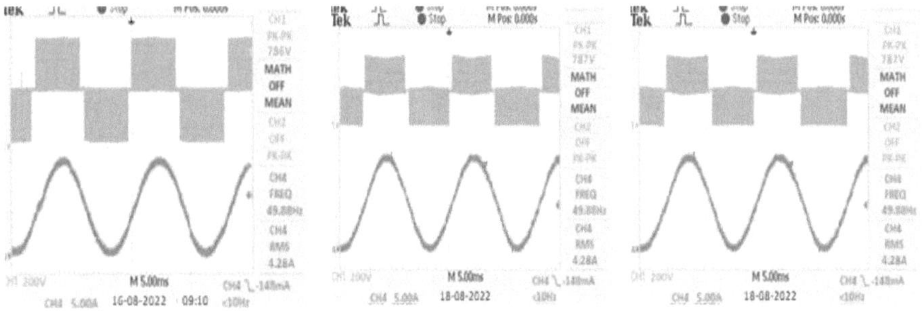

(a) (b) (c)

FIGURE 6.8 Output characteristics V_{out} (top) I_{out} (bottom): (a) HERIC, (b) H6-family, and (c) proposed topology.

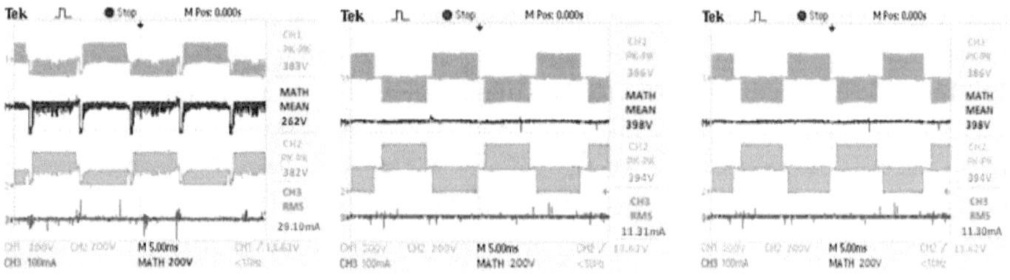

(a) (b) (c)

FIGURE 6.9 Common-mode characteristics-V_{an} (top), $2V_{CM}$ (middle), V_{bn} (bottom): (a) HERIC, (b) PN-NPC, and (c) proposed topology.

clamping topologies due to improved clamping circuit and corresponding modulation scheme. The THD in terms of output load current is captured by the FLUKE 434 Series II power Analyzer [24]. The %THD of various topologies are as follows with magnitude of 1.42%, 0.93%, 0.89%, which are placed in Figures 6.8a and 6.9b. And it is evident that all discussed HERIC, H6 and proposed topologies have less harmonic current distortion such as follows as per the IEEE1547 standards. And compared to simulation, experimental %THD is relatively higher due to grid affects and its parameters modelling in real time environment [28].

The HIOKI 3197 power analyzer is used for efficiency measurements. The European efficiency is computed using the below equation (6.21).

$$\eta_{\text{EU}} = 0.03\eta_{5\%} + 0.06\eta_{10\%} + 0.13\eta_{20\%} + 0.10\eta_{30\%} + 0.48\eta_{50\%} + 0.2\eta_{100\%} \quad (6.21)$$

Figure 6.10 gives the overall efficiency measurements of non-NPC and NPC based TL-PVITs. And the calculated European efficiency of the un-clamping and clamping

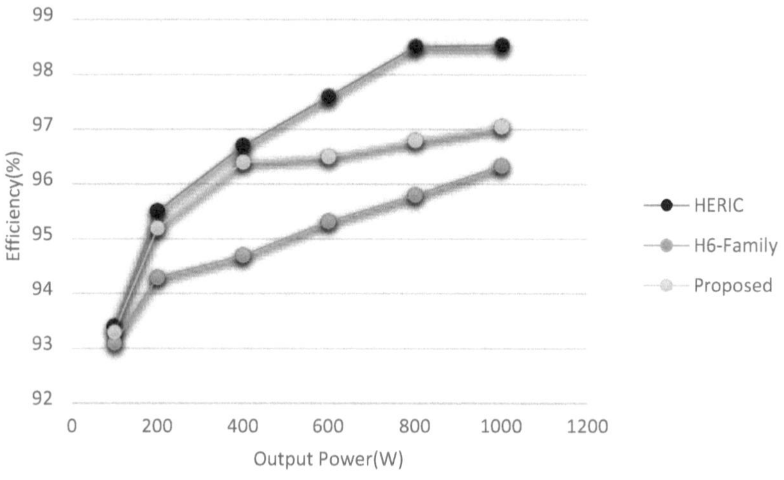

FIGURE 6.10 Efficiency analysis.

TABLE 6.4
Performance Parameter Analysis of Non-NPC and NPC Topologies

	Non-NPC	NPC	
Topology	**HERIC**	**H6-family**	**Proposed**
Output voltage	Unipolar	Unipolar	Unipolar
No. of switching count	6	8	7
No. of diodes	0	0	6
CMV	Floating	Constant	Constant
Leakage current (mA_{rms})	29.10	11.31	11.30
THD_i (%)	1.39	1.02	0.89
European efficiency (%)	98.89	96.98	97.01

based TL-PVITs are 98.89%, 96.98% and 97.01% respectively. However, the proposed configuration efficiency is registered lower among the conventional TL-PVITs topologies due to more active elements in the grid cycle. Nonetheless, the maximum proposed MRB-based configuration efficiency is relatively higher than the conventional H6-family topology because higher device total count in power freewheeling modes. And efficiency analysis is purely depending on the total switching count of power devices and thermal module selection such as device operating conditions. Here all topologies are modelled under one thermal condition to calculate losses through PSIM environment. And Table 6.4 is highlighted with an overall performance comparison of various topologies. Therefore, results with practical setup confirmed that the MRB-based TL-PVIT merged the features of both low-level loss non-NPC method (HERIC) and CM-leakage-current-decrement (H6-family) of the NPC clamping method. However, the proposed MRB-TL-PVIT is a more attractive solution for high-efficiency PV applications.

6.9 CONCLUSION

In this chapter, various MOSFETT-based non-NPC and NPC transformerless PV inverter circuits have been investigated thoroughly. It confirms that NPC topologies (H6-family) have better CMV performance as compared with non-NPC topologies (H5, HERIC) due to lack of clamping ability during the freewheeling periods. Based on that here a MOSFET-based topological circuit is proposed to combine the features of low-loss non-NPC (HERIC) method and CM-leakage-current-elimination of the NPC clamping method. The proposed circuit is able to reach the voltage at constant during freewheeling periods, which results in a very low leakage current as compared with non-NPC topologies. However, the proposed topology has less efficiency as compared with universal HERIC topology but relatively higher efficiency than H6-family topology because of a higher total device count in freewheeling periods. In addition to that, the proposed topology makes it possible to support the reactive power capability of the grid, which makes the proposed topology follow the quality of future trend inverters. Hence, proposed MRBCT is extremely suitable for both high efficiency PV applications as well as electric vehicle applications [22–28].

REFERENCES

[1] G. Carli and S. S. Williamson, "Technical Considerations on Power Conversion for Electric and Plug-in Hybrid Electric Vehicle Battery Charging in Photovoltaic Installations," *IEEE Transactions on Power Electronics*, vol. 28, pp. 5784–5792, 2013

[2] E. Afshari, G. Reza Moradi, A. Ramyar, R. Rahimi, B. Farhangi, and S. Farhangi, "Reactive Power Generation for Single-Phase Transformerless Vehicle-to-Grid Inverters: A Review and New Solutions," In: *IEEE Transport Electrification Conference and Expo (ITEC)*, pp. 69–76, June 22–24, 2017.

[3] E. Afshari, B. Farhangi, Y. Yang and S. Farhangi, "A Low-Voltage Ride-Through Control Strategy for Three-Phase Grid-Connected PV Systems," Presented at the *IEEE Power and Energy Conference at Illinois (PECI)*, Illinois, USA, 23–24 Feb, 2017.

[4] M. C. Kisacikoglu, M. Kesler and L. M. Tolbert, "Single-Phase On-Board Bidirectional PEV Charger for V2G Reactive Power Operation," *IEEE Transactions on Smart Grid*, vol. 6, pp. 767–775, 2015.

[5] B. Farhangi, "A Novel Modified Deadbeat Controller for Vehicle to Grid Application," In: *The 6th Power Electronics, Drives Systems & Technologies Conference (PEDSTC)*, 3–4 Feb, 2015, pp. 47–52.

[6] K. S. Tey and S. Mekhilef, "A Reduced Leakage Current Transformer-Less Photovoltaic Inverter," *Renewable Energy*, vol. 86, pp. 1103–1112, 2016.

[7] W. Li, Y. Gu, H. Luo, W. Cui, X. He and C. Xia, "Topology Review and Derivation Methodology of Single-Phase Transformerless Photovoltaic Inverters for Ground Leakage Current Suppression". *IEEE Transactions on Industrial Electronics,* vol. 62, pp. 4537–4551, 2016.

[8] M. Shayestegan et al., "An Overview on Prospects of New Generation Single-Phase Transformerless Inverters for Grid-Connected Photovoltaic (PV) Systems," *Renewable and Sustainable Energy Reviews*, vol. 82, pp. 515–530, 2018.

[9] S. Kirthiga and N. M. Jothi Swaroopan, "Highly Reliable Inverter Topology With a Novel Soft Computing Technique to Eliminate Leakage Current in Grid-Connected Transformerless Photovoltaic Systems," *Computer Electrical Engineering*, vol. 68, pp. 192–203, 2018.

[10] DIN VDEV0126-1-1, *Automatic Disconnection Device Between a Generator and the Public Low-Voltage Grid.* 2006, DIN VDE.

[11] Y. Mei, S. Hu, L. Lin, W. Li, X. He and F. Cao, "Highly Efficient and Reliable Inverter Concept-Based Transformerless Photovoltaic Inverters with Tri-Direction Clamping Cell for Leakage Current Elimination," *IET Power Electronics*, vol. 9, no. 8, pp. 1675–1683, 2016.

[12] T. K. S. Freddy, J. H. Lee, H. C. Moon, K. B. Lee and N. Abd Rahim, "Modulation Technique for Single-Phase Transformerless Photovoltaic Inverters With Reactive Power Capability," *IEEE Transactions on Industrial Electronics*, vol. 64, no. 9, pp. 6989–6999, 2017.

[13] L. Zhou, F. Gao and T. Xu, "A Family of Neutral-Point-Clamped Circuits of Single-Phase PV Inverters: Generalized Principle and Implementation," *IEEE Transactions on Power Electronics*, vol. 32, no. 6, pp. 4307–4319, 2016.

[14] L. Zhang, K. Sun, Y. Xing and M. Xing, "H6 Transformerless Full-Bridge PV Grid-Tied Inverters," *IEEE Transactions on Power Electronics*, vol. 29, no. 3, pp. 1229–1238, 2014.

[15] R. Hasan, S. Mekhilef, M. Seyedmahmoudian and B. Horan, "Grid-Connected Isolated PV Microinverters: A review," *Renewable and Sustainable Energy Reviews*, vol. 67, pp. 1065–1080, 2017.

[16] B. Chen, B. Gu, L. Zhang, Z. U. Zahid, J.-S. Lai, Z. Liao and R. Hao, "A High Efficiency MOSFET Transformerless Inverter for Non-Isolated Micro-Inverter Applications," *IEEE Transactions on Power Electronics*, vol. 30, no. 7, pp. 3610–3622, 2015.
[17] M Islam and S Mekhilef, "Efficient Transformerless MOSFET Inverter for a Grid-Tied Photovoltaic System" *IEEE Transactions on Power Electronics*, vol. 31, no. 9, pp. 6310–6316, 2016.
[18] D. Schmidt, D. Siedle and J. Ketterer, "Inverter for Transforming a DC Voltage into an AC Current or an AC Voltage," EP Paten: 1369985, Dec. 10, 2003
[19] M. Victor, K. Greizer and A. Bremicker, "Method of Converting a Direct Current Voltage From a Source of Direct Current Voltage, More Specifically From a Photovoltaic Source of Direct Current Voltage, into a Alternating Current Voltage," U.S. Patent: 20050286281A1, Apr. 23, 1998.
[20] Z. Özkan and A. M. Hava, "Classification of Grid Connected Transformerless PV Inverters with a Focus on the Leakage Current Characteristics and Extension of Topology Families," *Journal of Power Electron*, vol. 15, no. 1, pp. 256–267, 2015.
[21] T. Kerekes, R. Teodorescu et al., "A New High-Efficiency Single-Phase Transformerless PV Inverter Topology," *IEEE Transactions on Industrial Electronics*, vol. 58, pp. 184–191, 2011.
[22] T. K. S. Freddy, N. A. Rahim, W. P. Hew and H. S. Che, "Comparison and Analysis of Single-Phase Transformerless Grid-Connected PV Inverters," *IEEE Transactions on Power Electronics,* vol. 29, no. 10, pp. 5358–5369, 2014
[23] L. Zhang, S. Kai, L. Feng, H. Wu and Y. Xing, "A Family of Neutral Point Clamped Full Bridge Topologies for Transformerless Photovoltaic Grid Tied Inverter," *IEEE Transactions on Power Electronics*, vol. 28, pp. 730–739, 2013.
[24] Z. Ahmad and S. N. Singh, "An Improved Single Phase Transformerless Inverter Topology for Grid Connected PV System With Reduce GLC and Reactive Power Capability," *Solar Energy,* vol. 157, pp. 133–146, 2017
[25] K. Sateesh kumar, A. Kirubakaran and N. Subrahmanyam, "Bi-Directional Clamping Based H5, HERIC and H6 Transformerless Inverter Topologies with Improved Modulation Technique," In: *Pros IEEE International conference on power electronics, smart grid and Renewable energy(PESGRE 2020) Cochin*, India, pp. 1–6, 2020.
[26] M. N. H. Khan, Y. P. Siwakoti, L. Li and F. Blaabjerg, "H-Bridge Zero-Voltage Switch Controlled Rectifier (HB-ZVSCR) Transformerless Mid-Point-Clamped Inverter for Photovoltaic Applications" *IEEE Journal of Emerging and selected Topics in Power Electronics*, vol. 99, p. 1, 2019.
[27] PSIM User's Guide, 2007. https://www.psim europe. com.
[28] IEEE Std: 1547.1-2005, *IEEE Standard Conformance Test Procedures for Equipment Interconnecting Distributed Resources with Electric Power Systems.* 2015, pp. 442–449. IEEE, New York.

7 Battery Technology for Hybrid and Battery Vehicles

Sundara Subramanian Karuppasamy
National Taipei University of Technology)

N. Jeyaprakash
China University of Mining and Technology

Che-Hua Yang
National Taipei University of Technology

7.1 INTRODUCTION

Electric vehicles are considered as the future vehicles of transportation which get actuated by the electrochemical reactions that take place in the electrodes [1,2]. Since these vehicles utilize the energy produced in an electrochemical way for movement, they can overcome the drawbacks of the traditional IC engine vehicles. The major advantages of these electric vehicles are (i) Emission-free driving, (ii) Compactness, (iii) Less maintenance, (iv) Noise free operation, (v) Fuel cost and so on [3,4]. Based on the power source, electric vehicles are classified into five categories. They are

1. Battery electric vehicles (BEV),
2. Hybrid electric vehicles (HEV),
3. Plug-in hybrid electric vehicles (PHEV),
4. Fuel cell electric vehicles (FCEV)
5. Extended-range electric vehicles (EREV).

BEV utilizes the power generated only from the incorporated battery packs. The nominal voltage delivered by the battery pack is considered as an important criterion in deciding the vehicle's performance. A larger battery pack contributes for higher driving range [5]. HEV and PHEV employ two or more energy sources for vehicle movement. In most cases, the IC engine and battery packs are preferred as the energy source. The major difference between the HEV and PHEV is that in HEVs, the installed battery pack is charged by means of regenerative braking which is inbuilt into the vehicle whereas an external source (power grid) is used to charge the battery pack installed in the PHEVs [6,7]. FCEV generates the power from the fuel cell and

 DOI: 10.1201/9781003354901-7

transmits it to the motor for the vehicle's movement whereas the EREV is considered as the BEV updated version that incorporates the IC engine which is used to energize only the battery pack in the electric vehicles. The power from this IC engine is solely utilized in recharging the pack and not used in actuating the vehicle [8–10].

Mostly all types of electric vehicles need batteries for generating electricity. These batteries are made of electrochemical cells which transfer the electric charges. These cells are named as electrodes (anodes and cathodes). In the anode, the oxidation reaction takes place where the anodic material is oxidized to give out protons and electrons. In the presence of electrolyte, the anions reach the reducing electrode via electrolyte and the electrons flow through the circuit to produce electricity. This electricity drives the electric motors which are responsible for the electric vehicle movement [11,12]. A single cell (cathode and anode) cannot deliver the nominal voltage needed by the autonomous vehicle. Hence, to meet the nominal voltage requirements, these cells are packed to form a battery pack which consists of a greater number of cathodes and anodes. Since more electrodes are combined to form a battery pack, it can face many drawbacks like heating of the pack, and uneven delivery of voltage that can lead to battery pack failure. To overcome these drawbacks, these electrodes are arranged in a predefined fashion which is termed as battery pack configuration [13].

Depending on the requirements, the battery pack configuration is categorized into three major types as follows.

i. series configuration,
ii. parallel configuration,
iii. series-parallel configuration.

In most cases, the electrodes are arranged in any one of the aforementioned configurations to deliver efficient performance. Since the battery pack consists of many electrodes along with electrolytes and permeable membranes, there are many chances of failure to occur. Some of the factors that lead to battery pack failures are (i) pack's temperature rise, (ii) uneven charging and deep discharging rate, (iii) uneven charging, (iv) one or more cell failure, (v) improper cooling and so on. Hence, there is a need for some units which monitors the battery pack's performance and delivers the optimal voltage without any failures [14].

Battery Management System (BMS) supervises the battery pack by means of diagnosing failures that are going to occur and notifying users about the actions to be made. Apart from the diagnostic function, there are other functions which are performed by the battery management system. They are (i) monitoring the pack's condition, (ii) computing the SOC and SOH, (iii) delivering optimal voltage required for driving by means of cell balancing in case of any cell failures, (iv) recording the data such as battery pack temperature, operating voltage, for future purpose and so on [15]. The main scope of this chapter is to address the working principle of lead acid and lithium-ion batteries along with their cell chemistries. The three battery pack configurations followed by the electrolyte types have been discussed. Moreover, the need for BMS and its topology along with BMS modules and the thermal management methods followed by pack recycling strategies were presented in a detailed manner.

7.2 TYPES OF BATTERIES

This section elaborates on the major two battery types used in electric vehicles. In general, a battery cell has two electrodes and an electrolyte. In anode, oxidation takes place whereas the cathode experiences a reduction reaction in the presence of an electrolyte.

7.2.1 Lead Acid Batteries

Lead acid battery (LAB) belongs to the category of secondary batteries employed for electric vehicles. It uses lead (metallic lead) as the anode material whereas the lead oxide plates are preferred as the cathode material. The aqueous H_2SO_4 solution serves as the electrolytic medium for these kinds of batteries [16]. The chemical reactions that take place in the lead acid batteries are described as follows:

- *Anode reaction*: $Pb + HSO_4^- \rightarrow PbSO_4 + H^+ + 2e^-$
- *Cathode reaction*: $PbO_2 + 3H^+ + HSO_4^- + 2e^- \leftrightarrow PbSO_4 + 2H_2O$
- *Overall reaction*: $PbO_2 + Pb + 2H_2SO_4 \leftrightarrow 2PbSO_4 + 2H_2O$

The sulfate ions react with the anodic lead to form lead sulfate followed by the release of protons and electrons. The H^+ ions react with lead oxide in the presence of sulfate ions to form the $PbSO_4$ at the cathode as represented in Figure 7.1. The nominal cell voltage is found to be 2.048 V [17].

Advantages:

1. Cost efficient.
2. Greater discharge currents.

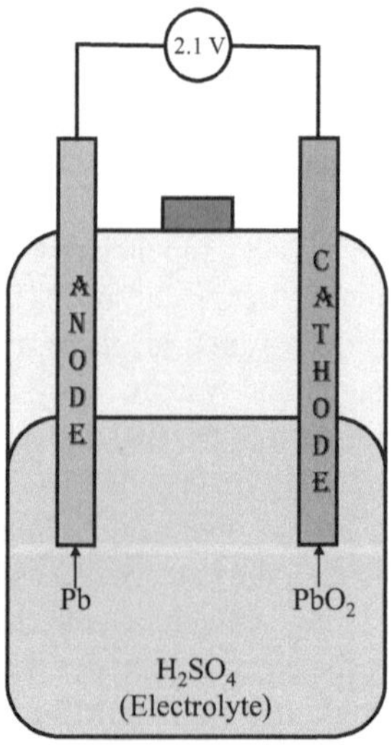

FIGURE 7.1 Schematic illustration of lead acid battery.

3. Higher specific power.
4. Better performance in both high and low temperatures.

Drawbacks:

1. Slow charging and the weight-to-energy ratio is poor.
2. Very low specific energy.
3. Greater environmental hazard due to the lead deposition.
4. Repeated deep discharge will lead to minimal battery life.

7.2.2 Lithium Ion Batteries

Nowadays, Li^+ ion battery (Figure 7.2) is mostly implemented in all electric vehicles due to its greater nominal cell voltage (3.6 V). Since this battery involves the transfer of Li ions, a wide range of materials can be adapted for both anode and cathode [18]. $LiNi_{0.33}Mn_{0.33}Co_{0.33}O_2$, $LiMn_2O_4$, $LiFePO_4$, and lithium nickel cobalt aluminum oxides are preferred as the cathode materials whereas the hard carbon, lithium titanate, natural and synthetic graphite are used as the anode materials. $LiPF_6$ salt solution is the electrolyte material [19]. The cell chemistries behind the Li ion battery are as follows:

- *Anode reaction*: $LiC_6 \leftrightarrow C_6 + Li^+ + e^-$
- *Cathode reaction*: $CoO_2 + Li^+ + e^- \leftrightarrow LiCoO_2$
- *Overall reaction*: $LiC_6 + CoO_2 \leftrightarrow C_6 + LiCoO_2$

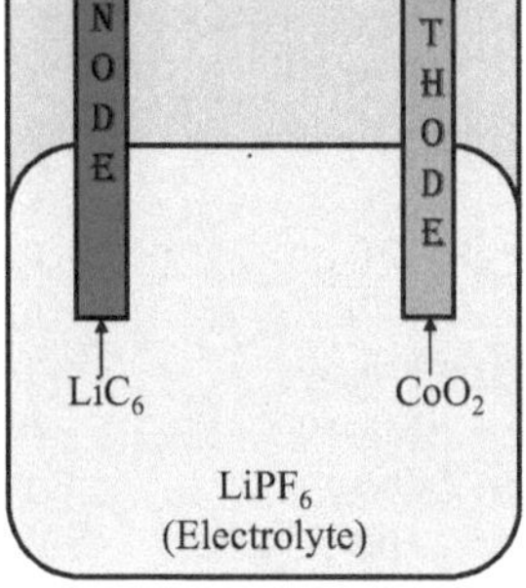

FIGURE 7.2 Schematic illustration of lithium-ion battery.

Advantages:

1. Greater energy density.
2. Lower self-discharge rate.
3. Less maintenance.
4. Higher nominal cell voltage.
5. Better load characteristics.
6. Greater number of charge and discharge cycles.
7. Rechargeable.

Disadvantages:

1. High cost.
2. Aging behavior.
3. Battery management system (BMS) needed.

The different battery parameters were compared among various batteries and listed in Table 7.1 [20, 21].

7.3 ELECTROLYTES

Electrolytes play an important role in transferring the ions from anode to cathode. It serves as the catalyst for the batteries to generate electricity [22]. In general, electrolytes can be either solid or liquid or gas. These three types are explained as follows:

TABLE 7.1
Battery Parameters Comparison Among Different Battery Types

Battery Parameters	Lead Acid	Nickel-Based Batteries	Lithium-Based Batteries
Gravimetric energy density (Wh/kg)	35–40	50–70	150–200
Energy density (Wh/L)	80–90	100–140	250–400
Gravimetric power density (W/kg)	285	200	260
Efficiency (%)	85	85	90
Rated capacity (kWh)	1.29	26.4	55
Life	200–300 cycles	300–600 cycles	600–1,000 cycles
Self-discharge rate (%/month)	5	20–30	3–10
Nominal cell voltage (V)	2.0	1.2	3.2–3.7
Charging efficiency (%)	50–95	65–90	80–90
Charging temperature (°C)	–20 to 50	0–45	0–45
Discharging temperature (°C)	–20 to 50	–20 to 65	–20 to 60
Cost	Cheap	Expensive	Moderate
Maintenance required	High	Low	Moderate
DoD (approx.)	20% for 500 cycles	20% for 2500 cycles	20% for 300 cycles

7.3.1 Solid-State Electrolytes

As the name suggests, the solid-state electrolytes are solid ionic conductors that allow the protons to pass through. These electrolytes act as the electron insulating material which restricts the flow of electrons inside the battery [23]. The major advantages of using the solid-state electrolytes are safety, no leakage of toxic solvents into the cell while operation, non-volatile, thermally stable, ease of processability, fewer self-discharge and so on. These electrolytes have better ionic conductivity, long cycle life, greater electrochemical stability, excellent compatibility, faster ions transfer, etc. All solid-state electrolyte and quasi solid-state electrolyte are the two major categories of solid-state electrolytes [24]. The former is further categorized into inorganic solid electrolytes and solid polymer electrolytes. Table 7.2 lists the ionic conductivity and electrochemical stability of various electrolyte types.

7.3.2 Liquid Phase Electrolytes

Liquid phase electrolyte contains acids or solutions containing dissolved salts where the ions transfer in a fast manner than the solid electrolytes. The cations and anions are uniformly dispersed thereby promoting ions transfer [25,26]. But there is a possibility of leakage of toxic solvents into the cell which minimizes the conducting behavior of electrodes. Some of the liquid phase electrolytes are hydroxides of sodium, potassium, calcium, magnesium, acids like sulfuric acid, nitric acid, hydrochloric acids, solutions containing lithium salts, ethers, lithium tetrafluoroborate, lithium trifluoro methane sulfonate, etc. [27,28].

7.3.3 Gas Phase Electrolytes

These electrolytes are termed as the future electrolytes which overcome the drawbacks of solid and liquid phase electrolytes. Gas phase electrolytes have superior ionic conductivity and can be implemented in operating temperatures [29]. The battery containing the gas phase electrolytes is reported as the clean energy process where the toxic solvents can be avoided and more eco-friendly operation. Liquefied gas electrolyte suits as the best example of this kind [30].

TABLE 7.2
Solid-State Electrolytes for Li-Ion Batteries

Electrolyte	Ionic Conductivity (S/cm)	Electrochemical Stability (V)
NASICON	2.14×10^{-2}–7.25×10^{-4}	0–7
Garnet	1×10^{-3}–8.1×10^{-4}	0–9
Perovskite	2×10^{-4}–8×10^{-5}	1–4.5
LISICON	1.03×10^{-5}–1.3×10^{-6}	0–9
LiPON	2×10^{-6}–6.7×10^{-6}	0–5.5
Li_3N	1.1×10^{-4}–7×10^{-5}	1.74–2.8
Sulfide	1.03×10^{-2}–6.5×10^{-5}	0–5
Argyrodite	1×10^{-2}–4.6×10^{-7}	0–7

7.4 BATTERY PACK

A single cell cannot meet the voltage requirements to run/accelerate the electric vehicle. Hence, many cells are combined to provide the sufficient voltage/energy needed to accelerate the electric vehicle [31,32]. Considering the practical requirements, the cells are arranged in modules to meet the demand of the electric powertrain. Figure 7.3 shows the battery pack components of electric vehicles [33]. In each battery pack, electric wires, fuses, switches, and sensors are used along with some mechanical components. Electric wires provide linkage between each cell/module whereas the fuses/switches are used as the detaching components. Inside the battery modules, the cells can be arranged in any one of the following configurations: (i) series configuration, (ii) parallel configuration, and (iii) mixed (series-parallel) configuration [34].

i. *Series configuration*: Here, the positive terminal of one cell is linked with the negative terminal of the next cell which forms the series configuration. The overall voltage of the pack is the sum of the nominal voltage of each cell connected in the series configuration. Series configuration results in higher nominal voltage with lower current but there is a need for the converter in order to achieve minimal voltage [35].
ii. *Parallel configuration*: On the other hand, a parallel configuration can be obtained by connecting all the positive terminals of each cell together and similarly by connecting all the negative terminals together in the pack. One of the major advantages of this type of configuration is that in a battery pack, if one cell didn't work, the other cells are still able to supply the energy to run the electric vehicle. Higher current draw and greater voltage drop are reported as the limitations of this configuration [36].
iii. *Mixed configuration*: Series and parallel configurations are not able to meet the load requirements in practical cases. Hence, researchers introduced a new type of configuration called as mixed configuration which combines

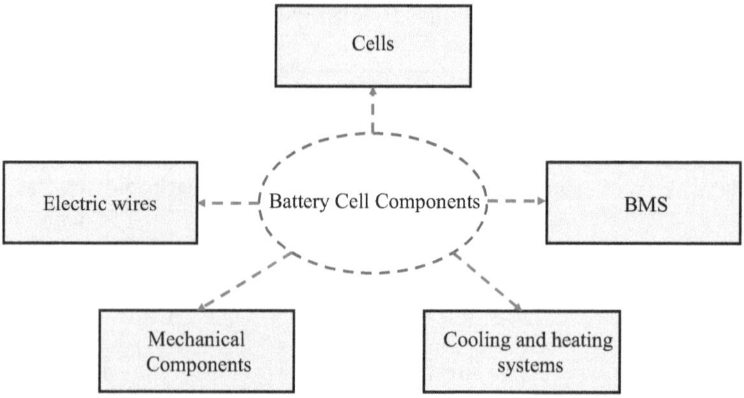

FIGURE 7.3 Battery pack components.

the advantages of the aforementioned configurations. This mixed configuration combines all the advantages reported in both configurations for eliminating the limitations of both combinations [37].

7.5 BATTERY MANAGEMENT SYSTEM

In electric vehicles, the BMS supervises the incorporated pack's performance. It diagnoses the battery failure and reports to the user before the failure happens [38]. This section deals with various aspects of BMS.

7.5.1 Definition and Need for BMS

The BMS minimizes the risk of module damage by receiving relevant information from the battery modules. This system ensures optimal power production from the pack to actuate the vehicle. It monitors the charging and discharging phenomenon so that the battery pack is maintained safely to provide the desired output voltage [39]. The need for BMS in terms of battery pack is as follows:

i. To improve the reliability of the pack.
ii. To increase the lifetime of the pack.
iii. To secure each and every cell from failures.
iv. To enhance the efficiency of the battery pack.
v. To make sure the battery modules are ready for usage.

7.5.2 Key Function of the BMS

i. Charging and discharging control
To deliver efficient performance, there should be stableness while charging and discharging the cells. More failures in the battery pack are reported due to the unstable charging and discharging. Hence, the major goal of any BMS is to monitor the charging and discharging levels to avoid failures [40]. Normally, two stages of charging namely constant current and constant voltage are employed [41]. The BMS monitors these two stages.
ii. Determining the SOC
The second main function of the BMS is to monitor the state of charge of the modules. By tracking the state of charge, the BMS can notify the user about the charging/discharging behaviors [42]. This can be done in three ways: (i) Voltmeter for measuring the voltage drop in the modules, (ii) Coulombic counting approach where the charge and discharge currents are measured and then notified by the user, and (iii) combining the Coulombic counting and voltmeter approach [43].
iii. Determining the SOH
The battery health is monitored in terms of operating voltage, service temperature for efficient performance and to enhance the working life of the battery packs [44].

iv. Cell balance
Some cells in the battery pack become weaker and deliver minimal nominal voltage than the normal cells. In such cases, the weaker cell voltage is said to be compensated by the rest of the cells in the battery pack for extending the battery life and to avoid failure. This phenomenon is referred as cell balancing [45]. The BMS should incorporate any one of the three cell balancing methods for the efficient performance. The three cell balancing methods are active cell balancing, passive cell balancing and charge shunting [46,47].

v. Logbook and communication protocol
The BMS should maintain the initial/standard working conditions data like a logbook. If the battery cell or modules deviates from the standard condition, the BMS should store the deviated data. It should compare the deviated data with the standard data and take necessary actions so that the user can see the actions taken by the battery management system via communication protocol incorporated with the BMS. This communication protocol notifies the user about the prevailing conditions in the battery pack [48,49].

7.5.3 Modules in the BMS

The schematic representation of the BMS is shown in Figure 7.4 [50]. This section deals with a brief description of each component of the BMS.

7.5.3.1 Data Acquisition

For the effective operation of the battery pack, the battery parameters need to be maintained as per the standard parameter values. The operating battery parameters are compared with the standard parameters in the case of any failures. Hence, the

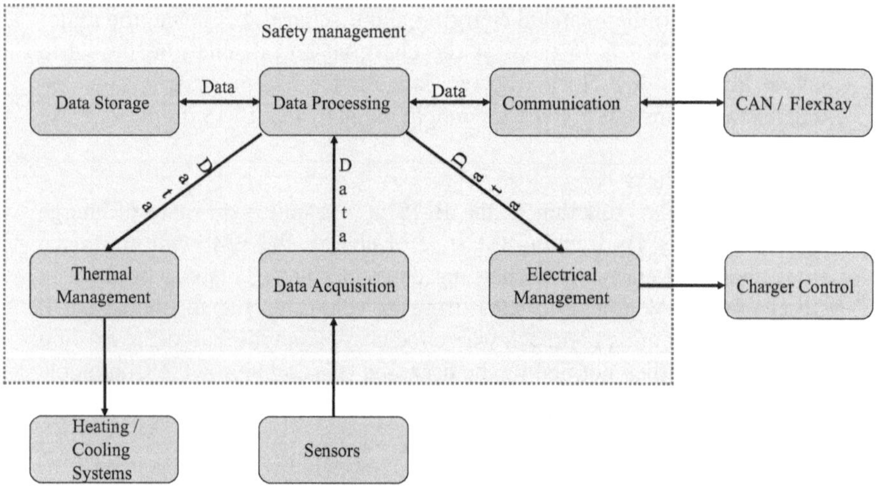

FIGURE 7.4 Modules in the battery management system.

BMS has a built-in data acquisition system that stores the battery parameters for future reference [51]. Also, these parameters are compared with the standard values that aid in making decisions by the BMS. These decisions are solely focused on fulfilling the energy demand of the powertrain in electric vehicles.

7.5.3.2 Data Processing and Storage

The acquired data in the BMS can be used for further estimation of relevant parameters that help in monitoring the performance of the battery pack in electric vehicles. This data is used to compute the SOC, SOH, remaining charge in the pack and so on. The data acquired are processed by certain algorithms that aid in predicting the parameters. After predicting these parameters, the values are stored for further reference. The stored values are used to notify the user about the prevailing condition of the battery pack or they can be used to make decision if the modules encounter failures. Thus, data acquisition, processing and storage in the BMS help in monitoring the prevailing conditions so that failures can be avoided [52].

7.5.3.3 Electrical Management

The electrical management module in the BMS monitors the electrical parameters such as the nominal voltage supplied by the pack. This module controls the charging and discharging process as per the powertrain demand. It tells the user about the current and voltage delivered by the battery pack and avoids deep discharging from the battery. Moreover, if any cell in the battery module fails to deliver the required voltage, the electrical management module is capable of performing the cell balancing phenomenon and delivering the required power to the electric vehicle [53].

7.5.3.4 Thermal Management

As the name implies, the thermal management module monitors the thermal parameters of the installed pack. For the effective functioning of the pack and to deliver the required nominal voltage, certain temperature is said to be maintained in the working environment. There will be a reduction in the voltage (power) delivered by the battery cell if the operating temperature is increased above the preferable range. Hence, cooling the battery pack to attain the preferred service temperature is more important for safety and durability. The BMS monitors the thermal parameters and allows the user or itself to take necessary actions for maintaining the service temperature [54]. Also, thermal management is elaborated in the upcoming section.

7.5.3.5 Safety Management

Since electric vehicles use the battery as the energy storage/supply device, these batteries are very sensitive to voltage and current fluctuations. These fluctuations are said to minimize the battery pack's life and performance. Hence, an earlier diagnosis of these fluctuations is required to avoid failures. Here, the safety management module monitors both operating (discharging) and charging voltage/current in order to protect the system from failure. Based on the acquired/processed data, this module takes necessary actions to eliminate the shutdown and notifies the user about the condition of the installed pack [55].

7.5.3.6 Communication

The important module in the BMS is the communication module. Other modules predict or diagnose the failure in the early stages but the communication module is responsible for notifying the user about the steps or processes taken by the battery management system [56]. Certain protocols are implemented to do the purpose of communication. Some of them are listed below:

i. *CAN*: Controller area network protocol.
ii. *BACnet*: Building automation and control networks.
iii. FlexRay protocol
iv. *UART*: Universal asynchronous receiver/transmitter protocol.
v. *AUTOSAR*: Automotive open system architecture protocol.

These communication protocols not only notify the user about the actions taken but also notifies the other units in the electric vehicle about the prevailing conditions of the battery pack [57].

7.5.4 Topology of the BMS

For better understanding about various topologies used in the electric vehicle, first the tiers are needed to be explained. Three tiers describe the topologies. They are cell monitoring unit, module management unit and pack management unit. These three tiers are brief explained as given below:

i. *Cell monitoring unit (CMU)*: This tier is the lowest level among the three tiers which is connected to each and every cell. This unit/tier computes the fundamental cell parameters. It does the cell level balancing in case of any failure.
ii. *Module management unit (MMU)*: This tier is considered as the middle level which is responsible for intercell balancing. The MMU monitors a series of cell monitoring units (more cells—up to 12 cells) and combines them to form a module.
iii. *Pack management unit (PMU)*: This tier is the highest level which monitors a group of module management units (more modules). This unit calibrates the battery pack parameters such as the pack's voltage, pack's temperature, pack's current and aids in safety management. Moreover, this unit notifies the user and other systems in the electric vehicle about the working condition.

After defining the tiers, the BMS topology is classified into four categories [50]. These categories are discussed as follows:

i. *Centralized and nondistributed BMS*: In the centralized topology, all the three units (CMU, MMU, and PMU) are combined in a unit system called the central control unit as in Figure 7.5. This is more complex as it requires

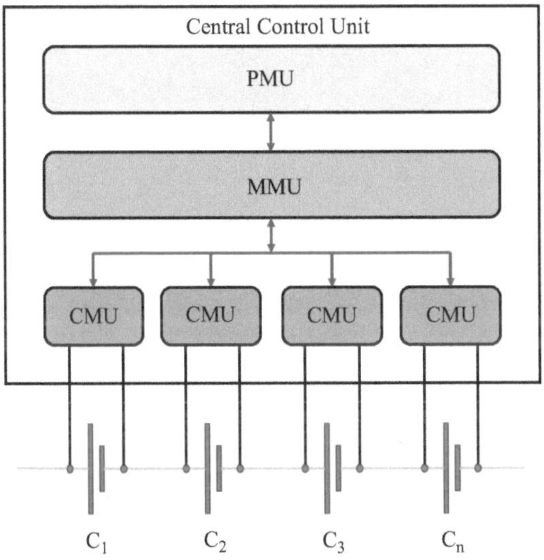

FIGURE 7.5 Centralized BMS topology.

more electrical wiring within the battery cells and is difficult to maintain. Due to its complexity, this type of topology is not suitable for large electric vehicles and holds good for low-capacity electric vehicles (bicycles) where the number of cells is limited [50].

ii. *Nondistributed and master-slave topology with distinct CMU*: Here the central control unit is termed as the master unit. In this type of topology, the master unit controls the MMU and PMU whereas the slave/CMU is connected to each battery cell as represented in Figure 7.6. The intercell balancing is performed either by the master or by the slave unit. This type is more flexible than the previous topology but it is difficult to maintain if the number of battery packs is more [50].

iii. *Nondistributed and master-slave topology with MMU*: This type of topology is most commonly implemented in electric vehicles of all kinds. Here, the master unit is combined with the PMU and this combination is solely to perform the PMU functions. On the other hand. the slave unit combines both the CMU and MMU as shown in Figure 7.7. This type of topology segregates (combines) the battery cells into distinct battery modules. Major advantages of this type are higher flexibility, easy to maintain, safe, high degree of replaceability, and easy to assemble [50].

iv. *Distributed software topology*: This topology is reported to be highly advanced and could be implemented in the future. This kind does not have any master (central) or slave unit. Each subsystem is said to initiate the functions from the three tiers and these functions in turn monitor one or more battery cells. It provides greater flexibility than the rest of the topologies but it is more complex. Moreover, to implement this type of topology, there is a requirement of highly advanced electronic control units, smart sensors and ultramodern RTOS (Real Time Operating Systems) systems [58].

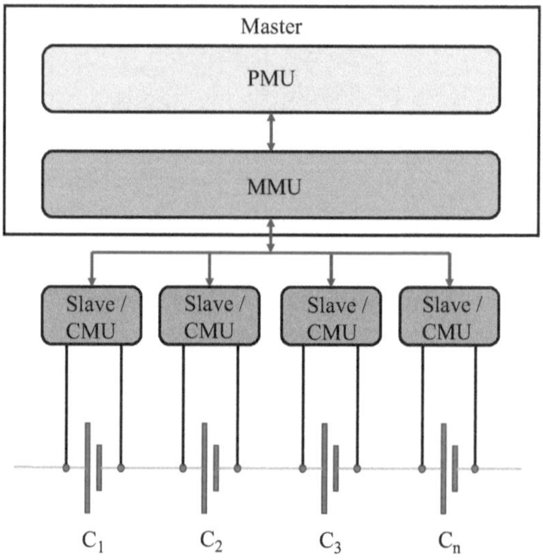

FIGURE 7.6 Master-slave topology with distinct CMU where C represents the battery cells.

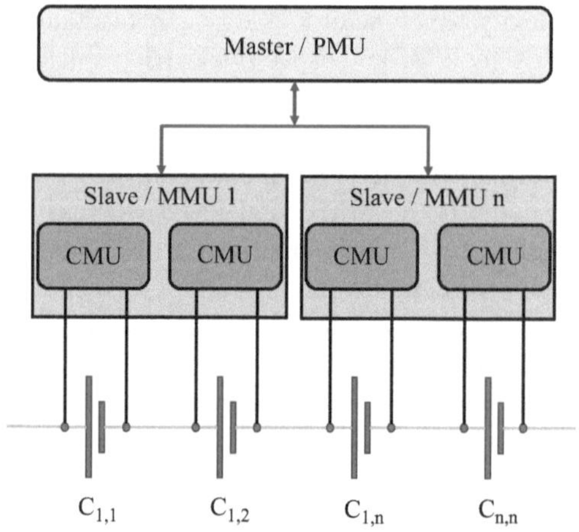

FIGURE 7.7 Master-slave topology with distinct MMU where C represents the battery cells.

7.5.5 Thermal Management for Batteries

Temperature variations are reported to be the main cause for most battery failures. To overcome this, certain cooling methods are employed to attain the nominal service temperature range. The thermal management systems are implemented for this purpose [59,60]. The major objectives are as follows:

1. To minimize the heat produced in the battery pack.
2. To reduce the temperature rise between each battery cells.

3. Helps in delivering the nominal voltage.
4. Act as a safeguard for the whole battery pack.
5. To control the humidity inside the pack.

Owing to the thermal cycle options, this system is categorized as depicted in Figure 7.8 [61].

7.5.5.1 Air Cooling Systems

As the name tells, the air cooling system uses the cooling medium as the air. This system is cost-efficient (considering the design, fabrication and maintenance cost), simple in design and direct contact between the cooling medium and battery pack. Based on the air it uses, this system is classified into two types namely cabin air and ambient air cooling systems. The atmospheric air is used as the cooling medium in this system. The ambient air reaches the battery pack through blower and after cooling the battery pack, it dissipates as exhaust air [62]. On the other hand, cabin air cooling uses the preconditioned air provided by the installed HVAC system in the electric vehicle [63]. In recent years, the electric vehicle uses the latest version of cabin air cooling system. This latest version has two HVAC systems, one is for preconditioning the cabin air and the other is for preconditioning the ambient air supplied to the battery pack for cooling [61].

7.5.5.2 Dielectric Liquid-Based Cooling Systems

This system uses liquid media to cool the battery packs. Here also, there exists a direct contact between the pack and liquid medium. Mostly dielectric liquid is preferred as the cooling medium in this system. Dielectric liquids are fluids that show

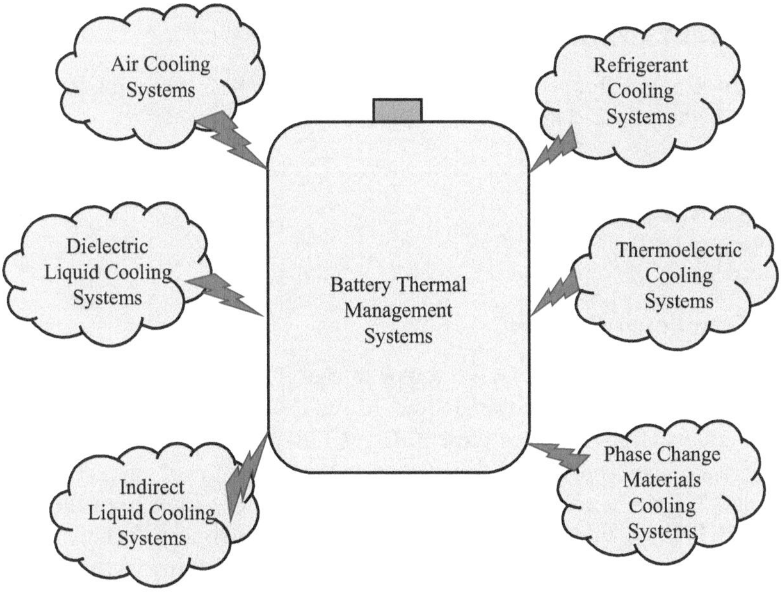

FIGURE 7.8 Battery thermal management systems.

greater resistance toward electrical breakdown. It minimizes or prevents the heating of battery packs. Though it reduces the heat to a greater extent, the maintenance and ease to assemble is very difficult for this kind of cooling systems [64].

7.5.5.3 Indirect Liquid Based Cooling Systems

This system is mostly implemented in electric vehicles as it overcomes the drawbacks of the aforementioned cooling systems. The liquid is said to be circulated between the battery pack and heat sink. This liquid absorbs the heat inside the battery pack and delivers it into the heat sink. This system is easier to maintain and safe for the battery pack [61].

7.5.5.4 Refrigerant-Based Cooling Systems

Refrigerant-based cooling system is similar to indirect liquid-based cooling systems where a refrigerant is circulated between the battery cells. The refrigerant will absorb the heat to a greater extent and has many advantages compared to the aforementioned types. Some of the commonly used refrigerants are R744, R134a and R1234yf [65–67].

7.5.5.5 Thermoelectric and Phase Change Material-Based Cooling Systems

This system holds good for limited load capacity electric vehicles as it uses thermoelectric materials and phase change materials to provide sufficient cooling to the pack. Peltier element acts as the thermoelectric cooler which dissipates the heat from one side to the other whereas the phase change material absorbs heat by changing its phase [68]. Some of the phase change materials used in electric vehicles are paraffin, lauric acid, $LiNO_3$, $CaCl_2$, water and so on [69–71].

7.6 BATTERY CHARGING METHODS

In electric vehicles, three types of charging methods have been implemented to charge the battery pack. The three charging methods are

i. Conductive charging
ii. Inductive charging
iii. Portable battery pack

7.6.1 Conductive Charging

Conductive charging (AC and DC) is a type of charging that has direct connectivity between the battery and the power source. In AC charging, there are various levels (level 1 and level 2) and modes present. Level 1 charging is mostly like the home charging where the electric vehicle is connected to the on-board charger (grounded receptacle—NEMA 5-15R). This level charges the battery pack from the existing AC network [72]. On the other hand, the level 2 charger energizes the battery via EV supply equipment. Level 2 charging is referred as the fast charging where the electric vehicle is connected to the on-board charger (EV supply equipment—single phase 7kW or 3-phase 21kW) [73]. The DC charging system charges the battery pack through off-board charger. Here the charging time is said to be rapid and within 20 minutes, 80% of the battery pack can be charged by means of a DC charger [74,75].

7.6.2 Inductive Charging

This type of charging is carried out with the help of induction coils. These coils are reported to generate the electromagnetic field (waves) from the charging base (transmitter). The generated electromagnetic field is then converted into the electric current with the help of receiver which is attached to the battery pack. This electric current is responsible for charging the battery pack without any connections [76,77]. The major advantage of this type of charging is wireless which means that there is no direct connection between the electric vehicle and the charger. But inductive charging is often termed as slow charging and also this type is more expensive. Hence, it holds good for electric vehicles with limited capacity.

7.6.3 Portable Battery Pack

This type is considered as the advanced charging method which reduces the anxiety of finding the nearest charging stations available. Here, the battery packs are designed in such a way that it can be taken from the stations (portable battery packs) and directly implemented in the electric vehicle. This holds good for long distance driving and the charged battery pack from the station is directly installed which eliminates the charging time [78,79].

7.7 BATTERY RECYCLING METHODS

Each battery pack has a predefined service lifetime depending on its capacity. When a battery pack reach its service lifetime, it can be recycled to some extent to recover the valuable elements. There are many methods which are implemented for recycling battery packs. This section deals with the five most efficient methods to recycle battery packs [80].

7.7.1 Pyrometallurgical Recovery

Pyrometallurgical process involves the separation of valuable elements from the battery pack in the presence of heat. In this process, the valuable elements are recovered after processing in a furnace (high-temperature type). After this process, the recovered elements can be of pure metals or alloys which can be a suitable feed for further processing [81,82]. In pyrometallurgical recovery, there are four steps to recover the valuable elements. These steps are briefly discussed as below:

i. *Calcining*: In calcining, the battery pack's components are heated to a level before melting them. This process removes moisture, any vaporous impurities and made it suitable for the next process.
ii. *Roasting*: Here, thermal gas-solid reactions such as redox, sulfation, and chlorination reaction are involved. The oxygen in the air reacts with the

calcined products in the presence of heat and forms oxide. These oxides dissipate or are collected separately thereby leaving the elements alone.

iii. *Smelting*: The roasted products undergo smelting process where high temperature is used to melt the impurities in the roasted product which is removed as liquid form leaving the pure metal alone.

iv. *Refining*: Refining is like smelting process where the roasted products are melted at very high temperatures to form a liquid melt pool. This melt pool is refined at various stages in order to extract the pure metals.

7.7.2 Physical Material Separation

As the name suggests, this recovery method uses physical ways for recovering valuable elements. A mixture of battery components is prone to physical media such as magnets, sieves, shakers, filters, etc. During this process, each physical media will segregate valuable elements/components that can be used for further processing [83,84].

7.7.3 Hydrometallurgical Metals Recovery

In this process, the valuable elements are recovered by means of leaching the battery components with leaching reagents. The leaching reagents react with the components and recover the valuable elements by chemical reaction [85,86]. Some of the commonly used leaching reagents are sulfuric acid, hydrochloric acid, sodium carbonate, sodium cyanide, ammonium hydroxide and so on.

7.7.4 Direct Recycling

In this approach, the cathode and anode materials are directly removed from both electrodes and can be implemented in other battery packs after undergoing relevant processing stages. This type of recycling avoids the buying of new electrode materials so that the old materials can be used effectively [87,88].

7.7.5 Bioleaching

This process is similar to the hydrometallurgical metal recovery process in which instead of using the leaching reagents, some microorganisms (bacteria) are used to remove the valuable elements [89]. Some acidophilic bacteria under specified conditions can react with certain forms of metal (metal sulfides, metal oxides) and convert them into valuable metals. Examples of bacteria used in bioleaching are Acidiphilium cryptum, Acidithiobacillus ferrooxidans, Leptospirillum ferrooxidans, Metallosphaera species, Sulfobacillus, and so on [90–92].

7.8 CONCLUSION

In most electric vehicles, batteries serve as the primary source of energy for actuating it. To supply the desired voltage required for long distance transportation, the battery cells are arranged in a predefined pattern to deliver the optimal voltage. Moreover, many factors influence the rated output voltage which is delivered to the motor for movement. To monitor the installed pack and to avoid these influences, each vehicle has BMS that supervises the entire pack. This chapter explained the cells' chemistries and the pack's configuration which is employed in most electric vehicles. Further, the need for BMS along with its topological characteristics is briefly discussed. Also, the thermal management strategies that are employed in cooling the battery packs are elaborated. In addition, the pack's charging types along with the battery recycling methods that are used to recover the valuable elements from the pack are detailed in this chapter.

ACKNOWLEDGMENT

The authors like to acknowledge the editors of this book for providing an opportunity to write this section.

REFERENCES

[1] Mi C, Masrur MA. *Hybrid Electric Vehicles: Principles and Applications with Practical Perspectives.* John Wiley & Sons; 2017.

[2] Patel N, Bhoi AK, Padmanaban S, Holm-Nielsen JB, editors. *Electric Vehicles: Modern Technologies and Trends.* Springer; 2021.

[3] Chan CC, Chau KT. *Modern Electric Vehicle Technology.* Oxford University Press on Demand; 2001.

[4] Wakefield EH. History *of the Electric Automobile: Hybrid Electric Vehicles.* SAE International; 1998.

[5] König A, Nicoletti L, Schröder D, Wolff S, Waclaw A, Lienkamp M. An overview of parameter and cost for battery electric vehicles. *World Electric Vehicle Journal.* 2021;12(1):21.

[6] Hannan MA, Azidin FA, Mohamed A. Hybrid electric vehicles and their challenges: A review. *Renewable and Sustainable Energy Reviews.* 2014;29:135–50.

[7] Clement-Nyns K, Haesen E, Driesen J. The impact of charging plug-in hybrid electric vehicles on a residential distribution grid. *IEEE Transactions on power systems.* 2009;25(1):371–80.

[8] Tanç B, Arat HT, Baltacıoğlu E, Aydın K. Overview of the next quarter century vision of hydrogen fuel cell electric vehicles. *International Journal of Hydrogen Energy.* 2019;44(20):10120–8.

[9] Sorlei IS, Bizon N, Thounthong P, Varlam M, Carcadea E, Culcer M, Iliescu M, Raceanu M. Fuel cell electric vehicles-A brief review of current topologies and energy management strategies. *Energies.* 2021;14(1):252.

[10] Tate ED, Harpster MO, Savagian PJ. *The Electrification of the Automobile: From Conventional Hybrid, to Plug-in Hybrids, to Extended-Range Electric Vehicles.* SAE International; 2008.

[11] Liu W, Placke T, Chau KT. Overview of batteries and battery management for electric vehicles. *Energy Reports.* 2022;8:4058–84.

[12] Fichtner M. Recent research and progress in batteries for electric vehicles. *Batteries & Supercaps.* 2022;5(2):e202100224.

[13] Ramkumar MS, Reddy C, Ramakrishnan A, Raja K, Pushpa S, Jose S, Jayakumar M. Review on Li-Ion Battery with Battery Management System in Electrical Vehicle. *Advances in Materials Science and Engineering.* 2022;9:56–64.

[14] See KW, Wang G, Zhang Y, Wang Y, Meng L, Gu X, Zhang N, Lim KC, Zhao L, Xie B. Critical review and functional safety of a battery management system for large-scale lithium-ion battery pack technologies. *International Journal of Coal Science & Technology*. 2022;9(1):36.

[15] Vaideeswaran V, Bhuvanesh S, Devasena M. Battery management systems for electric vehicles using lithium ion batteries. *Innovations in Power and Advanced Computing Technologies*. 2019;1:1–9.

[16] Ruetschi P. Review on the lead-acid battery science and technology. *Journal of Power Sources*. 1977;2(1):3–120.

[17] Moseley PT, Rand DA, Garche J. Lead-acid batteries for future automobiles: Status and prospects. *Lead-Acid Batteries for Future Automobiles*. 2017;601–18.

[18] Etacheri V, Marom R, Elazari R, Salitra G, Aurbach D. Challenges in the development of advanced Li-ion batteries: A review. *Energy & Environmental Science*. 2011;4(9):3243–62.

[19] Aurbach D, Talyosef Y, Markovsky B, Markevich E, Zinigrad E, Asraf L, Gnanaraj JS, Kim HJ. Design of electrolyte solutions for Li and Li-ion batteries: A review. *Electrochimica acta*. 2004;50(2-3):247–54.

[20] Bukhari SM, Maqsood J, Baig MQ, Ashraf S, Khan TA. Comparison of characteristics--lead acid, nickel based, lead crystal and lithium based batteries. In: *2015 17th UKSim-AMSS International Conference on Modelling and Simulation (UKSim)*. IEEE, 2015, pp. 444–450.

[21] Muslimin S, Nawawi Z, Suprapto BY, Dewi T. Comparison of batteries used in electrical vehicles. In: *5th FIRST T1 T2 2021 International Conference (FIRST-T1-T2 2021)*. Atlantis Press, 2022, pp. 421–425.

[22] Huang J, Guo Z, Ma Y, Bin D, Wang Y, Xia Y. Recent progress of rechargeable batteries using mild aqueous electrolytes. *Small Methods*. 2019;3(1):1800272.

[23] Zhao Q, Stalin S, Zhao CZ, Archer LA. Designing solid-state electrolytes for safe, energy-dense batteries. *Nature Reviews Materials*. 2020;5(3):229–52.

[24] Gambe Y, Sun Y, Honma I. Development of bipolar all-solid-state lithium battery based on quasi-solid-state electrolyte containing tetraglyme-LiTFSA equimolar complex. *Scientific Reports*. 2015;5(1):1–4.

[25] Fan X, Wang C. High-voltage liquid electrolytes for Li batteries: Progress and perspectives. *Chemical Society Reviews*. 2021;50(18):10486–566.

[26] Zhang SS. A review on the separators of liquid electrolyte Li-ion batteries. *Journal of Power Sources*. 2007;164(1):351–64.

[27] Zuo X, Fan C, Liu J, Xiao X, Wu J, Nan J. Lithium tetrafluoroborate as an electrolyte additive to improve the high voltage performance of lithium-ion battery. *Journal of The Electrochemical Society*. 2013;160(8):A1199.

[28] Parimalam BS, Lucht BL. Reduction reactions of electrolyte salts for lithium ion batteries: LiPF6, LiBF4, LiDFOB, LiBOB, and LiTFSI. *Journal of The Electrochemical Society*. 2018;165(2):A251.

[29] Rustomji CS, Yang Y, Kim TK, Mac J, Kim YJ, Caldwell E, Chung H, Meng YS. Liquefied gas electrolytes for electrochemical energy storage devices. *Science*. 2017;356(6345):eaal4263.

[30] Yang Y, Yin Y, Davies DM, Zhang M, Mayer M, Zhang Y, Sablina ES, Wang S, Lee JZ, Borodin O, Rustomji CS. Liquefied gas electrolytes for wide-temperature lithium metal batteries. *Energy & Environmental Science*. 2020;13(7):2209–19.

[31] Arora S, Shen W, Kapoor A. Review of mechanical design and strategic placement technique of a robust battery pack for electric vehicles. *Renewable and Sustainable Energy Reviews*. 2016;60:1319–31.

[32] Saw LH, Ye Y, Tay AA. Integration issues of lithium-ion battery into electric vehicles battery pack. *Journal of Cleaner Production*. 2016;113:1032–45.

[33] Perner A, Vetter J. Lithium-ion batteries for hybrid electric vehicles and battery electric vehicles. In: Scrosati B, Garche J, Tillmetz W, editors. *Advances in Battery Technologies for Electric Vehicles*. Woodhead Publishing, 2015, pp. 173–190.
[34] Xiong R, Sun W, Yu Q, Sun F. Research progress, challenges and prospects of fault diagnosis on battery system of electric vehicles. *Applied Energy*. 2020;279:115855.
[35] Xiong R, Sun F, Gong X, He H. Adaptive state of charge estimator for lithium-ion cells series battery pack in electric vehicles. *Journal of Power Sources*. 2013;242:699–713.
[36] Yong JY, Ramachandaramurthy VK, Tan KM, Mithulananthan N. A review on the state-of-the-art technologies of electric vehicle, its impacts and prospects. *Renewable and Sustainable Energy Reviews*. 2015;49:365–85.
[37] Ding N, Prasad K, Lie TT. The electric vehicle: A review. *International Journal of Electric and Hybrid Vehicles*. 2017;9(1):49–66.
[38] Conte FV. Battery and battery management for hybrid electric vehicles: A review. *E & I Elektrotechnik und Informationstechnik*. 2006;123(10):424–31.
[39] Rahimi-Eichi H, Ojha U, Baronti F, Chow MY. Battery management system: An overview of its application in the smart grid and electric vehicles. *IEEE Industrial Electronics Magazine*. 2013;7(2):4–16.
[40] Lipu MH, Hannan MA, Karim TF, Hussain A, Saad MH, Ayob A, Miah MS, Mahlia TI. Intelligent algorithms and control strategies for battery management system in electric vehicles: Progress, challenges and future outlook. *Journal of Cleaner Production*. 2021;292:126044.
[41] Li Y, Li K, Xie Y, Liu J, Fu C, Liu B. Optimized charging of lithium-ion battery for electric vehicles: Adaptive multistage constant current-constant voltage charging strategy. *Renewable Energy*. 2020;146:2688–99.
[42] Ali MU, Zafar A, Nengroo SH, Hussain S, Junaid Alvi M, Kim HJ. Towards a smarter battery management system for electric vehicle applications: A critical review of lithium-ion battery state of charge estimation. *Energies*. 2019;12(3):446.
[43] Singirikonda S, Obulesu YP. Battery modelling and state of charge estimation methods for energy management in electric vehicle: A review. *IOP Conference Series: Materials Science and Engineering*. 2020;937(1):012046.
[44] Noura N, Boulon L, Jemeï S. A review of battery state of health estimation methods: Hybrid electric vehicle challenges. *World Electric Vehicle Journal*. 2020;11(4):66.
[45] Xing Y, Ma EW, Tsui KL, Pecht M. Battery management systems in electric and hybrid vehicles. *Energies*. 2011;4(11):1840–57.
[46] Uzair M, Abbas G, Hosain S. Characteristics of battery management systems of electric vehicles with consideration of the active and passive cell balancing process. *World Electric Vehicle Journal*. 2021;12(3):120.
[47] Cao J, Schofield N, Emadi A. Battery balancing methods: A comprehensive review. In: *2008 IEEE Vehicle Power and Propulsion Conference*. IEEE, 2008, pp. 1–6.
[48] Hu, R. Battery management system for electric vehicle applications (2011), *Electronic Theses and Dissertations*. 5007.
[49] Zhou Z, Gong J, He Y, Zhang Y. Software defined machine-to-machine communication for smart energy management. *IEEE Communications Magazine*. 2017;55(10):52–60.
[50] Hauser A, Kuhn R. High-voltage battery management systems (BMS) for electric vehicles. In: Scrosati B, Garche J, Tillmetz W, editors. *Advances in Battery Technologies for Electric Vehicles*. Woodhead Publishing, 2015, pp. 265–282.
[51] Brandl M, Gall H, Wenger M, Lorentz V, Giegerich M, Baronti F, Fantechi G, Fanucci L, Roncella R, Saletti R, Saponara S. Batteries and battery management systems for electric vehicles. In: *2012 Design, Automation and Test in Europe Conference and Exhibition (DATE)*. IEEE, 2012, pp. 971–976.

[52] Hannan MA, Hoque MM, Hussain A, Yusof Y, Ker PJ. State-of-the-art and energy management system of lithium-ion batteries in electric vehicle applications: Issues and recommendations. *IEEE Access*. 2018;6:19362–78.

[53] Sabri MF, Danapalasingam KA, Rahmat MF. A review on hybrid electric vehicles architecture and energy management strategies. *Renewable and Sustainable Energy Reviews*. 2016;53:1433–42.

[54] Kim J, Oh J, Lee H. Review on battery thermal management system for electric vehicles. *Applied Thermal Engineering*. 2019;149:192–212.

[55] Lu L, Han X, Li J, Hua J, Ouyang M. A review on the key issues for lithium-ion battery management in electric vehicles. *Journal of Power Sources*. 2013;226:272–88.

[56] Zhu X, Zhang H, Cao D, Fang Z. Robust control of integrated motor-transmission powertrain system over controller area network for automotive applications. *Mechanical Systems and Signal Processing*. 2015;58:15–28.

[57] Quigley, C. Electronic system integration for hybrid and electric vehicles. *IET Hybrid Vehicle Conference*, 2006;125–139.

[58] Chakraborty S, Lukasiewycz M, Buckl C, Fahmy S, Chang N, Park S, Kim Y, Leteinturier P, Adlkofer H. Embedded systems and software challenges in electric vehicles. In: *2012 Design, Automation and Test in Europe Conference and Exhibition (DATE)*. IEEE, 2012, pp. 424–429.

[59] Tete PR, Gupta MM, Joshi SS. Developments in battery thermal management systems for electric vehicles: A technical review. *Journal of Energy Storage*. 2021;35:102255.

[60] Xia G, Cao L, Bi G. A review on battery thermal management in electric vehicle application. *Journal of Power Sources*. 2017;367:90–105.

[61] Huber C, Kuhn R. Thermal management of batteries for electric vehicles. In: Scrosati B, Garche J, Tillmetz W, editors. *Advances in Battery Technologies for Electric Vehicles*. Woodhead Publishing, 2015, pp. 327–358.

[62] Peng Q, Du Q. Progress in heat pump air conditioning systems for electric vehicles-a review. *Energies*. 2016;9(4):240.

[63] Zhang Z, Wang J, Feng X, Chang L, Chen Y, Wang X. The solutions to electric vehicle air conditioning systems: A review. *Renewable and Sustainable Energy Reviews*. 2018;91:443–63.

[64] Patil MS, Seo JH, Lee MY. A novel dielectric fluid immersion cooling technology for Li-ion battery thermal management. *Energy Conversion and Management*. 2021;229:113715.

[65] Junqi D, Yibiao W, Shiwei J, Xianhui Z, Linjie H. Experimental study of R744 heat pump system for electric vehicle application. *Applied Thermal Engineering*. 2021;183:116191.

[66] Yibiao WA, Junqi DO, Shiwei JI, Huang L. Experimental comparison of R744 and R134a heat pump systems for electric vehicle application. *International Journal of Refrigeration*. 2021;121:10–22.

[67] Li W, Liu R, Liu Y, Wang D, Shi J, Chen J. Performance evaluation of R1234yf heat pump system for an electric vehicle in cold climate. *International Journal of Refrigeration*. 2020;115:117–25.

[68] Lu M, Zhang X, Ji J, Xu X, Zhang Y. Research progress on power battery cooling technology for electric vehicles. *Journal of Energy Storage*. 2020;27:101155.

[69] Kim GH, Gonder J, Lustbader J, Pesaran A. Thermal management of batteries in advanced vehicles using phase-change materials. *World Electric Vehicle Journal*. 2008;2(2):134–47.

[70] Jilte R, Afzal A, Panchal S. A novel battery thermal management system using nano-enhanced phase change materials. *Energy*. 2021;219:119564.

[71] Ianniciello L, Biwolé PH, Achard P. Electric vehicles batteries thermal management systems employing phase change materials. *Journal of Power Sources*. 2018;378:383–403.

[72] Dericioglu C, Yiri K E, Unal E, Cuma MU, Onur B, Tumay M. A review of charging technologies for commercial electric vehicles. *International Journal of Advances on Automotive and Technology.* 2018;2(1):61–70.
[73] Habib S, Khan MM, Hashmi K, Ali M, Tang H. A comparative study of electric vehicles concerning charging infrastructure and power levels. In: *2017 International Conference on Frontiers of Information Technology (FIT).* IEEE, 2017, pp. 327–332.
[74] Arancibia A, Strunz K. Modeling of an electric vehicle charging station for fast DC charging. In: *2012 IEEE International Electric Vehicle Conference.* IEEE, 2012, pp. 1–6.
[75] Botsford C, Szczepanek A. Fast charging vs. slow charging: Pros and cons for the new age of electric vehicles. In: *International Battery Hybrid Fuel Cell Electric Vehicle Symposium.* World Electric Vehicle Association, 2009, pp. 1–9.
[76] Wu HH, Gilchrist A, Sealy K, Israelsen P, Muhs J. A review on inductive charging for electric vehicles. In: *2011 IEEE International Electric Machines and Drives Conference (IEMDC).* IEEE, 2011, pp. 143–147.
[77] Khaligh A, Dusmez S. Comprehensive topological analysis of conductive and inductive charging solutions for plug-in electric vehicles. *IEEE Transactions on Vehicular Technology.* 2012;61(8):3475–89.
[78] Gong J, Wang Y, Zhao A, Zhao Z. Optimal sizing of portable modular batteries for electric vehicles. *Journal of Cleaner Production.* 2020;277:123868.
[79] Lin BM, Suen SH, Jang JS. Promotion strategy of electric scooters in Taiwan. *World Electric Vehicle Journal.* 2009;3(1):69–72.
[80] Harper G, Sommerville R, Kendrick E, Driscoll L, Slater P, Stolkin R, Walton A, Christensen P, Heidrich O, Lambert S, Abbott A. Recycling lithium-ion batteries from electric vehicles. *Nature.* 2019;575(7781):75–86.
[81] Makuza B, Tian Q, Guo X, Chattopadhyay K, Yu D. Pyrometallurgical options for recycling spent lithium-ion batteries: A comprehensive review. *Journal of Power Sources.* 2021;491:229622.
[82] Liu C, Lin J, Cao H, Zhang Y, Sun Z. Recycling of spent lithium-ion batteries in view of lithium recovery: A critical review. *Journal of Cleaner Production.* 2019;228:801–13.
[83] Li J, Wang G, Xu Z. Environmentally-friendly oxygen-free roasting/wet magnetic separation technology for in situ recycling cobalt, lithium carbonate and graphite from spent LiCoO2/graphite lithium batteries. *Journal of Hazardous Materials.* 2016;302:97–104.
[84] Li X, Zhang J, Song D, Song J, Zhang L. Direct regeneration of recycled cathode material mixture from scrapped LiFePO4 batteries. *Journal of Power Sources.* 2017;345:78–84.
[85] Golmohammadzadeh R, Faraji F, Rashchi F. Recovery of lithium and cobalt from spent lithium ion batteries (LIBs) using organic acids as leaching reagents: A review. *Resources, Conservation and Recycling.* 2018;136:418–35.
[86] Chagnes A, Pospiech B. A brief review on hydrometallurgical technologies for recycling spent lithium-ion batteries. *Journal of Chemical Technology & Biotechnology.* 2013;88(7):1191–9.
[87] Shi Y, Chen G, Liu F, Yue X, Chen Z. Resolving the compositional and structural defects of degraded LiNi x Co y Mn z O2 particles to directly regenerate high-performance lithium-ion battery cathodes. *ACS Energy Letters.* 2018;3(7):1683–92.
[88] Dunn JB, Gaines L, Sullivan J, Wang MQ. Impact of recycling on cradle-to-gate energy consumption and greenhouse gas emissions of automotive lithium-ion batteries. *Environmental Science & Technology.* 2012;46(22):12704–10.
[89] Roy JJ, Cao B, Madhavi S. A review on the recycling of spent lithium-ion batteries (LIBs) by the bioleaching approach. *Chemosphere.* 2021;282:130944.
[90] Cai X, Tian L, Chen C, Huang W, Yu Y, Liu C, Yang B, Lu X, Mao Y. Phylogenetically divergent bacteria consortium from neutral activated sludge showed heightened potential on bioleaching spent lithium-ion batteries. *Ecotoxicology and Environmental Safety.* 2021;223:112592.

[91] Moazzam P, Boroumand Y, Rabiei P, Baghbaderani SS, Mokarian P, Mohagheghian F, Mohammed LJ, Razmjou A. Lithium bioleaching: An emerging approach for the recovery of Li from spent lithium ion batteries. *Chemosphere*. 2021;277:130196.

[92] Moosakazemi F, Ghassa S, Jafari M, Chelgani SC. Bioleaching for recovery of metals from spent batteries-a review. *Mineral Processing and Extractive Metallurgy Review*. 2022;2:1–1.

8 Challenges and Limitations of Electric Vehicles

Venkata Charan Kantumuchu
Bradley University

8.1 INTRODUCTION

The history of the automotive industry could go back as far as the 1800s. The first car was made back in the 1800s with gas as the fuel, and since then many developments have taken place in the automotive industry such as mass production, multiple cylinder engines, and climate-controlled cabinets. Today's cars are far more advanced, and they not only cover the comfort aspect such as leather heated seats or Dolby audio systems, but they also have superior safety functionalities like lane departure warning and adaptive cruise control, collision prevention and many more such features. In recent years the focus has shifted toward electric cars. With more and more people concerned about the emissions from the gas vehicles, electric vehicles have been viewed as a good alternate mode of transport. The emissions caused by the transportation sector are so significant, according to an article published by Li et al., in recent years, the US transportation industry has been responsible for nearly 30% of the country's total greenhouse gas emissions, over 50% of the emissions of carbon monoxide and nitrogen oxides, and approximately 25% of hydrocarbon emissions. Additionally, this sector has been accountable for about 75% of the petroleum consumption in the US [1]. According to Muratori et al. 2021 Prog. Energy 3 022002, electric vehicles (EVs) offer several benefits when compared to internal combustion engine vehicles (ICEVs) powered by fossil fuels, including but not limited to zero emissions from the tailpipe, independence from petroleum, better fuel efficiency, reduced maintenance costs, and an enhanced driving experience (such as smoother acceleration, reduced noise, and convenient recharging options at home and public stations) [2]. The electric cars, unlike gas and steam engines, had their advantages, such as not requiring a manual effort to start, such as a hand crank on gasoline engines, and not having to shift gears [3]. According to the Department of Energy, the first electric car was made in 1890 in Des Moines, Iowa, United States, with a top speed of 14 miles per hour [4]. After more than 100 years the technology has advanced and today, we have several automotive companies bringing their efforts together to make the most efficient yet powerful electric cars. Today we have companies like Tesla, Rivian, Fisker, Lucid and many others that are focusing on the EV market. To encourage the use of low/zero emission cars the state of California has

DOI: 10.1201/9781003354901-8

even announced rebates for drivers up to $7,000 [5]. The global sales trends of EVs suggest the rapid growth of this market. According to an article published by CNBC, the EV sales hit an astounding 6.6 million units in 2021 and in 2022, two million EVs were sold in the first quarter of 2022 alone, which is a 75% increase compared to the first quarter of 2021 [6]. The growth of the EV market has been increasing exponentially lately, and it is of pivotal importance to consider the challenges and limitations that surface with this growth. The challenges with EVs could range anywhere from more mining of minerals such as lithium for making batteries to the long wait times at charging stations in populated cities that cause severe inconvenience.

This chapter intends to uncover many such challenges and limitations that emerge with more EVs on the roads and the potential solutions to these problems. The following challenges and limitations will be covered through the remainder of the chapter, while outlining the possible solutions to each of these problems.

8.2 LITHIUM-ION BATTERIES AND THE ENVIRONMENTAL EFFECTS

EVs in the recent times have been relying on lithium-ion batteries (LIBs) as their energy source. These batteries have several advantages over internal combustion engines (ICEs) such as lower cost, maintenance, zero emissions and sustainability. For this reason, lithium is considered to be a suitable metal for making EV batteries. Due to their high energy density, high power density, and long cycle life, LIBs have found widespread usage in various applications such as portable electronics, EVs, and grid storage [7]. With these many vast applications, lithium has been in high demand and according to USGS (2022) report, the cost of spot lithium metal prices has increased from approximately $77,000 per ton in January to about $97,000 per ton in July in China, which is approximately a 26% increase in a period of 6 months. The figure below shows the various markets for lithium [8].

As seen from the above figure, two thirds of the lithium is used toward batteries and the use of batteries for EVs is growing exponentially. It is estimated that a typical EV battery contains 10 kg or 22 pounds of lithium. This number could be higher in higher performing EVs. For example, a Tesla roadster is estimated to have 6,831 individual Li-ion batteries that roughly weigh about 900 pounds [9], while this number is about three times that of a Hummer EV. So, the amount of lithium required for these batteries and the growing number of EVs certainly transpire into a concern for EVs and their battery management system. The battery capcity plays a major role in the amount of lithium consumed. The battery capacities of several EVs are compared in two different tables along with their respective ranges (Table 8.1).

It is expected that the EV passenger market will account for 30% of the market in the next 10 years [10]. With such rapid and significant growth, there would be challenges associated with the mining and extraction of the materials required for making EV batteries. Besides the challenges with mining, there are other issues such as disposal and storage. According to an article published on the disposal of lithium batteries, there has been a 38% increase in fire damages either through transportation or material recovery facilities due to the presence of lithium batteries in the waste [11].

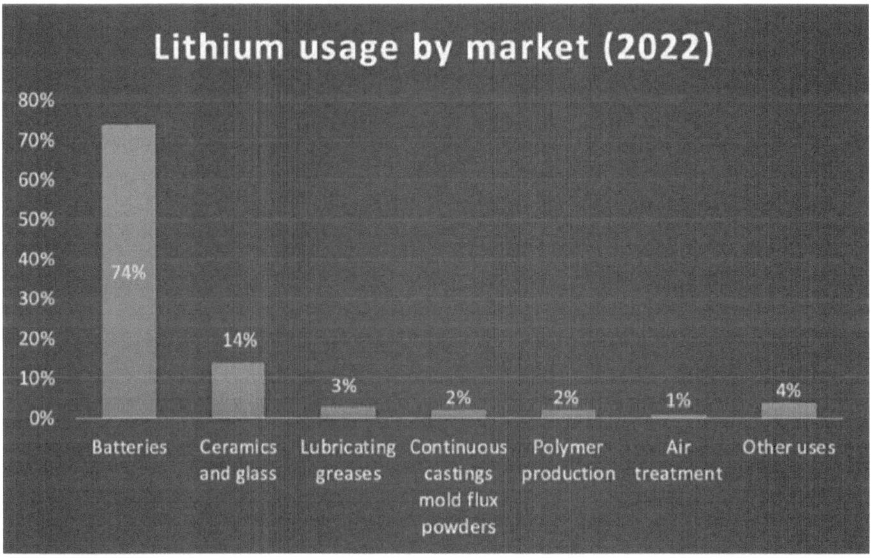

FIGURE 8.1 Showing lithium usage by each market.

TABLE 8.1
A List of Some Popular EVs With Low Ranges

Vehicle Model	Battery Capacity (KWh)	Drivetrain	Range—EPA
2022 Porsche Taycan Turbo S 21″	93.4	AWD	201 mi
2022 Porsche Taycan 19″	79.2	RWD	200 mi
2022 Porsche Taycan 4S 19″	79.2	AWD	199 mi
2022 Audi e-tron S Sportback 21″	95	AWD	185 mi
2022 Audi e-tron S 21″	95	AWD	181mi
2021 Hyundai IONIQ Electric	38.3	FWD	170 mi
2022 Nissan LEAF S	40	FWD	149 mi
2022 MINI Cooper SE	32.6	FWD	114 mi
2022 Mazda MX-30	35.5	FWD	100 mi

This has a drastic effect on the environment, and with the exponentially increasing use of lithium batteries, this effect will only get worse. This, combined with mining rare earth metals, will certainly remain a challenge.

A potential solution to this problem is the efficient handling of lithium batteries toward the end of their life cycle. This is possible through effective recycling methods. Recycling materials is a great option for reducing mining since it would not only prevent used batteries from causing dreadful environmental effects but also cause less mining since the recycled batteries can satisfy the growing needs of lithium-dependent industries. According to an article, it is estimated that about 60% of lithium demand for LIBs produced in China can be met through recycling by 2050 [12].

Not just for lithium, but recycling can help reduce the mining of other metals as well. "Based on the projected demand, our findings suggest that recycling could potentially lower primary demand compared to total demand in 2040 by roughly 25% for lithium, 35% for cobalt and nickel, and 55% for copper" [13]. Other solutions include but are not limited to using alternate types of batteries such as hydrogen batteries and batteries that use new technology that provide longer ranges and more durability. Hydrogen fuel cells though promising, have their own set of problems such as hydrogen storage but newer technologies should be able to overcome these limitations [14].

8.3 INADEQUATE CHARGING STATIONS

Gas vehicles today have sufficient infrastructure in place to support both long distance and internal city commute. The engines are being designed to produce more and more gas mileage with little to no loss of performance. The demand for better mileage cars is not only a requirement from the consumers but also from the regulatory bodies. According to the standards set by Corporate Average Fuel Economy (CAFÉ), the requirements for gas mileage are expected to be at 54.5 miles per gallon (mpg) [15]. Today, compact cars such as Nissan Sentra are able to deliver 29 MPG in the city and 39 MPG on the highway [16]. Similarly, cars like Hyundai Elantra could produce mileage of up to 42 mpg on the highway [17]. So, gas cars can produce excellent gas mileage. For example, it is estimated that a Toyota Camry may go up to 600 miles with a full tank of gas, while the commercially available EVs are still struggling to maintain a range of upwards of 350 miles on a single charge. This would result in stopping more frequently on a long trip. Add this to the charging time, and it would be even more painful to go for a longer commute. So, in order to support charging the exponentially growing number of EVs, the charging infrastructure needs to grow proportionally. As discussed before, it is claimed that global EV sales will increase by 41% in 2020. This number is only growing year over year; in 2021, the number of electric cars sold reached a record high of 6.6 million, which is almost twice as much as compared to 2020 [18]. According to a report, "Between 2018 and 2020, the count of EVs on US roads rose by 87.5%, while the number of public charging outlets increased by roughly 31%" [19]. According to [20] a significant percentage of EVs are charged at homes and workplaces, while only approximately 10% of charging is done at charging stations outside work and home. This certainly turns out to be a problem when traveling long distances and having to detour and wait before it is one's chance to charge. On top of this, the charging speeds slow down when multiple cars charge at the same time. Tesla, on their website has stated that the charging speed is impacted by the number of cars charging at that time [21]. In the figure below, it is evident that there are far more gas stations compared to the EV chargers in the US (2020), and with the exponential potential growth that is expected in the EV market, the demand for more charging stations is only reemphasized.

A potential solution to this problem is to increase the number of charging stations and ensuring that consumers can drive without the fear of being stranded at an inadvertent place. So, building more charging stations is both necessary and imperative. Even the US government is supporting this initiative. According to a note released by the Whitehouse, "The vision of creating a national network of 500,000 charging

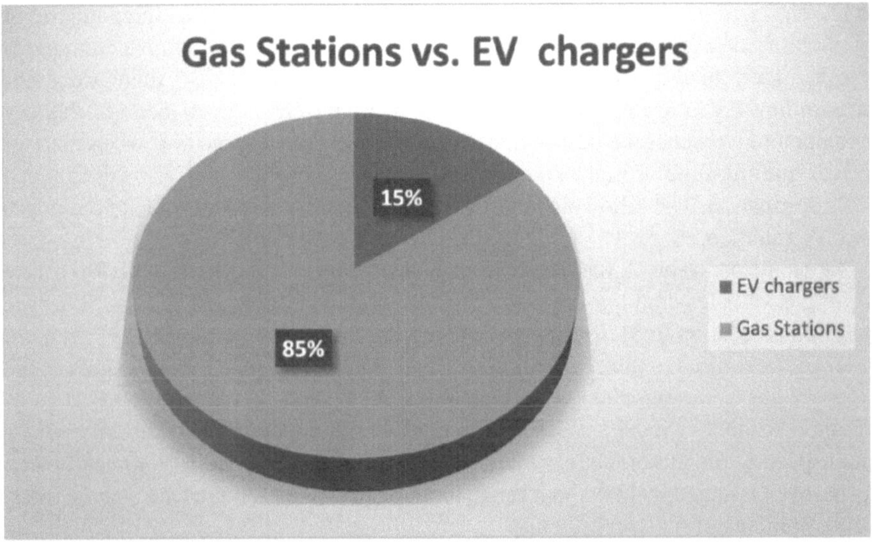

FIGURE 8.2 Showing the number of gas stations compared to EV chargers.

stations will be supported by a transformational investment of $15 billion as part of President Biden's American Jobs Plan" [22]. Even from a consumer standpoint, having more charging stations provides the confidence and encourages consumers to incline toward the EVs. According to [23] increasing the number of charging stations has more impact than increasing the financial incentives to consumers, "Providing an extra charging station per 100,000 residents had a roughly twofold impact on the electric vehicle (EV) market share compared to offering an additional $1,000 in consumer financial incentives". Another solution may be to repurpose the existing gas stations to accommodate the EV charger stations, so that more business and employment may be created at the gas stations due to longer wait times of charging vs. filling gas.

8.4 INSUFFICIENT RANGE

As part of the problem discussed above, another common challenge that EVs have is the range issues. Often referred to as the range anxiety, range issues with EVs are another major complaint. According to Henry et al. (2016) "If the availability and compatibility of charging stations cannot be guaranteed to a prospective buyer, they are unlikely to buy an electric vehicle (EV)" [24]. So, the range issues have a major impact on the sales of the EVs. In fact, it is such a crucial factor that according to Adnan et al., the most influential factor in the process of making an EV purchase decision is the range anxiety, which is simply the fear of running out of charge before a charging station can be reached [25]. According to an article published by Zhengming and Renran (2021), "The issue of range anxiety has been extensively debated as one of the major apprehensions surrounding the adoption and usage of battery electric vehicles (BEVs), and its impact on user experience" [26]. According to [27], the range anxiety could be twofold and may include: (i) the user perception of EV chargers' infrastructure vs. conventional gas stations and (ii) the potential of the inability of a user to reach a charger station before the battery drains completely and then the probability of a charge port available upon arrival. So, there are issues surrounding EV charging such as the number of chargers, the chance of a battery dying before it reaches the destination, the availability of a charge port when the user arrives and the time it might take to charge before there is enough power to reach one's destination. The following tables give us a quick overview of some of the popular EVs and their ranges

As seen from Table 2, the best EVs still have a range in the vicinity of 500 miles, while gas vehicles could go upwards of 600 miles. While the EVs in the 500-mile range could cost up to $170k, the gas cars could cost a mere $26k [28, 29] for more mileage. According to [30], range is also affected by several other factors, such as the gross weight of the vehicle, tyre radius, moment of inertia, wind area, coefficient of rolling resistance, windage coefficient, reduction ratio, battery rated capacity, maximum speed of the motor, battery rated quantity of electricity, and motor peak power. A survey conducted by [31] about range anxiety among a group revealed some interesting opinions of the users, such as

- More than 20% of the participants chose charging stations that were closer than the current gas stations.
- 50% of the participants believe placing charging stations just as close the gas stations are today, will solve the range anxiety issues.
- On average though, the participants believed that the distance between the charging stations should be 0.12 km lower than the distance between the gas stations

TABLE 8.2
A List of Some Popular EVs with High Ranges

Vehicle Model	Battery Capacity (KWh)	Drivetrain	Range—EPA
2022 Lucid Air Dream Edition Range 19″	118	AWD	520 mi
2022 Lucid Air Dream Edition Range 21″	118	AWD	481 mi
2022 Lucid Air Grand Touring 21″	112	AWD	469 mi
2022 Tesla Model S LR AWD 19″	100*	AWD	405 mi
2022 Tesla Model S LR AWD 21″	100*	AWD	375 mi
2022 Tesla Model 3 Long Range AWD 18″	80*	AWD	358 mi
2022 Porsche Taycan Turbo S 21″	93.4	AWD	201mi
2022 Audi e-tron S Sportback 21″	95	AWD	185 mi
2021 Hyundai IONIQ Electric	38.3	FWD	170 mi
2022 Nissan LEAF S (40 kWh)	40	FWD	149 mi

Note: Battery capacity for Tesla models is estimated and not official.

This survey falls in part between the adequacy of charge stations and the range issue topics. Now coming to the solutions, there are a few potential solutions to the range problem.

- Increase the number of fast chargers available to the public
- Develop battery technology with more power density
- Ensure charging stations are only used for charging and not for other purposes such as parking
- Redesign EVs to reduce drag and improve aerodynamics
- Use lighter but safer materials to reduce the weight of the cars for longer mileage
- Battery optimization through improved technological advances by identifying wastes
- Scaling the current production of chargers to reduce the costs of EV charging infrastructure

8.5 AFFORDABILITY

The affordability of EVs has always been a topic of interest. For example, one of the cheapest gas cars available on the market, the Nissan Versa (2022 model), is priced at $15,580 [32] on their website. On the other hand, Nissan has also priced one of the cheapest EVs available on the market, the Nissan Leaf (2022 model), at $28,040 [33]. There is a price variation of $13,000 between the two cheapest gas and EV models available on the market, made by the same company. According to Progressive [34], an insurance company in the United States, "The cost of insurance for an electric vehicle may be higher compared to that of a regular gas-powered car due to the former's higher price and the more intricate equipment it employs, leading to more expensive repairs or replacements in the event of an accident. Consequently, individuals with comprehensive and collision coverage may have to pay higher rates". According to an article published by Benjamin Preston, "The cost of insuring an electric car can be higher compared to a conventional gasoline-powered car. For instance, when compared to a Honda Insight hybrid, the Chevrolet Bolt EV may cost an additional $184 annually to insure, and a Tesla Model 3 could cost $318 more to insure than an Audi A4, which runs on gasoline" [35]. Apart from these, one factor that is often considered during the purchase of EVs is the tax credits. The federal and sometimes some states could offer tax credits for purchasing clean vehicles (EVs) but these are applied only under certain conditions. One such condition is the cap of EVs beyond which the tax credit would not apply. For example, according to the US Department of Energy, some manufacturers such as Chevrolet, GMC and Tesla have reached their cap of 200,000 EVs assembled in North America and are therefore not eligible for the clean vehicle tax credit [36].

Let us now consider how buying an EV could still be a good option.

- Though the initial cost of an EV may be higher, the cost of ownership is relatively low. Based on a consumer report [37], "The costs of owning an electric vehicle over its lifetime were significantly lower than those of all comparable ICEVs, with most electric vehicles providing savings of approximately $6,000 to $10,000".

- According to an article published by Idaho National laboratory, EVs can travel twice as much compared to a gas vehicle for the same cost [38].
- According to [39], "Electric vehicle (EV) owners typically spend around 60% less on annual fuel expenses than drivers of traditional gasoline-powered vehicles".

8.6 NOT REALLY "CLEAN"

EVs are said to have an advantage over combustion engines because of the zero emissions. The performance of electric vehicles in terms of greenhouse gas emissions compared to ICEVs is a subject of significant debate among scholars and practitioners. There are many variables that can affect this performance, and the extent to which EVs are better in this regard is still a matter of discussion [40]. A debate that often arises with EVs is whether they are clean energy vehicles. Many people advocate the use of EVs because they are essentially zero emission vehicles. According to [41], "EVs have the potential to significantly decrease carbon emissions and air pollution at a local level". This is one big factor that is promoting the sales of EVs as many people who care for the environment and the future generations like to see reduced carbon emissions. According to [42], "A global approach to decreasing carbon emissions and achieving a cleaner environment in numerous countries is to encourage the adoption of electric and hybrid vehicles". However, on the flipside there is an argument about whether EVs are really zero carbon emissions, and this is because the energy produced to charge these EVs could still be produced by burning fossil fuels. According to [43], "Electric vehicles (EVs) and HPs eliminate the use of fossil fuels and the emission of greenhouse gases from tailpipes and on-site sources, but they do contribute to emissions from the generation of electricity". If the energy required to charge the EVs comes from burning fossil fuels or not "clean" energy such as solar or wind energy, then though the EVs are zero emission, they are not completely "clean" energy vehicles. According to [44], the CO_2 produced by an electric car depends on the source of energy that is feeding it. For example, while a clean energy state like California would only produce 100 gm of greenhouse gases per mile, a fossil-fuel dependent state like Minnesota produces three times as much of greenhouse gases as California. In some countries power grids are energized by clean energy, while in many others they are still dependent on fossil fuels. According to [45], Most electricity grids worldwide still rely on fossil fuels like coal or oil, which are used to generate the energy needed to charge EVs. BP's statistical review of world energy has reported that primary energy use from fossil fuels has decreased from 85% 5 years ago to 82% last year, according to a recent article.

Though there is research published about the claims that the energy used to fuel the EVs is from fossil fuels, the emissions caused by the EVs are relatively low. According to the United States Environmental Protection Agency, even when the emissions from fossil fuels to produce electricity are considered, this may still be less than the total emissions from gas vehicles. According to the EPA, "Taking into consideration the electricity emissions, studies indicate that an electric vehicle generally produces less greenhouse gases (GHGs) compared to an average new gasoline car" [46]. However, the production of more electricity from renewable sources such as wind energy and solar energy may be considered an alternative to burning fossil fuels.

8.7 SERVICEABILITY

It is not uncommon to see gas-powered vehicles run out of gas or have a mechanical breakdown issue and need assistance. In such cases, a can of gas or a trained person can help get the car moved. This is typically referred to as roadside assistance. According to [47], "Vehicle breakdown, flat tire, running out of fuel, or getting stranded are some of the scenarios covered under roadside assistance. In the case of gas-powered cars, these services can also be used to obtain gasoline when stranded at the roadside. Depending on the specific service, a mechanic or service provider closest to your location can be summoned to deliver the amount of fuel you require" [48]. For EVs, this concept is new and unheard of. The options could include a way to charge your vehicle enough to get to the nearest charging station or tow your car to the closest EV charger. Though this may not sound like a daunting problem, it is the result of range anxiety and failing to reach a designated point with not enough charge left.

The good news with EVs is that they have significantly lesser mechanical parts meaning the chances of having a mechanical breakdown are minimized considerably. According to [3], "Compared to a gasoline-powered vehicle, which comprises hundreds of moving parts, an electric vehicle is much simpler, with only one moving part—the motor. This disparity in the number of moving parts is significant since the electric vehicle necessitates less frequent maintenance and is consequently more dependable". And one of the solutions for the EVs stopping in the middle of the roads due to lack of electricity may be addressed with portable mobile chargers just like a can of gas for gas vehicles. According to [49], "Roadside assistance for electric vehicles is provided by portable charging solutions".

8.8 CYBER RISK

We live in a smart age. Many things around us are smart already, like the watches we wear, the phones we use, the televisions we watch, and sometimes the clothes we put-on. It is no wonder that cars, which are a big part of our lives, have also become smart in this process. Many of these devices interact with each other. For example, a smart watch can interact with your phone, tablet, and even a payment kiosk. According to the Google Wear OS website, "In specific countries or areas, it is possible to utilize the Google Wallet app on your smartwatch for contactless payments at places where Google Pay is recognized" [50]. Similarly, cars can interact with your smart devices. This ability of interaction of devices (and people) is often referred to as IoT (Internet of Things). According to [51], "The swift progress and integration of smart and IoT (Internet of Things) technologies have enabled diverse prospects for technological innovations across various facets of life". According to [52], "The IoT has led to numerous transformations, such as smart cities, smart homes, pollution control,

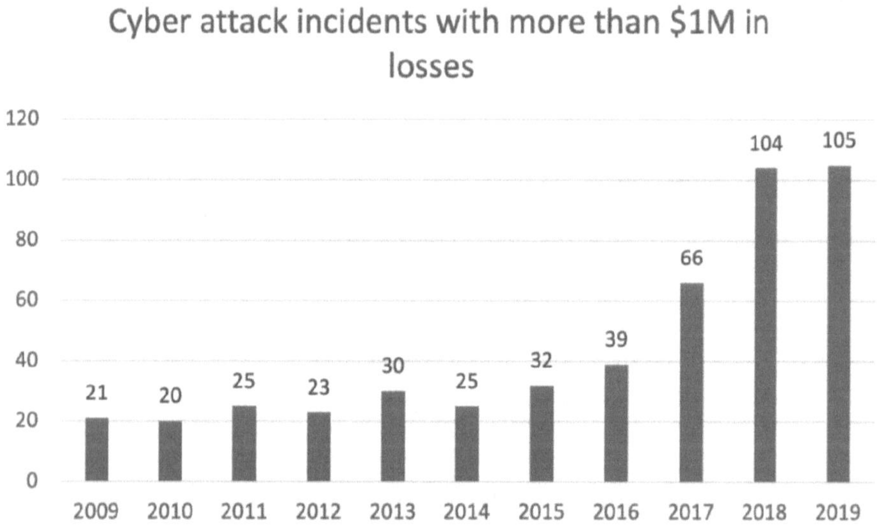

FIGURE 8.3 Showing the cyber-attacks with more than $1 M in losses.

energy conservation, smart transportation, and smart industries". There is a risk that comes with this smart technology. Cyber-attacks happen from time to time, and they can happen anywhere such as government agencies and tech industries. The following picture shows the losses incurred in excess of $1 M from 2009 to 2019. The graph shows a clear raise in the number of incidents and the number has grown fivefold in a span of 10 years and this is seriously concerning.

When this technology is used for EVs, it is referred to as Internet of Electric Vehicles (IoEV). According to [53], IoEV uses many communication technologies to communicate, and it is therefore vulnerable to severe cyber-attacks such as DoS, false data injection and modification. The smart technology makes EVs an easy target for hackers. According to [54], "With the increasing intelligence and connectivity of vehicles, their vulnerability to cyber-attacks is also rising. There have been instances where smart or autonomous vehicles were successfully hacked, highlighting the feasibility of such attacks on various occasions and platforms". According to an article published by the Security magazine in April 2022, "In the previous month, a tech security expert aged 19, managed to hack into 25 Tesla cars in over a dozen nations using a third-party software application called TeslaMate" [55]. This news could be concerning for several EV users because EVs have not completely penetrated the market yet and we already have hackers hacking into some of the world's secured electric cars. With the smart features that come with the IoT, the risk of being monitored, watched and robbed could increase significantly.

A potential solution to this problem is to research more advanced cybersecurity methods and tools that would minimize/eliminate the probability of cyber-attacks. According to [56], "To safeguard EV charging infrastructure, it is crucial to establish comprehensive cybersecurity guidelines based on well-founded research". Similarly, it is also important to protect EVs at public charging stations as EVs could be vulnerable to cyber-attacks at public charging stations. According to [57], by using a residual generation technique, an attack resilient EV charging system may be designed, which can detect false data injection under the denial of service (DOS) attack. There is existing technology to support the security of EVs. According to [58], "The technology required for secure implementation of a connected EV charging infrastructure is already available, with lessons drawn from the Internet and industrial control systems. Cryptographic algorithms such as AES-256, protocols like TLS, digital signatures, and certificates constitute the key tools for achieving this objective". Another solution could be regulating the production of EV chargers and making sure that protocols are in place to confirm compliance.

8.9 GRID PROBLEMS

EVs may be charged at public chargers and at home. Regardless of where they are charged, the act of drawing power by multiple electric vehicles in a given time frame may result in the overloading of the power grid. According to [59], "The increase in electric vehicle usage may place an excessive burden on the electric grid unless charging is primarily carried out during the day". This statement could be true since it is a common practice to charge EVs during night and have them ready for a busy day the next morning. According to [60], "This implies that almost 80 percent of EV charging occurs at the owner's residence during the night when the vehicle is not required, and sufficient time is available for a complete charge". This is very similar to charging your phone overnight when you need it the least. But what could be affected in the power distribution system because of charging EVs overnight? According to [61], "The EV load has an impact on several key components of the power grid, including transformers, transmission lines, and switchgear protective devices". But these problems may also depend on the time of the day that these vehicles get charged. According to [62], "If a large number of individuals purchase electric vehicles and attempt to charge them right after returning home from work, similar to the present scenario, it could potentially overwhelm the system or compel utilities to supply more power than their present production capacity". This could mean that households may be asked to charge their vehicles at different times of the day to decrease the load on the grid which may not be feasible for everyone. This simply means a lot of planning needs to go into managing the load on the power grids and not just urge people to buy more EVs. According to [63], "The adoption of EVs may result in grid limitations and higher transmission and distribution peaks, necessitating expensive measures like peaker plants, unplanned grid upgrades, and other solutions to address the issue".

When we look at the solutions to these problems, there are some interesting solutions such as using shared EVs instead of privately-owned EVs for better grid management [64]. Another solution could be a smart charging technology that enables communication between the vehicle and the charger, allowing for optimized charging. According to [65], "Smart charging, also referred to as V1G charging, is a process where an electric vehicle and a charging device establish a data connection. This connection allows you to efficiently regulate the charging of your EV by linking it to the power grid. Smart charging leverages the available energy during a specific timeframe to make the best use of it". The power grids may just need to produce more energy to keep up with the increasing demand and this is not an impossible task. According to [66], "Currently, Americans drive approximately three trillion miles annually, which is equivalent to half a light-year. If they drove all these miles in vehicles similar to the Tesla Model 3, an additional 750-terawatt hours (tWh) of electricity per year would be required. This represents a rise of roughly 20% in power demand compared to the US's 2020 generation of 4,000 tWh. However, this demand is not expected to occur until approximately 2050". So, all the evidence suggests that EVs will have little to no effect on the grids in terms of overloading them.

8.10 LONG CHARGE TIMES

EVs are notorious for long charging times. This is one of the biggest drawbacks of EVs that several non-EV users complain about. While it might take a typical gasoline car 3–5 minutes to fill up, it could take 15–30 minutes for an EV to charge, even at a high-speed charge station. EVs such as trucks could wait longer. According to [67] "Although drivers have become used to refilling their gas tanks in a matter of minutes, charging an EV typically takes much longer. The amount of time it takes to charge an EV to 80% capacity varies depending on the size and specifications of the battery, but at the fastest charging stations currently available, it usually takes at least 30 minutes". Though EVs have several advantages, like being almost maintenance-free, emissions-free, and having a reduced cost of ownership over time, the charging speeds are still a cause for concern for potential buyers. The charge speeds worsen when using chargers at home that plug into the wall. According to [68], "The charging time for electric vehicles can significantly differ, ranging from as little as 20 minutes at a public charging station to several hours when using a home charging station. Due to this variability, charging wait times and access to charging infrastructure are commonly reported sources of concern among potential EV buyers". Research done on the behavior of EV drivers revealed an interesting aspect that EV drivers prefer both fast and slow charging depending on their driving needs, but the study also suggests that fast charging is essential during a long commute [69]. Technology today is yet to support making an affordable EV that has long range and short charging times. For example, Chevvy Bolt EUV, which is one of the most affordable electric vehicles available in the market with a maximum range of 247 miles, charges at four miles per hour with a 120 V charger and takes 7 hours for a full charge on a 240 V charger and takes about 30 minutes to get a range of 95 miles with a DC fast charge [70]. To wait for 30 minutes just to get a range of 95 miles with the fastest charger is perhaps several times slower than a gas car which can be filled in 3–5 minutes. These are some genuine concerns that need to be addressed if the market share of EVs needs to rise.

On the other hand, if we look at the possible solutions to this problem, there are several technological advances that are taking place that could help with reducing the charging times. According to [71], "The engineers at Purdue University have created a new charging station cable that is currently pending a patent. This cable has the capability of fully recharging specific electric vehicles in less than 5 minutes, which is comparable to the time it takes to fill a gas tank". The students have designed a charging cable that would remove up to 24.22 KW of heat and deliver 4.6 times more current. Such technological advancements can make it possible to cut down on the charge times and enable EVs to compete with gas cars in all aspects. According to [72], it may be soon possible to charge your EV up to 90%, in 10 minutes which is perhaps faster than charging your phone. "We've significantly increased the amount of energy that can go into a battery cell in a short amount of time", says Dufek. "At present, it is possible to charge batteries up to 90% within 10 minutes without encountering issues such as lithium plating or cathode cracking". As more consumers move toward the EV market, the scalability of the battery and charging technology exponentially increases and more efficient batteries and charging technologies may be available to us.

8.11 OTHER CONCERNS

There are also other concerns besides what has been discussed already. EVs have been reported to have issues in freezing weather conditions such as batteries dying down. EVs are also notorious for the strange things that they could do such as autonomous driving, and this has caused several accidents with consumers paying little to no attention toward the driving conditions. According to [73], as many as 400 accidents have been reported in the US in a span of 11 months, with cars involving automated tech. This is a big number considering the safety features many of these vehicles have including ADAS. Of these 400, 273 are reported to be Tesla cars that have superior technology. Though the idea of driving cars that use electricity to run is an exciting and innovative idea, this is far from reality in many third world countries. There are many nations in the world that still struggle to produce enough electricity to keep their economy running. This would get only get worse in summers when the need for electricity grows several folds. According to [74], "On the other hand, electric vehicle sales are not yet picking up in other developing and emerging countries as only a limited number of models are available and are still considered too expensive for most consumers". Another concern with EVs is that it may not be the only car you want to have in your house. This is because when electricity is not available to charge your car, during difficult times such as severe weather conditions, you may be left with no car to drive. Besides this there are other risks with EVs such as the risk of catching fire. In a recent catastrophe, that took place in Florida, several EVs caught fire after they got flooded in the water. According to [75], At least 11 electric vehicles caught fire during the recent hurricane Ian which caused excessive flooding. This has not only caused concerns about EVs but also how they might be risky to drive later after the dust settles. There could be other severe concerns that have not been discussed here but may be important.

CONCLUSION

Though there are risks with EVs and how they could be concerning in several aspects such as low range, not enough charge stations, affordability, service issues and others, there are several positive points to be considered such as zero emissions and perhaps net zero emissions with further green energy sources used to charge these EVs. With advanced technology and scalability, EVs could be the only transport system that would sustain itself as we exhaust current fossil fuels at a rapid pace, to satisfy the needs of the technologically advanced world that is craving for more and more energy every day.

FUTURE WORK

Future work could entail addressing all the challenges discussed in the chapter. This would mean eliminating issues such as burning of fossil fuels to produce electricity required for EVs and resorting to wind, and solar energy for harnessing more power. We need more powerful battery packs to address range issues. Strategic planning needs to be done for charging infrastructure as part of mitigating range anxiety. We need to focus on reducing costs through methods such as scalability. We also need to look into minimizing the cyber risks and attacks by developing robust security systems. Finally, we need to look for alternative battery materials that could produce more power (energy-dense) and yet be environmentally friendly. The future work would involve upgrading the current technologies to completely replace all vehicles using fossil fuels and minimize the carbon footprint. This would include all vehicles such as heavy trucks, semis, agricultural, mining, and all other vehicles that are used for both commercial and non-commercial purposes.

REFERENCES

[1] Li, S., L. Tong, J. Xing and Y. Zhou. The Market for Electric Vehicles: Indirect Network Effects and Policy Design. *Journal of the Association of Environmental and Resource Economists* 4, no. 1 (2017): 89–133. doi:10.1086/689702.

[2] Muratori, M., M. Alexander, D. Arent, M. Bazilian, P. Cazzola, E. M. Dede, J. Farrell, et al. The Rise of Electric Vehicles-2020 Status and Future Expectations. *Progress in Energy* 3, no. 2 (2021): 022002. doi:10.1088/2516-1083/abe0ad.

[3] https://avt.inl.gov/sites/default/files/pdf/fsev/HistoryOfElectricCars.pdf.

[4] Matulka, R. *The History of the Electric Car* (2014). Retrieved November 14, 2022, from https://www.energy.gov/articles/history-electric-car#:~:text=Here%20in%20the%20U.S.%2C%20the,spark%20interest%20in%20electric%20vehicles.

[5] Clean Vehicle Rebate Program (CVRP). (nd). Retrieved November 14, 2022, from https://ww2.arb.ca.gov/sites/default/files/movingca/cvrp.html

[6] Frangoul, A. *Ev Sales to Hit All-Time High in 2022, IEA Says, But More Work Needed to Put World on Net-Zero Path* (2022). CNBC. Retrieved February 21, 2023, from https://www.cnbc.com/2022/09/23/electric-vehicle-ev-sales-set-to-hit-an-all-time-high-in-2022-iea-says.html

[7] Liu, Y., R. Zhang, J. Wang and Y. Wang. Current and Future Lithium-Ion Battery Manufacturing. *IScience* 24, no. 4 (2021): 102332. doi:10.1016/j.isci.2021.102332.

[8] https://pubs.usgs.gov/periodicals/mcs2022/mcs2022-lithium.pdf

[9] Eberhard, M. *A Bit About Batteries: Electric Cars, Solar and Clean Energy* (2010). Retrieved November 14, 2022, from https://www.tesla.com/pt_PT/blog/bit-about-batteries

[10] Rietmann, N., B. Hügler and T. Lieven. Forecasting the Trajectory of Electric Vehicle Sales and the Consequences for Worldwide CO_2 Emissions. *Journal of Cleaner Production* 261 (2020): 121038. doi:10.1016/j.jclepro.2020.121038.

[11] Graveson, G. (2020). *Preventing Battery Fires.* Retrieved November 14, 2022, from https://www.veolia.co.uk/press-releases/preventing-battery-fires

[12] Qiao, D., G. Wang, T. Gao, B. Wen and T. Dai. Potential Impact of the End-of-Life Batteries Recycling of Electric Vehicles on Lithium Demand in China: 2010-2050. *Science of The Total Environment* 764 (2021): 142835. doi:10.1016/j.scitotenv.2020.142835.

[13] https://earthworks.org/wp-content/uploads/2021/09/UTS-EV-battery-metals-sourcing-20210419-FINAL.pdf

[14] Baum, Z. *A Greener Future: Lithium-Ion Batteries and Hydrogen Fuel Cells* (2022). CAS. Retrieved November 14, 2022, from https://www.cas.org/resources/cas-insights/sustainability/lithium-batteries-hydrogen-fuel-cells

[15] https://nepis.epa.gov/Exe/ZyPDF.cgi/P100EZ7C.PDF?Dockey=P100EZ7C.PDF

[16] Nissan. *Nissan Sentra Sedan* (2022). Nissan USA. Retrieved November 15, 2022, from https://www.nissanusa.com/vehicles/cars/sentra.html

[17] Hyundai. *Hyundai Elantra|Compact Sedan. Hyundai Elantra|Compact Car* (2023). Hyundai USA. Retrieved November 14, 2022, from https://www.hyundaiusa.com/us/en/vehicles/elantra

[18] IEA. *Electric Vehicles—Analysis* (2022). IEA. Retrieved November 14, 2022, from https://www.iea.org/reports/electric-vehicles

[19] McNutt, K. Charging Stations Not Keeping up with Growth of EVS. *The Journal Record* (2022). Retrieved November 15, 2022, from https://journalrecord.com/2022/09/29/charging-stations-not-keeping-up-with-growth-of-evs/

[20] Hardman, S., A. Jenn, G. Tal, J. Axsen, G. Beard, N. Daina, E. Figenbaum, et al. A Review of Consumer Preferences of and Interactions with Electric Vehicle Charging Infrastructure. *Transportation Research Part D: Transport and Environment* 62 (2018): 508–23. doi:10.1016/j.trd.2018.04.002.

[21] https://www.tesla.com/support/supercharger

[22] The United States Government. *Fact Sheet: Biden Administration Advances Electric Vehicle Charging Infrastructure* (2022). The White House. Retrieved November 15, 2022, from https://www.whitehouse.gov/briefing-room/statements-releases/2021/04/22/fact-sheet-biden-administration-advances-electric-vehicle-charging-infrastructure/

[23] LaMonaca, S. and L. Ryan. The State of Play in Electric Vehicle Charging Services: A Review of Infrastructure Provision, Players and Policies. *Renewable and Sustainable Energy Reviews* 154 (2022): 111733. doi:10.1016/j.rser.2021.111733.

[24] Bonges, H. A. and A. C. Lusk. Addressing Electric Vehicle (EV) Sales and Range Anxiety through Parking Layout, Policy and Regulation. *Transportation Research Part A: Policy and Practice* 83 (2016): 63–73. doi:10.1016/j.tra.2015.09.011.

[25] N. Adnan, S. M. Nordin, I. Rahman, P. M. Vasant and A. Noor, A Comprehensive Review on Theoretical Framework-Based Electric Vehicle Consumer Adoption Research. *International Journal of Energy Research* 41, no. 3 (2017): 317–335,.

[26] file:///Users/charan/Downloads/1071181321651243.pdf

[27] Pevec, D., J. Babic, A. Carvalho, Y. Ghiassi-Farrokhfal, W. Ketter and V. Podobnik. Electric Vehicle Range Anxiety: An Obstacle for the Personal Transportation (R)Evolution? In: *2019 4th International Conference on Smart and Sustainable Technologies (SpliTech)*, 2019. doi:10.23919/splitech.2019.8783178.

[28] Elfalan, J. Track Tested: 2022 Lucid Air Dream Range Breaks 3 Seconds to 60. *Edmunds* (2022). Retrieved November 15, 2022, from https://www.edmunds.com/car-news/track-tested-2022-lucid-air-dream-range.html#:~:text=The%20Dream%20Range%20variant%20of,(before%20federal%20tax%20credits)

[29] https://www.toyota.com/camry/2022/

[30] Pan, C., W. Dai, L. Chen, L. Chen and L. Wang. Driving Range Estimation for Electric Vehicles Based on Driving Condition Identification and Forecast. *AIP Advances* 7, no. 10 (2017): 105206. doi:10.1063/1.4993945.

[31] https://arthurcarvalho.info/publications/SpliTech.pdf

[32] https://www.nissanusa.com/vehicles/cars/versa-sedan.html
[33] https://www.nissanusa.com/shopping-tools/build-price?models=nissan-leaf
[34] Progressive. *Electric Car Insurance: Is it More Expensive?* (nd). Retrieved November 15, 2022, from https://www.progressive.com/answers/car-insurance-electric-vehicles/#:~:text=Insurance%20for%20an%20electric%20car,carry%20comprehensive%20and%20collision%20coverage
[35] Preston, B. *Electric Vehicles May Cost More to Insure Than Gasoline-Powered Cars: Consumer Reports* (2021). Retrieved November 15, 2022, from https://www.consumerreports.org/car-insurance/electric-vehicles-cost-more-to-insure-than-gasoline-powered-a6372607024/
[36] U.S Department of Energy. *Electric Vehicles With Final Assembly in North America.* (n.d.). Alternative Fuels Data Center. Retrieved November 15, 2022, from https://afdc.energy.gov/laws/electric-vehicles-for-tax-credit
[37] https://advocacy.consumerreports.org/wp-content/uploads/2020/10/EV-Ownership-Cost-Final-Report-1.pdf
[38] https://avt.inl.gov/sites/default/files/pdf/fsev/compare.pdf
[39] Lindwall, C. Electric vs. Gas Cars: Is it Cheaper to Drive an EV? *NRDC* (2022). Retrieved November 15, 2022, from https://www.nrdc.org/stories/electric-vs-gas-it-cheaper-drive-ev
[40] Franzò, S. and A. Nasca. The Environmental Impact of Electric Vehicles: A Novel Life Cycle-Based Evaluation Framework and Its Applications to Multi-Country Scenarios. *Journal of Cleaner Production* 315 (2021): 128005. doi:10.1016/j.jclepro.2021.128005.
[41] https://theicct.org/sites/default/files/publications/EV-equity-feb2021.pdf
[42] Zhenyu, S., J. Hu, H. Lin, Z. Chen, J. Zhu, L. Yang, Z. Hu, X. Chen, X. Wu. Lagrangian Observation of the Kuroshio Current by Surface Drifters in 2019. *AGRIS* (nd). Retrieved November 15, 2022, from https://agris.fao.org/agris-search/search.do?recordID=CH20221104435
[43] Knobloch, F., S. Hanssen, A. Lam, H. Pollitt, P. Salas, U. Chewpreecha, M. A. J. Huijbregts and J.-F. Mercure. NET Emission Reductions from Electric Cars and Heat Pumps in 59 World Regions Over Time. *Nature Sustainability* (2020). Retrieved November 16, 2022, from https://www.ncbi.nlm.nih.gov/pmc/articles/PMC7308170/
[44] Biello, D. Electric Cars Are Not Necessarily Clean. *Scientific American* (2016). Retrieved November 16, 2022, from https://www.scientificamerican.com/article/electric-cars-are-not-necessarily-clean/
[45] Sahelirc. Are Electric Cars 'Green'? The Answer Is Yes, But It's Complicated. *CNBC* (2021). Retrieved November 16, 2022, from https://www.cnbc.com/2021/07/26/lifetime-emissions-of-evs-are-lower-than-gasoline-cars-experts-say.html
[46] https://www.epa.gov/greenvehicles/electric-vehicle-myths
[47] Rivelli, E., and L. Ward. Roadside Assistance: What Is it and Do I Need It? (2020). Retrieved November 16, 2022, from https://www.coverage.com/insurance/auto/roadside-assistance/
[48] Sharbono, C. *The Best Out of Gas Roadside Assistance Service.* Mach 1 Services (2021). Retrieved November 16, 2022, from https://www.mach1services.com/out-of-gas-roadside-assistance/
[49] Schweber, B. Portable Charging Solutions Offer Roadside Assistance for EVS. *EDN* (2020). Retrieved November 16, 2022, from https://www.edn.com/portable-charging-solutions-offer-roadside-assistance-for-evs/
[50] https://support.google.com/wearos/answer/12059876?hl=en
[51] Nižetić, S. P. Šolić, D. López-de-Ipiña González-de-Artaza and L. Patrono. Internet of Things (IOT): Opportunities, Issues and Challenges Towards a Smart and Sustainable Future. *Journal of Cleaner Production* (2020). Retrieved November 16, 2022, from https://www.ncbi.nlm.nih.gov/pmc/articles/PMC7368922/

[52] Kumar, S., P. Tiwari and M. Zymbler. Internet of Things Is a Revolutionary Approach for Future Technology Enhancement: A Review. *Journal of Big Data* (2019). Retrieved November 16, 2022, from https://journalofbigdata.springeropen.com/articles/10.1186/s40537-019-0268-2
[53] Y. F., L. Ben Azzouz, W. Trojet and L. A. Saidane. Cyber Security Issues of Internet of Electric Vehicles. In: *2018 IEEE Wireless Communications and Networking Conference (WCNC)*. 2018, pp. 1–6, doi:10.1109/WCNC.2018.8377181.
[54] Mousavian, S., M. Erol-Kantarci, L. Wu and T. Ortmeyer. A Risk-Based Optimization Model for Electric Vehicle Infrastructure Response to Cyber Attacks. *IEEE Transactions on Smart Grid* 9, no. 6 (2018): 6160–69. doi:10.1109/tsg.2017.2705188.
[55] Nawy, R. Electric Vehicles are Taking Over Hackers Are Waiting. *Security Magazine RSS* (2022). Retrieved November 16, 2022, from https://www.securitymagazine.com/articles/97461-electric-vehicles-are-taking-over-hackers-are-waiting
[56] Johnson, J., B. Anderson, B. Wright, J. Quiroz, T. Berg, R. Graves, J. Daley, K. Phan, M. Kunz, R. Pratt, T. Carroll, L. R. ONeil, B. Dindlebeck, P. Maloney, D. O'Brien, D. Gotthold, R. Varriale, T. Bohn and K. Hardy. *Cybersecurity for Electric Vehicle Charging Infrastructure* (Technical Report) (2022). Retrieved November 17, 2022, from https://www.osti.gov/biblio/1877784
[57] Anvaripour, M., S.M. Mahdi Alavi, M. Hayes and M. Saif. Cybersecurity of Electric Vehicles Charging Systems By Using Residual Generation Technique (2022). doi:10.36227/techrxiv.20387397.v1.
[58] https://arxiv.org/pdf/2105.02905.pdf
[59] Powell, S., G. V. Cezar, L. Min, I. M. L. Azevedo and R. Rajagopal. *Nature Energy* (2022). doi:10.1038/s41560-022-01105-7.
[60] Osaka, S. It's Common to Charge Electric Vehicles at Night That Will Be a Problem. *The Washington Post* (2022). Retrieved November 17, 2022, from https://www.washingtonpost.com/climate-environment/2022/09/22/its-common-charge-electric-vehicles-night-that-will-be-problem/
[61] Rizvi, S. A. A., A. Xin, A. Masood, S. Iqbal, M. U. Jan and H. Rehman. Electric Vehicles and Their Impacts on Integration into Power Grid: A Review. In: *2018 2nd IEEE Conference on Energy Internet and Energy System Integration (EI2)*, October 2018. doi:10.1109/ei2.2018.8582069.
[62] Brown, A. *Electric Cars Will Challenge State Power Grids*. The Pew Charitable Trusts (2020). Retrieved November 17, 2022, from https://www.pewtrusts.org/en/research-and-analysis/blogs/stateline/2020/01/09/electric-cars-will-challenge-state-power-grids
[63] Trabish, H. K. *The Rise of Evs Could Overwhelm the Grid, But PG&E Has a Better Plan*. Utility Dive (2019, June 13). Retrieved November 17, 2022, from https://www.utilitydive.com/news/the-rise-of-evs-could-overwhelm-the-grid-but-pge-has-a-better-plan/555984/
[64] Brinkel, N., T. AlSkaif and W. van Sark. Grid Congestion Mitigation in the Era of Shared Electric Vehicles. *Journal of Energy Storage* 48 (2022): 103806. doi:10.1016/j.est.2021.103806.
[65] Virta. Myth Buster: Electric Vehicles Will Overload the Power Grid: Virta. *Virta Global* (2021). Retrieved November 17, 2022, from https://www.virta.global/blog/myth-buster-electric-vehicles-will-overload-the-power-grid
[66] Templeton, B. EVS Won't Overload the Power Grid, in Fact EVS (and Ice) Are Its Salvation. *Forbes* (2022). Retrieved November 17, 2022, from https://www.forbes.com/sites/bradtempleton/2022/09/12/evs-wont-overload-the-power-grid-in-fact-evs-and-ice-are-its-salvation/?sh=696159aa49c5
[67] Stone, M. Will Charging Electric Cars Ever Be as Fast as Pumping Gas? *Environment* (2021). Retrieved November 17, 2022, from https://www.nationalgeographic.com/environment/article/will-charging-electric-cars-ever-be-as-fast-as-pumping-gas

[68] Service, P. N. (n.d.). Electric Vehicles Could Fully Recharge in Under 5 Minutes With New Charging Station Cable Design. *Purdue University News*. Retrieved November 17, 2022, from https://www.purdue.edu/newsroom/releases/2021/Q4/electric-vehicles-could-fully-recharge-in-under-5-minutes-with-new-charging-station-cable-design.html
[69] Ashkrof, P., G. H. de Almeida Correia and B. van Arem. Analysis of the Effect of Charging Needs on Battery Electric Vehicle Drivers' Route Choice Behaviour: A Case Study in the Netherlands. *Transportation Research Part D: Transport and Environment* 78 (2020): 102206. doi:10.1016/j.trd.2019.102206.
[70] *2022 Bolt EUV and Bolt EV Specifications* (nd). Retrieved November 17, 2022, from https://media.chevrolet.com/media/us/en/chevrolet/2022-bolt-euv-bolt-ev.detail.html/content/Pages/news/us/en/2021/feb/0214-boltev-bolteuv-specifications.html
[71] Wiles, K. Electric Vehicles Could Fully Recharge in Under 5 Minutes With New Charging Station Cable Design. *The Persistent Pursuit* (2022). Retrieved November 17, 2022, from https://stories.purdue.edu/electric-vehicles-could-fully-recharge-in-under-5-minutes-with-new-charging-station-cable-design/
[72] ACSpressroom. Super-Fast Electric Car Charging, With a Tailor-Made Touch. *EurekAlert!* (nd) Retrieved November 17, 2022, from https://www.eurekalert.org/news-releases/960985
[73] Press, T. A. *Nearly 400 Car Crashes in 11 Months Involved Automated Tech Companies Tell Regulators*. NPR (2022). Retrieved November 17, 2022, from https://www.npr.org/2022/06/15/1105252793/nearly-400-car-crashes-in-11-months-involved-automated-tech-companies-tell-regul
[74] IEA. *Executive Summary—Global EV Outlook 2022—Analysis* (nd). IEA. Retrieved November 17, 2022, from https://www.iea.org/reports/global-ev-outlook-2022/executive-summary
[75] Weise, E. Something Surprising Can Cause Electric Vehicles to Catch on Fire. Here's What Experts Want You to Know. *USA Today* (2022). Retrieved November 17, 2022, from https://www.usatoday.com/story/money/cars/2022/10/26/electric-vehicle-fires-florida-flooding-what-happened/10553207002/

9 Battery Management System for Electric Vehicles

Suresh Goka
National Institute of Technology, Warangal

Syed Quadir Moinuddin
King Faisal University, Al Hofuf

Ashok Kumar Dewangan
National Institute of Technology, Delhi

Muralimohan Cheepu
Super-TIG Welding Company Limited

Venkata Charan Kantumuchu
Electrex Inc Hutchinson

NOMENCLATURE

BMS	Battery management system
EV	Electric vehicle
HEV	Hybrid electrical vehicle
FCEV	Fuel cell electric vehicle
DER	Distributed energy resources
BEV	Battery electric vehicle
CAN	Controller area network
ASIC	Application-specific integrated circuit
SOH	State of health
SOC	State of charge
TOU	Time of use
T & D	Transmission and distribution
PHEV	Plug-in hybrid electric vehicle
IEA	International energy agency
LIB	Lithium-ion battery
PCM	Phase change materials
IEA	International energy agency

DOI: 10.1201/9781003354901-9

9.1 INTRODUCTION

In recent years, contamination of air and more reliance on renewable fuels has increased global warming and drastic climatic change due to carbon emission from conventional transportation. To overcome this, numerous initiatives were taken to develop alternative transportation vehicles with less or zero emissions, green [1, 2]. One such alternative vehicle is electric vehicles (EVs) with some alternative sources [3–5], considered as future transportation systems and classified into four categories, namely, hybrid electrical vehicles (HEVs), fuel cell electric vehicles (FCEVs), plug-in hybrid electric vehicles (PHEVs), and fully battery electric vehicles (BEVs) [6]. All these four categories are electrically propelled with the help of a battery which is considered the heart of EVs. The advances in the use of batteries as propellants are the capacity to store energy, less emission, the ability to recharge, etc. Battery as an electric propellant requires a continuous breakthrough in the direction of EVs. However, batteries have been used in several billion electronic devices such as toys, watches, cellphones, etc. The primary performance indicator of the battery that attracts industries and consumers are energy, power, lifetime, safety, cost, etc. Further, these indicators have been sub-classified as specific energy and power, energy and power density, fast charging, cycle and calendar life, safety with respect to mechanical, electrical and thermal, and cost per energy content. The key performance indicators vary depending on the end application. In the market, three types of traction batteries are available such as nickel-metal hydride type, lead-acid type and lithium-ion type batteries. Lithium-ion batteries have more benefits than the other two types. Nevertheless, the performance of battery is depending on the effective battery management system. For different applications, different types of modules of energy storage from flywheel to various power storage chemistries are in employment for supporting the storage of power. EVs and smart grid are two technologies that are growing greatly and will be benefitted in developing some advanced components and infrastructure facilities.

A smart grid is a typical concept involving electricity grids to deliver electrical energy using computer technology, control and communications at a low cost provided with more reliability. Its benefits and market potential assessment were described [7] to understand the storage of energy for the electricity grid. The design of experiments and energy storage devices were thoroughly studied by the authors. The US Department of Energy has proven a number of features, such as consumer engagement, incorporating all generation and storage alternatives, and creating markets and services for new products.

The energy storage devices such as Li-ion batteries, advanced lead-acid batteries and flywheels are identified as potential solutions for energy applications. Whereas power quality, electric service reliability, voltage support, and frequency/area regulations are suitable for power applications [8]. Li-ion and sodium-sulphur have been found to have the capacity to store a lot of energy and discharge it for a long time, measured in hours [9].

9.2 BACKGROUND

In this competitive world with evolving technologies rapidly, the most prominent energy storage devices are batteries. With the fantastic performance of exclusive lithium-ion batteries, Boeing 787 has declared that the battery technology has been growing rapidly by developing cells with higher energy density and power. In addition, to make enhance performance, it is also important to manufacture a battery within low-cost limits and more specifically reliable and safe in handle or usage.

Based on earlier analysis [10], In America, the vehicles manufactured using internal combustion (IC) engines are caused to air pollution. Airborne pollution contains 62% carbon monoxide, 32% nitrogen oxides, 28% lead, 27% volatile organic compounds, and 18% suspended particles, among other things. Carbon Dioxide (CO_2), which is released from many sources such as industries/factories, automobiles, and other traditional vehicles, is the primary cause of the greenhouse effect in the universe. Here 25% of the CO_2 is released into the atmosphere because of traditional vehicles only. As a growing factor of population, the usage of vehicles also increased which implies that battery usage also tremendously increased and hence controlling the greenhouse effect is required obviously. In this way, BMS have a prominent role in the automobile sector.

9.3 BATTERY MANAGEMENT SYSTEM

It is the most common type, recording the operational parameters of the battery, such as current, voltage, internal temperature, and so on, as well as the ambient temperature. The recording is happening both when it is charging and while it is charging. In pure electric vehicles (EVs), the battery is the only power source. Therefore, the battery management system must consist of a protection system, which is a system to signal the battery to get ready to produce maximum power when requires, and an additional system that can enhance the life of the battery. Apart from all of the above-mentioned systems, a system of controlling the charging and in addition a system of managing the thermal issues is also required. Here the protection system that monitors the controlling circuits and other devices generates an alarm whenever extra loads/parameters disturb the entire circuit. For safety in operation, it disconnects the battery from the loop consisting of the overloads.

9.3.1 Need for BMS in EVs

The integration of the energy storage systems like batteries into EVs depends on some parameters viz. cost, lifetime, power delivery and environmental safety. The following Figure 9.1 reveals the need for BMS in detail.

There are various types of batteries are available. Some of the names mentioned here like nickel-metal hydride, sodium-sulphur, lithium-ion, lead-acid, etc. depend upon their potential, storage capability, energy consumption, heat dissipation and some other parameters, these batteries will be incorporated into thermal managing systems. Among the various types that exist, lithium-ion batteries carry more roles in thermal systems and hence further information in terms of performance and significance is discussed in detail.

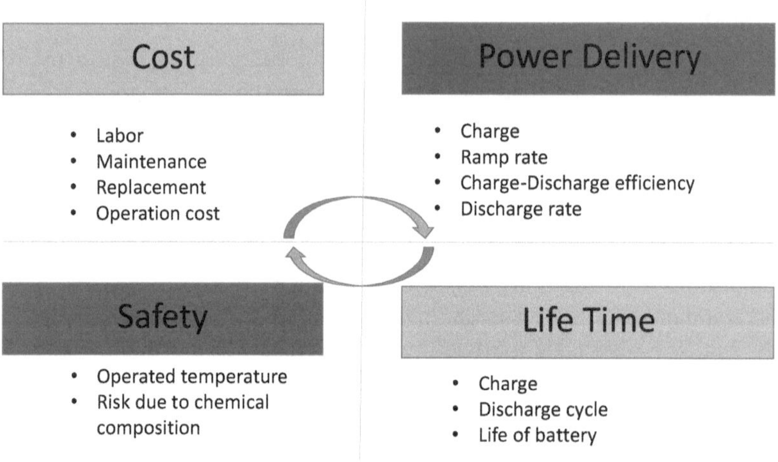

FIGURE 9.1 Need for BMS.

9.3.2 Features of BMS

For proper management of the battery, it should be controlled with good features. The primary features are detailed below shown in Figure 9.2.

a. Discharge control

   The BMS's primary goal is to operate the battery within a range of parameters that are safe, particularly in terms of current and voltage. The BMS should need to keep guard the cell against abnormal consequences during the discharging process. In case the cell functions out of its boundary limits, there may occurrence of huge damage to the circuit and the system.

b. Charging control

   The process of charging is one of the important features of batteries. Frequently, the batteries get damaged because of the improper charging process.

c. Determination of state of charge (SOC)

   SOC measures the capacity of a battery as a percentage of its total capacity under identical conditions. It is comparable to the gasoline gauge used in cars with IC engines. By managing the charging and discharging of the batteries, SOC estimate is crucial for monitoring and improving their performance. A precise SOC estimation is able to determine the lasting of an EV go without recharging [11]. Typically, a battery's usable capacity is between 100% and 5% SOC, and its runtime is determined by how long it takes to drain it from 100% to 5%. Direct estimation of the SOC is not possible. If I_b (τ) represents the charging current, then equation (9.1) definition of the charge provided to the battery applies. SOC is then determined by the [12]

$$\int_{t_0}^{t} I_b(T)\mathrm{d}(T) \tag{9.1}$$

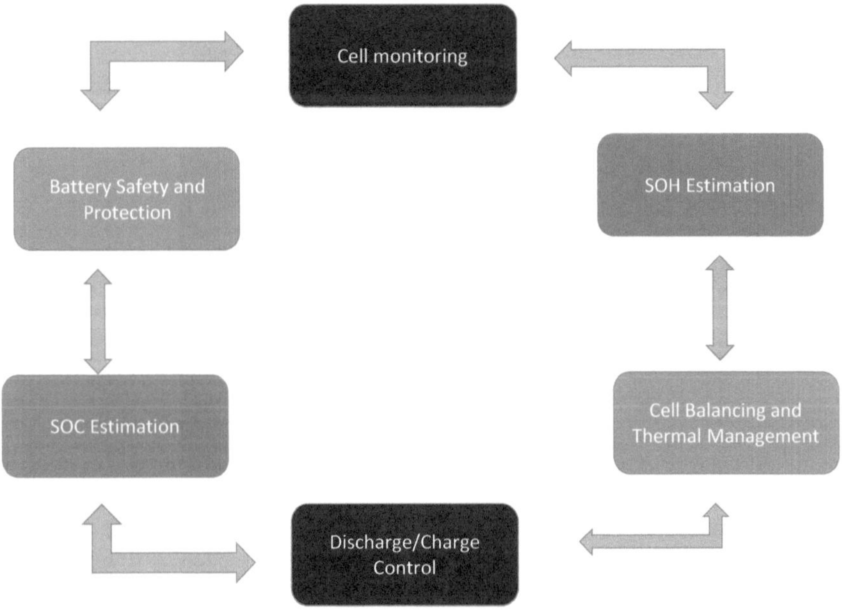

FIGURE 9.2 Features of BMS.

$$SOC(T) = \int_{t_0}^{t} \frac{I_b dT}{Q} \times 100\% \tag{9.2}$$

d. Determination of state of health

SOH determines the degree of deterioration, of the batteries. SOH depends on the depth-of-charge function. SOH estimation is more difficult to define simply than SOC. It can be stated as follows: It indicates how many cycles are left until the battery's useful life is over.

$$SOH = \frac{\text{Current capacity}}{\text{Initial capacity}} \times 100 \tag{9.3}$$

e. Cell balancing

A BMS's most important task is cell balancing. A BMS must attempt to retain an equal charge state in discrete cell. The SOC impact of imbalance during charging and discharging circumstances [13] is depicted in Figure 9.3.

f. Logbook function

BMS functions as a log book because it compares the new battery to the old battery using the SOH, which is a relative word. This measurement should thus be possible to compare with the same cell as it ages and save a record of the original circumstances of a cell. As a result, the MEMS' log book feature would save such crucial information in memory.

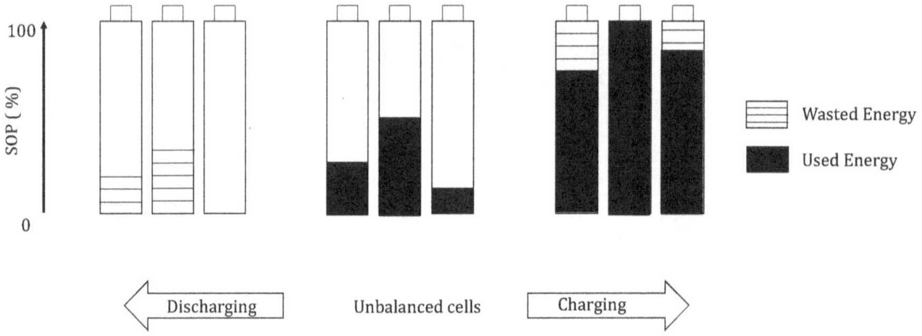

FIGURE 9.3 SOC imbalances on a battery pack (4-cell) [13].

g. Communications

Along with much more information for service and maintenance, BMS also offers information on health SOC. Typically, a driver is not a technical advisor when operating a vehicle. Consequently, the data should be easy to understand. However, in order to conduct research or identify faults, battery service professionals and engineers require extremely exact technical information, and it is the responsibility of the BMS to give both of them the information they require. Therefore, a BMS should do all of these tasks to ensure the battery's safety.

9.3.3 Characteristics of BMS

Software and hardware combine the BMS architecture. To gather additional battery characteristics, such as temperature, voltage, and, current the hardware components are incorporated into the EV equipment. Real-time data acquisition with a high sample rate is advised to boost exactness and give the essential monitoring competencies deprived of any errors. The software component is in charge of actively controlling variables, internal and external communication, alerts and protection, computing different types of estimations, balancing the battery cells, and other tasks [14–19]. The block diagram of the BMS features is shown in Figure 9.4.

9.3.4 BMS Topologies

BMS topologies specify the system's cost, installation complexity, maintenance requirements, reliability, control structure, and communication architecture. The following describes the BMS topologies.

a. Centralized topology

The batteries are serially linked by a single master module, which allows for readings of each cell voltage, current, and other factors. The topology is economical, takes little upkeep, and offers decent accuracy [20]. The likelihood of short circuits and other types of mistakes, however, is increased

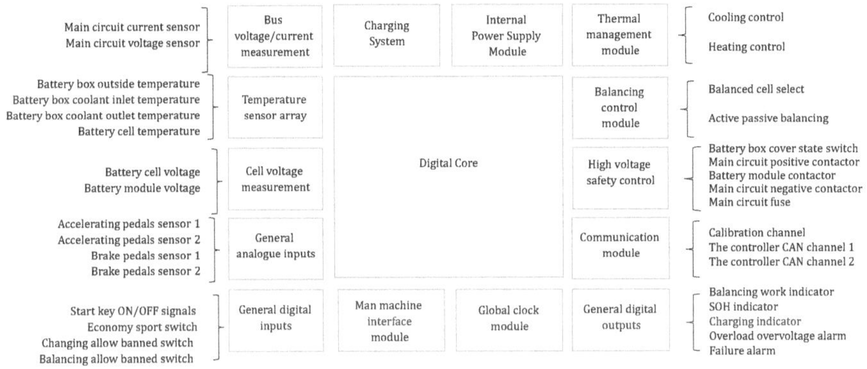

FIGURE 9.4 Basic framework of a BMS [14].

by long cables, the potential for mixing up or misconnecting inputs, and other factors. In addition, the flexibility and scalability of the overall system are limited because the maximum quantity of batteries is exactly set. Furthermore, the system as a whole will collapse if the master controller malfunctions, greatly threatening reliability [11].

b. Modularized topology

   Many alike modules are allied to the distinct batteries in this approach, just like the centralized BMS topology. Among the elements, one serves as a master element that controls the complete battery pack, whereas the additional modules collect data from the batteries and communicate to the master element. Amended scalability and flexibility are offered by this design. A further BMS module may be simply added to extend the scalability. The BMS's functionality can also be implemented separately via modules. The amount of inputs remains fixed, and the total cost of this architecture is high. Additionally, this design requires additional hardware and programming work.

c. Distributed topology

   Each cell in this architecture has a BMS module attached to it. The BMS controller manages the communication and overall metrics. This topology is trustworthy, durable, and economical. In comparison to centralized and modularized systems, this strategy is more adaptable and scalable. The flexibility and scalability are increased by the ability to add or remove cells. Moreover, thanks to short connection cables and improved interference immunity, great measurement precision is attained. This topology is pricey and complicated [20].

d. Decentralized topology

   The full functionality is locally, dependably, and autonomously provided by many equal decentralized BMS units under this architecture. There is no chance of a systemic failure because there is no central control device. Subsequently, installation, additional or removed inputs are possible because this architecture is adaptable and expandable. Growing decentralization also promotes scalability. This strategy has several benefits, but its growth necessitates new techniques [11].

9.3.5 Prerequisite of BMS

a. Electromagnetic interference (EMI)

Sensors are subject to EMI, which can lead to inaccurate readings and, in the worst situation, render data utterly worthless despite being observable. Therefore, EMI should be taken into consideration while designing all electrical and electronic components as well as other loads. Measures should be made to decrease EMI effects, like employing blocking capacitors, and common mode chokes as specified in the ISO7637-1 standard [21].

b. Contactors requirement

In a battery pack, the battery poles must be galvanically isolated. Contactors, that are used to eradicate influential DC currents during the incident of breakdown, can be utilized to accomplish this. In the case of DC currents, the system's capacity to shut off may also be considerably reduced If the electrical arc on the blades of the contactors is erroneously disengaged and extinguished [22]. Similar to how using magnetic arc extinguishers to drive away the arc from the blades in contact might significantly improve switch off efficiency. Similar to this, additional risk is presented by the blades' switch-on function. When blades close, they will bounce, creating an electric arc that will heat the blades.

c. Redundancy

It is challenging to implement a BMS with a high level of severance brought on by the cost, the design process's time restrictions, and other problems. The ISO 26262 standard, however, requires compensation of the system wherein increases dependability. To accurately gauge cell voltages, the standard stipulates that there must be at least two distinct chips, and BMS must choose the most reliable signal from these two sources [22]. Similar to this, there should be redundancy in the BMS for further tasks such higher level processing in protection strategies, self-test mechanism implementation, central processing units, memory error correction, and other technologies.

d. Overall protection

A BMS must ensure that all of its components are protected from LIBs' potential for explosion as well as from short circuits, excessive charging and discharging, over- and undercurrent, overheating, pressure, leakage current, liabilities in connected devices, ground responsibilities, system or command catastrophes, etc. A BMS's protective circuit has to be effective and should be created with consideration for the battery or cell's type, size, connected device, intended use, etc. [23]

9.3.6 Applications of BMS

Figure 9.5 summarizes the application of five categories of electrical energy storage system-wise.

Distributed energy storage applications: The storage configured as a distributed energy resource (DER) may be suitable for all applications depending on location, as shown in Figure 9.6.

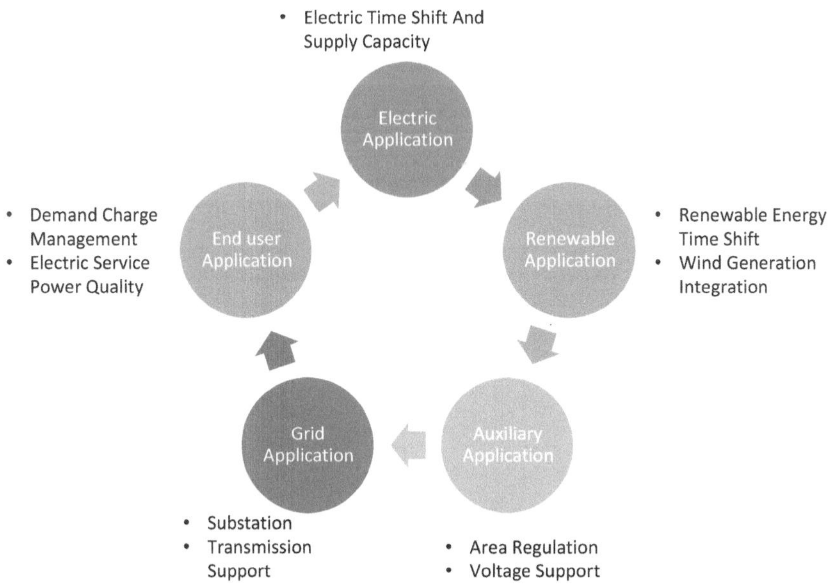

FIGURE 9.5 Electrical energy storage.

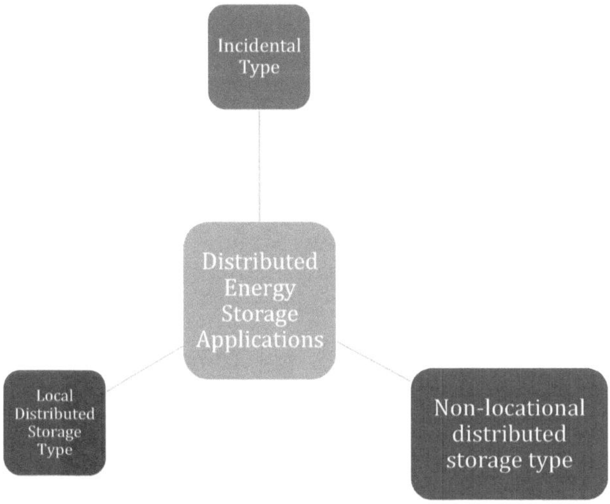

FIGURE 9.6 Distributed energy storage applications.

9.4 BATTERY THERMAL MANAGEMENT SYSTEM

The high expenses and low energy density of today make it imperative to utilize existing cells to their fullest potential. The operating temperature of lithium-ion batteries has a considerable impact on their electrochemical effectiveness, longevity, and safety. The safe working temperature range for lithium-ion batteries is between 20°C and 50°C. The lithium-ion battery pack's utmost temperature difference should be kept to 5°C. The lifespan and range of EVs can be increased by extending battery life and promoting appropriate operating conditions. Due to their operating conditions, including their state of charge (SOC), current, and extreme temperatures, lithium-ion batteries inherently deteriorate over time. Due to speedy charging during performance driving for high cell currents, the cells have considerable heat losses. Heat is predominantly generated in a battery cell via activation interlayer kinetics, concentration ion transfer, and Ohmic Joule heating caused by charged particle motion. This last process is particularly for bigger cell sizes, considerable like the ones conferred in this paper. The difference in temperature inside a cell or module and its average or maximum temperature impact ageing. Low working temperatures can cause metallic lithium dendrite and rapid capacity fading, while excessive working temperatures can cause thermal runaway, resulting in fire or even explosion. The use of a Li-ion battery pack with a significant temperature differential will not be consistent, which causes early deterioration and possibly safety incidents. Different types of BTMS are shown in Figure 9.7.

9.4.1 Lithium-Ion Battery-Information

Among all the metals found to form nature, lithium is one of the metals with the lightest and greatest energy density (per weight). In addition, lithium is also having the highest electrochemical potential. Hence, the usage of lithium as an anode/

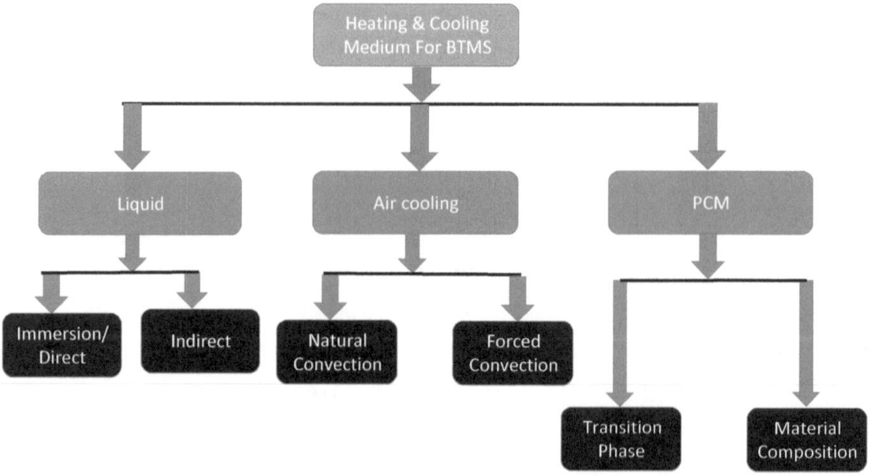

FIGURE 9.7 Classification of battery TMS.

cathode can able to result in high voltage, electrical capacity and energy density in rechargeable batteries. The first commercialized Li-ion rechargeable battery was in 1991 by the famous company namely Sony. Henceforth, it gains popularity and remained the best choice among rechargeable batteries. Though, during charging lithium is unstable inherently. That's the reason; lithium ions have been replaced with lithium metals. Despite lower energy density, lithium ions are safer when compared to lithium ions. Hence ion batteries have attention across various applications.

Li-ion batteries have some advantages over the other types are as almost no maintenance during their life cycle, no schedule cycles are required and hence no memory effect in the battery. That's why it is very much applicable to EVs. In addition to the aforementioned benefits, the batteries' self-discharge rate is less than half that of lead-acid and Ni-MH (nickel-metal hydride) batteries. On the other hand, certain precautions have to be taken during discharging and charging of any rechargeable batteries.

Regardless of the advantages of the batteries, drawbacks also exist certainly for some of them. Lithium ions are brittle in nature. Hence, they require a protective inbuilt device to have a safe operation and life. The protective device comes under a battery management system that will monitor the cell temperature and it controls the maximum levels of currents for charging and discharging. Each cell's peak voltage is really constrained by the BMS during charging, and it also prevents the cell's voltage from falling below the threshold level during discharging.

9.5 BATTERY MODELLING

Fundamentals of Battery Modeling an overview of the many behaviours displayed by the battery is given in this section.

a. Polarizations

Uncomfortable impedance develops during the substance responses in batteries. Polarization is the term used to describe these undesirable impedances. The slowest step in a process causes polarization to occur at the beginning. Due to focal differences known as fixation polarization, the dispersion also affects the response rate of reactants and objects. The ionic impediment of electrolytes, the electrical resistance between vibrant mass and current accumulators, and the contact resistance between vibrant mass and current establishments all promote Ohmic divergence. Protections R trans, L, R-series, and R trans S take these polarization-related misfortunes into account. I^2R errors are mistakes that can happen with any polarization. The battery's temperature rises as the problem intensifies and can be expressed as follows [24]:

$$T_{rise} = I^2 \ R \ R_{thermal} \tag{9.4}$$

where $R_{thermal}$ is the thermal resistance.

b. Charge recovery effect

The battery recovery effect abuses an internal event that is hidden from view. The velocity of artificial responses within the battery cannot keep up with the speed of outside action, such as when the battery's energy is swiftly dissipated. Thus, the location of the anodes within the battery is not exactly where the dynamic charges are converging near them. When the load on the battery is low or there is no load for an extended period of time, active charges adequate time to disseminate to the anodes therefore charge recovery occurs. As a result, a battery released at a fast pace has a lower usable limit than one released at a low rate [24].

c. Utilization factor

Typically expressed as Coulomb proficiency, voltage or vitality productivity, this is sometimes referred to as battery efficacy. Under the general phrase "vital effectiveness," both voltage and Coulomb competence are taken into account. Loss of vitality has an impact on the battery's capacity. Thus, when the usual release current raises, the capacity decreases. This suggests that the battery's release function is an extremely straight measurement of the release current's capacity value [25]. According to common definitions, energy efficiency is:

$$\text{Energy efficiency} = \frac{\text{Output energy}}{\text{Input energy}} \times 100 \tag{9.5}$$

where the input energy is used to charge the system to its starting condition and the output energy is used to discharge the system. Battery efficiency is influenced by energy loss (I^2R), indicating that efficiency declines as average dissipated current increases. [16].

9.6 BATTERY MODELS

Constructing an accurate battery model is crucial for the design, optimization, operation, and control, of the BMS. In Figure 9.8, many battery models are shown.

9.6.1 Battery Electric Model

The four primary battery electric concepts are electrochemical, reduced-order, equivalent circuit, and data-driven. Strength, physical characteristics, stress, and current of the battery are the primary characteristics of electrochemical models. These models are capable of precisely depicting the battery's electrochemical process. The representation of various characteristics, such as chemical composition, in real time is challenging. Due to the extensive use of partial differential equations, these models also have high computing costs. By making the right assumptions, reduced-order models could be used for estimating from full-order electrochemical models along with damage of evidence. In this method, computational overhead is minimized plus real-time applications may identify desirable parameters. Without any prior information, data-driven models represent the interconnection among the input and output

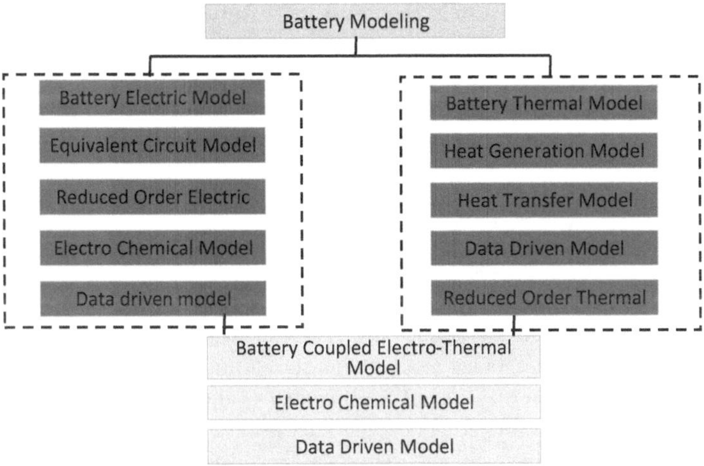

FIGURE 9.8 Classifications of battery modelling [26].

signals of batteries or the electric behaviours of batteries. The comprehensive test data should be given for improved model accuracy strong generalizability, and battery operating range. Machine learning, filtering, and stochastic models can be used to broadly classify the data-based models.

9.6.2 Battery Thermal Model

The thermal models used to gather battery thermal parameters include the data-driven model, heat transfer and generation model, and reduced-order thermal model. Activation, concentration, and Ohmic losses are frequently used in heat generation models to depict heat generation. According to [26], three well-liked methods have been applied frequently in real-time applications.

Heat conduction, convection, and radiation are used in battery heat transfer to determine the thermal characteristics of the batteries. 3D methods can identify high-heat hot spots, whereas 1-D heat transfer models could record thermal data in one direction only. However, various thermal models of reduced-order have intended to uphold battery thermal management. In addition, the LIB model is condensed from 1D to low-order linear model that resulted in predicting the temperature accurately [27].

9.6.3 Battery-Coupled Electro-Thermal Model

The thermal and electric characteristics of a battery are tightly coupled. Several linked electro-thermal models, such as coupled lump and distributed parameter models, have been developed in order to concurrently obtain battery electric behaviours (such as current, voltage), as well as thermal behaviours (such as surface/internal temperature). One method involved in creating a 3-D electro-thermal model to simulate battery conditions and calculate the heat output. A connected 3-D electro-thermal model was developed [26] to study the impact of various battery functions, such as coolant flow rate and discharge current, battery temperature. For battery modelling, hybrid and empirical methods have been applied. These models fit simply the data without taking into account the physical or chemical principles that are practical for real-time systems [27], giving no information about the specific battery properties.

9.7 RESEARCH ASPECTS/CASE STUDIES

The heat generated from the electric vehicle battery pack is one of the serious issues that affect the energy storage, durability, efficiency and life cycle of the vehicle battery package. Hence some techniques are required to manage the generated heat. There are many research articles available as shown in Figure 9.9, that explain the thermal management system. Hence the efficiency of the battery has enhanced and the battery life can be prolonged. The generated heat can able to manage using air, liquid, phase and other phase change materials. In addition, with some due research interest Nanofluids and refrigerants are also taken into consideration for thermal management systems for batteries of EV [28–32]. Most of the authors used liquid refrigerant system with maintaining the battery operating temperature [33–38].

The authors presented a numerical study through set of equations. They followed a model namely the Eulerian two-phase approach based on some assumptions such as neglecting viscous dissipation, radiation, considering properties as constant, etc. to simplify the modelling problem and easy to estimate the efficiency of the battery. These equations represent the continuity equation, momentum equation and energy equation and they are deployed for different phases, whereas the pressure is distributed by all three phases. The battery of the vehicle has a cooling module where nanofluids are the coolant flowing in the corrugated min channel. the conclusions given are that the movement of the suspended nanoparticles of the base fluid has a significant effect on the cooling ability of coolant that results in an inferior maximum

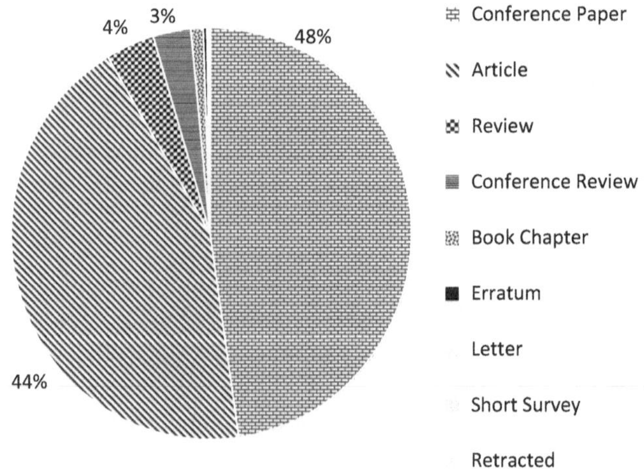

FIGURE 9.9 Scopus database [19].

temperature of the battery for highly concentrated nanofluids. Hence their proposed model could be used to optimize the cooling module used for the battery's thermal management used for electrical vehicles.

In our daily lives, the applications of EVs and hybrid vehicles are growing rapidly and hence it's quite common to practical use lithium-ion batteries (LIB) in everyday lives. It is evident that the performance of the battery pack has a huge impact on the battery's durability, ageing, stability, etc. That's why the thermal management system has become a prominent research zone in recent years. A bibliometric study of the TMS has surveyed the research done for the last two decades and summarized the information. According to their analysis, LIBs have rapidly expanded their operations in thermal management systems. Many countries across the globe have been utilizing these batteries. China is a country actively participating in research-oriented approaches. University-wise, more work has been progressing with Ontario Tech University from Canada and Tsinghua University from China is the leading body from a research point of view.

EVs are having the substantial potential to minimize GHG emissions and environmental pollution [39]. In 2011, the European Union proposed that the use of EVs in the transportation sector is an effective path to reduce the emission of greenhouse gas [40]. Later on, many countries China, Briton, Norway, France, etc. have announced down the line. Similarly, many nations across the world support the advancement of EVs [41]. The trend in the development of the international market, the number of electric vehicle usage will reach 1.3 million by the end of the year 2030 [42] as per the information from International Energy Agency (IEA)

The conventional vehicles that operate with combustion engines need to be replaced because of the raising pressure of emissions and shortage of energy. Hence most countries are looking forward to better replacements and exploring the possibilities of developing alternative vehicles. So EVs with battery set-up is good technology with well in function to eliminate emissions and diverse energy sources also.

For EVs, LIBs are very much suitable essentially. But temperature sensitivity is an acute issue. Thermal management in batteries can be improved by nanofluids, phase change materials and heat pipes introduced for the thermal management system. These thermal management systems of batteries are carried out based on various cooling methods employing air, liquid, phase change materials, heat pipe, etc. A review of different categories of cooling strategies utilized is summarized [11] and their comparison [43–44].

Hybrid battery thermal management systems are enhanced models of thermal management systems. When comparing the hybrid systems to normal thermal management systems, temperature rise can be decreased and uniformity of the temperature increased and improved consequently.

9.8 CHALLENGES AND PROSPECTS FOR BTMS

For EVs, the economical, lightweight design and ultimately the thermal performance are the three main factors. The accuracy in forecasting the temperature of the battery can be enhanced by a complete span of life, techno-economic behaviour and the driving atmosphere. Also, the chemistry of the battery can reproduce the electro-thermal properties of the battery accurately even at sub-cell levels. In addition, battery charging strategy should be improvised to control higher heat rates and hence temperature rises.

Multi-physical properties can be managed by employing self-adaptive control of the BTMS. Importantly, the cooling and pre-heating systems are to be considered. Various cooling systems can be incorporated provided with benefits like economical, good scalability and savings of energy consumption when compared to traditional cooling phenomena [18].

Battery performance can be enhanced by proper insulation of the heat at low temperatures especially. So-called phase change materials are having the ability. So avoiding such barriers lead to improvement in the performance of batteries and thermal management systems. Such problems can be resolved using PCM-based thermal management systems of batteries. In extreme environmental conditions, hybrid cooling systems perform well. But these stems have a little higher weight. Economical procedures are to be introduced for these battery thermal management systems using nanofluids. The synthetic procedures of manufacturing the nanoparticles are defective and also costs are considerably high. So need to be implemented economic methods.

While using nanoparticles, agglomeration may take place. Whereas the application of an external magnetic field makes different forms of linear chain-like structures of the nanoparticles that result in improved heat transfer coefficient. Similarly, a high concentration of nanoparticles can influence the stability, corrosion, sedimentation and drop of the pressure of the Nanofluids in the thermal management system. The nanofluids employed are to be purchased not only at the possible low costs but also to avoid the hazards cause globally.

9.9 CONCLUSIONS

A smart battery management system (BMS) is crucial in the understanding of the smart grid and intensification of the EVs sector. Proper circuits and algorithms are required to control and monitor the battery. In addition, reliability and safety are to be assured for the energy storage devices. With the successful research in the area of BMS, the applications are spread not only over EVs but also in electronic gadgets such as cell phones, laptops, etc. In EVs, the research is extended to inventing more accurate and practical algorithms to estimate the RUL, SOH, EOL, and SOC of the battery. The nanofluids, nanoparticles and heat pipes had a key role in augmenting the efficiency of battery thermal management systems. Nevertheless, the hybrid thermal systems had a more significant effect on the battery systems of EVs.

REFERENCES

1. M. Maskey, M. Patten, D. Vines and T. Maxwell, "Intelligent battery management system for electric and hybrid electric vehicles," *IEEE VTS 50th Veh. Technol. Conf. VTC 1999-Fall*, vol. 2, pp. 1389–1391, 1999, doi:10.1109/vetec.1999.780575.
2. H. Rahimi-Eichi, U. Ojha, F. Baronti and M. Y. Chow, "Battery management system: An overview of its application in the smart grid and electric vehicles," *IEEE Ind. Electron. Mag.*, vol. 7, no. 2, pp. 4–16, 2013, doi:10.1109/MIE.2013.2250351.

3. D. P. Mishra, A. Kumar Dewangan and A. Singh, "Recent Trends in Thermal and Fluid Sciences," In: *Lecture Notes in Mechanical Engineering*, Springer, 2022.
4. A. Shrivastava, D. Singh, A. K. Sharma and A. K. Dewangan, "Design and performance analysis of a wick-type solar distillation unit", In: Kannan Govindan, Harish Kumar and Sanjay Yadav (eds). *Advances in Mechanical and Materials Technology, Lecture Notes in Mechanical Engineering*, Springer, 2022, doi:10.1007/978-981-16-2794-1_99.
5. O. Khan, S. Mojahid Ul Islam, T. Hassan, H. Ahmad, M. Adib Ur Rahman and A. K. Dewangan, "Optimization of solar collector system based on different nanofluids", *Therm. Energy Syst. Des. Comput. Tech. Appl.*, vol. 1, p. 352, 2023.
6. K. Liu, K. Li, Q. Peng and C. Zhang, "A brief review on key technologies in the battery management system of electric vehicles," *Front. Mech. Eng.*, vol. 14, no. 1, pp. 47–64, 2019, doi:10.1007/s11465-018-0516-8.
7. J. Eyer, G. P. Corey and Sandia, "Energy storage for the electricity grid: Benefits and market potential assessment guide," In: *A Study DOE Energy Storage System Program* No. SAND2010-0815, p. 232, 2010, Available: https://www.ntis.gov/help/ordermethods.asp?loc=7-4-0#online
8. A. Khaligh and Z. Li, "Battery, ultracapacitor, fuel cell and hybrid energy storage systems for electric, hybrid electric, fuel cell and plug-in hybrid electric vehicles: State of the art," *IEEE Trans. Veh. Technol.*, vol. 59, no. 6, pp. 2806–2814, 2010, doi:10.1109/TVT.2010.2047877.
9. A. Abele, E. Elkind, J. Intrator and B. Washom, *2020 Strategic Analysis of Energy Storage in California*, Berkeley School of Law, University of California, 2011.
10. S. Dhameja, *Electric Vehicle Battery Systems*, Elsevier, 2002. doi:10.1016/B978-0-7506-9916-7.X5000-6.
11. C. Sangang, F. Chen and B. Li, "Lithium-ion battery management system for electric vehicles," *Int. J. Perform. Eng.*, vol. 14, pp. 3184–3194, 2018.
12. A. Fill and K. P. Birke, "Impacts of cell topology, parameter distributions and current profile on the usable power and energy of lithium-ion batteries," In: *Proceedings of the 2019 International Conference on Smart Energy Systems and Technologies (SEST)*, Porto, Portugal, 9–11 September 2019, pp. 585–592.
13. S. Hemavathi, "Overview of cell balancing methods for Li-ion battery technology," *Energy Stor.*, vol. 3, p. 203, 2021.
14. Z. B. Omariba, Z. Lijun and S. Dongbai, "Review on health management system for lithium-ion batteries of electric vehicles," *Electronics*, vol. 7, p. 72, 2018.
15. M. Naguib, P. Kollmeyer and A. Emadi, "Lithium-ion battery pack robust state of charge estimation, cell inconsistency and balancing: review," *IEEE Access*, vol. 9, pp. 50570–50582, 2021.
16. M. Abbas, S. K. Eung and K. K. Seul, "Comparative analysis of battery behavior with different modes of discharge for optimal capacity sizing and BMS operation," *Energies*, vol. 9, p. 812, 2016.
17. P. R. Tete, M. M. Gupta and S. S. Joshi, "Developments in battery thermal management systems for electric vehicles: a technical review," *J. Energy Stor.*, vol. 35, p. 102255, 2021, doi:10.1016/j.est.2021.102255.
18. S. Arora, "Selection of thermal management system for modular battery packs of electric vehicles: a review of existing and emerging technologies," *J. Power Sour.*, vol. 400, pp. 621–640, 2018, doi:10.1016/j.jpowsour.2018.08.020.
19. M. A. Abdelkareem *et al.*, "Thermal Management Systems Based on Heat Pipes for Batteries in EVs/HEVs," *J. Energy Stor.*, vol. 51, 2022, doi:10.1016/j.est.2022.104384.
20. B. André, K. Wolfgang, L. Stefan, T. Sven, P. Thilo Carsten and T. Scalable, "Decentralized battery management system based on self-organizing nodes," *Archit. Comput Syst.*, vol. 121, pp. 171–184, 2020.

21. P. Dost, V. Kipke and C. Sourkounis, "Direct active cell balancing with integrated cell monitoring," *IET Electr. Syst. Transp.*, vol. 9, pp. 244–250, 2019.
22. M. Lelie, B. Thomas, M. Knips, H. Nordmann, R. Florian, Z. Hendrik and D. Sauer, "Battery management system hardware concepts: An overview," *Appl. Sci.*, vol. 8, p. 534, 2018.
23. G. H. Min and J. I. Ha, "Active cell balancing algorithm for serially connected li-ion batteries based on power to energy ratio," In: *Proceedings of the IEEE Energy Conversion Congress and Exposition (ECCE)*, Cincinnati, OH, USA, 1–5 October 2017, pp. 1550–1558.
24. M. Hannan, H. Murshadul, A. Hussain, Y. Yushaizad and P. Ker, "State-of-the-art and energy management system of lithium-ion batteries in electric vehicle applications: Issues and recommendations," *IEEE Access*, vol. 6, pp. 19362–19378, 2018.
25. A. K. Dewangan, S. Q. Moinuddin, M. M. Cheepu, S. K. Sajjan and N. Dutt, "Thermal energy storage devices: opportunities, challenges and future scope," *Therm. Energy Syst. Des. Comput. Tech. App.*, vol. 1, p. 352, 2023.
26. K. Liu, K. Li, P. Qiao and Z. Cheng, "A brief review on key technologies in the battery management system of electric vehicles," *Front. Mech. Eng.*, vol. 14, pp. 47–64, 2019.
27. N. T. H. Dickson, M. Hannan, L. Hossain and P. Ker, "State of charge estimation for lithium-ion batteries using model-based and data-driven methods: a review," *IEEE Access* vol. 7, pp. 136116–136136, 2019.
28. S. Wiriyasart, C. Hommalee, S. Sirikasemsuk, R. Prurapark and P. Naphon, "Thermal management system with nanofluids for electric vehicle battery cooling modules," *Case Stud. Therm. Eng.*, vol. 18, p. 100583, 2020, doi:10.1016/j.csite.2020.100583.
29. M. Kalteh, A. Abbassi, M. Saffar-Avval, A. Frijns, A. Darhuber and J. Harting, "Experimental and numerical investigation of nanofluid forced convection inside a wide microchannel heat sink," *Appl. Therm. Eng.*, vol. 36, pp. 260–268, 2012, doi:10.1016/j.applthermaleng.2011.10.023.
30. A. K. Dewangan, A. Kumar and R. Kumar, "Nucleate boiling heat transfer of pure and quasi-azeotropic refrigerants from copper coated surfaces", *Appl. Therm. Eng.*, vol. 94, pp. 395–403, 2016, doi:10.1016/j.applthermaleng.2015.10.138
31. M. A. Abdelkareem *et al.*, "Battery thermal management systems based on nanofluids for electric vehicles," *J. Energy Stor.*, vol. 50, p. 104385, 2022, doi:10.1016/j.est.2022.104385.
32. M. Akbari, N. Galanis and A. Behzadmehr, "Comparative analysis of single and two-phase models for CFD studies of nanofluid heat transfer," *Int. J. Therm. Sci.*, vol. 50, no. 8, pp. 1343–1354, 2011, doi:10.1016/j.ijthermalsci.2011.03.008.
33. K. Khanafer and K. Vafai, "A critical synthesis of thermophysical characteristics of nanofluids," *Int. J. Heat Mass Transf.*, vol. 54, no. 19–20, pp. 4410–4428, 2011, doi:10.1016/j.ijheatmasstransfer.2011.04.048.
34. A. K. Dewangan, A. Kumar and R. Kumar, "Pool boiling of iso-butane and quasi azeotropic refrigerant mixture on coated surfaces", *Exp. Therm. Fluid Sci.*, vol. 85, pp. 176–188, 2017, doi:10.1016/j.expthermflusci.2017.02.028.
35. Y. Xuan and Q. Li, "Heat transfer enhancement of nanofluids," *Int. J. Heat Fluid Flow*, vol. 21, no. 1, pp. 58–64, 2000, doi:10.1016/S0142-727X(99)00067-3.
36. S. K. Verma, A. K. Tiwari and M. Tripathi, "An evaluative observation on impact of optical properties of nanofluids in performance of photo-thermal concentrating systems," *Sol. Energy*, vol. 176, pp. 709–724, 2018, doi:10.1016/j.solener.2018.10.084.
37. A. Ahmed, H. Baig, S. Sundaram and T. K. Mallick, "Use of Nanofluids in Solar PV/Thermal Systems," *Int. J. Photoenergy*, vol. 2019, pp. 1–17, 2019, doi:10.1155/2019/8039129.
38. A. K. Dewangan, A. Kumar and R. Kumar, "Experimental Study of Nucleate Pool Boiling of R-134a and R-410A on a Porous Surface", *Heat Transf. Eng.*, vol. 40, no. 15, pp. 1249–1258, 2019, http://doi.org/10.1080/01457632.2018.1460922.

39. A. G. Abo-Khalil *et al.*, "Electric vehicle impact on energy industry, policy, technical barriers and power systems," *Int. J. Thermofluids*, vol. 13, p. 100134, 2022, doi:10.1016/j.ijft.2022.100134.
40. D. Keiner, M. Ram, L. D. S. N. S. Barbosa, D. Bogdanov and C. Breyer, "Cost optimal self-consumption of PV prosumers with stationary batteries, heat pumps, thermal energy storage and electric vehicles across the world up to 2050," *Sol. Energy*, vol. 185, pp. 406–423, 2019, doi:10.1016/j.solener.2019.04.081.
41. P. Sun, R. Bisschop, H. Niu and X. Huang, "A review of battery fires in electric vehicles," *Fire Technol.*, vol. 56, no. 4, 2020. doi:10.1007/s10694-019-00944-3.
42. N. Kittner, I. Tsiropoulos, D. Tarvydas, O. Schmidt, I. Staffell and D. M. Kammen, "Electric vehicles," In: Martin Junginger and Atse Louwen (eds). *Technological Learning in the Transition to a Low-Carbon Energy System*, Elsevier, pp. 145–163, 2020. doi:10.1016/B978-0-12-818762-3.00009-1.
43. P. Kumar, D. Chaudhary, P. Varshney, U. Varshney, S. M. Yahya and Y. Rafat, "Critical review on battery thermal management and role of nanomaterial in heat transfer enhancement for electrical vehicle application," *J. Energy Stor.*, vol. 32, p. 102003, 2020, doi:10.1016/j.est.2020.102003.
44. S. S. Madani, M. J. Swierczynski and S. K. Kaer, "A review of thermal management and safety for lithium ion batteries," in *2017 Twelfth International Conference on Ecological Vehicles and Renewable Energies (EVER)*, pp. 1–20, 2017. doi:10.1109/EVER.2017.7935914.

10 Charging Infrastructure for EVs

S. Hari Prasadh and R. Gopalakrishnan
K.S. Rangasamy College of Technology

S. Sathish
Madras Institute of Technology

M. Ravichandran
K. Ramakrishnan College of Engineering

10.1 INTRODUCTION OF ELECTRIC VEHICLE CHARGING INFRASTRUCTURE

There are voluminous techniques to charge the E-vehicle, depending on location and needs. A well-designed E-vehicle ecosystem must have reliable electrical infrastructure at various points along the distribution grid to support the E-vehicle charging infrastructure, known as E-vehicle supply equipment. As a result, numerous E-vehicle charging infrastructure forms have been developed for a range of applications. E-vehicle supply equipment (EVSE), moreover known as E-vehicle charger specifications and guidelines, varies in each country and it depends on the E-vehicle models available on the bazaar features of the electrical grid. It emphasizes the necessity for a connected methodology to indigenous development and execution of E-vehicle charging networks and describes E-vehicle charging infrastructure's systematic ideas to meet the increased power consumption [1]. The outcome of the developments on the power systems and battery technology made the different the electric vehicles (EVs) in different segment with affordable price [2,3]. Though several research accomplishments are made on the advancement of EV technology, the anticipated progress is not achieved still now [4].The foremost issue in the employment of EV is non-existence of charging facilities and this need to focus in near future [5].

10.2 SYMPATHETIC THE ELECTRIC VEHICLE CHARGING INFRASTRUCTURE

The typical EV charging infrastructure comprises AC and DC charging circuits with the EV supply equipment under the monitoring of the battery management system. A typical scheme of such an EV charging infrastructure is shown in Figure 10.1. EVs

 DOI: 10.1201/9781003354901-10

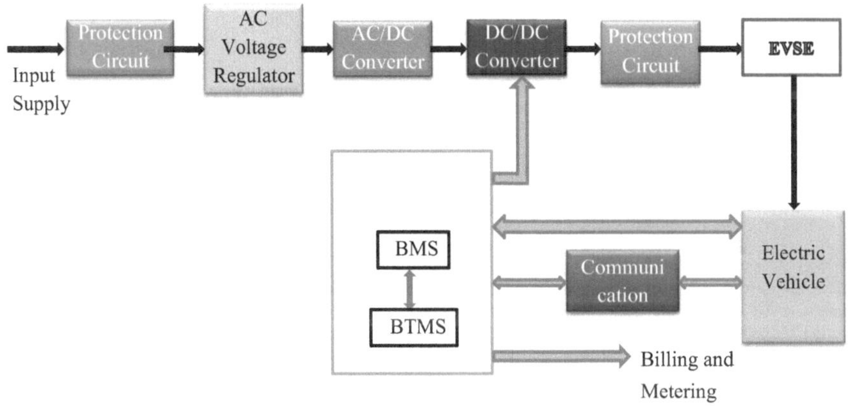

FIGURE 10.1 Scheme of EV charging infrastructure block diagram.

can be delivered by several means. The charging requirement of EVs varies with type as 2, 3, and 4 W depending upon location, power requirement, and the vehicle's ranking. In India, conductive charging, the conventional charging technology, is used. The EV supply equipment securely charges the EV by retrieving power from the confined electricity supply and then moving it through a wired connection and an intrinsic control system. The categorization of the power source of the EV because of the supply voltage and battery rating is shown in Table 10.1 [6]. The standard power range for Alternative current charging is from 3 to 22 kW and is suitable for E-2 wheelers, E-3 wheelers, four wheelers and other light commercial vehicles (LCVs) up to 1 tonne. High-power DC chargers, which are closer but require a higher electricity supply with added infrastructure, are being steadily integrated into the link and will be more prevalent soon.

10.2.1 Charging Techniques for EVs

EVs operate with a direct current power source directly supplied to the battery pack as part of the E-vehicle charging process. But, the available electrical distribution systems supply the power source as alternating current (AC). Hence it is required to convert the AC to direct current with the help of a converter as illustrated in Figure 10.2 [7]. Conductive charging is possible with alternating and direct currents

TABLE 10.1
Various Vehicle Required Power Sources

Vehicle Description	Battery Rating (kWh)	Supply Voltage (V)
Two wheeler	1–3.5	48–72
Three wheeler	3.5–8	48–60
Four wheeler	24	72
High power vehicle	Above 75	350–500

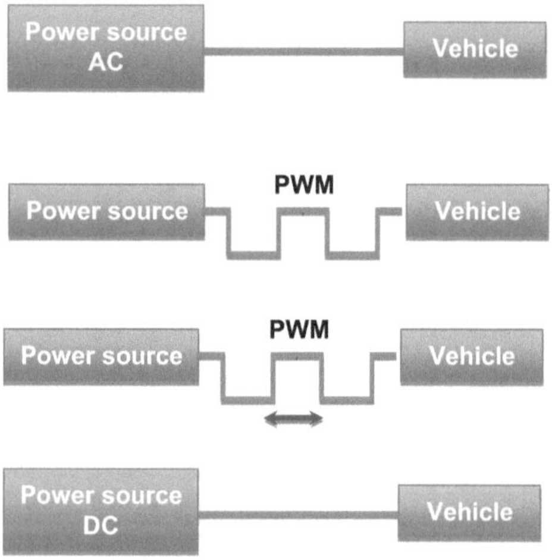

FIGURE 10.2 Charging techniques for electric vehicles.

when using EV supply equipment. The on-board charger of an electric car receives AC electricity and converts it to direct current. Direct current EV supply equipment transforms external power and distributes them directly to the battery. The method of charging is categorized in to four different methods and method 1–3 relates to alternate current charging, whereas method 4 relates to direct current charging. An EV can be connected to the conventional outlet using a cable and an Anderson plugs in methods 1 and 2, respectively [8]. As a result, method one is called "dumb charging" because there is no authorization of signals between the electric car and the equipment that supplies electricity to the vehicle, which is necessary. Method 2 contains a portable cable with built-in protection and control capabilities for its usage in a home charging process. Methods 3 and 4 have superior control systems due to a separate charger provided for use in commercial or public charging.

10.2.2 Power Rating

EVSE has different power evaluations or levels in light of charging necessities, which like this, decide the input power (I/P) provisions for charging groundwork. Table 10.2 categorizes the E-vehicle power level charging to the ordinary power level (energizing to 22–200 kW) [6].Though, worldwide E-vehicle has an estimated 500 kW manageable for heavy vehicles like trucks used for transportation. Typical power AC charging is adequate for EVs with two wheelers, three wheelers and high-powered EVs. Due to the ubiquity of light electric vehicles (LEVs) and low-voltage batteries in EVs, typical power DC charging stands out in India. For vehicles equipped with single-stage prepared chargers, single-stage alternate current chargers with the most ludicrously high power rating of 7 kW are sufficient. For EVs with better locally available chargers, three-stage AC chargers with a power rating of up to 22 kW are the norm. The conventional power scattering association can provide the input power source for routine power charging. High-power DC charging of 50 kW with battery

TABLE 10.2
E-Vehicle Supply Equipment Power Charging

Description	Power Limit, kw	Supply	Compatible EV Segments
Normal power rating	$P \leq 7$	AC, DC	Two wheeler and three wheeler
	$7 < P \leq 22$	AC, DC	
High power rating	$22 < P \leq 50$	DC	Light and heavy vehicles
	$50 < P \leq 200$	DC	

Note: Values may vary from country to country.

constraints between 30 and 80 kWh is used for high-voltage EVs. DC chargers on guard have power outputs between 25 and 60 kW. In any case, more potent DC chargers will be available soon. While high-power DC charging gives E-vehicles considerable flexibility, it requires a significant amount of power supply with the extra framework. For the majority of charging requirements, including slow or accelerated E-vehicle charging, standard power charging focuses are suitable.

10.2.3 Standards for Charging

The strategic E-vehicle charging standard used in India is named IS 17017, and it is composed of three parts and six sections [9]. The basic features of all E-vehicle charging systems are provided by IS-17017-Part-1. This standard should be followed while creating AC EVs supply equipment, whereas the AC connector should adhere to IS-17017-Part-2. Equipment used to provide AC and DC EVs must adhere to IS-17017-Parts 21 and 22. Also, parking spaces should be used by light EVs and E-cars (in the tradition of using inexpensive charging stations) in accordance with Indian regulations for AC EVSE.

DC charging hubs with an output power of 48–220 kW are specified to IS-17017-Part-23. For buses and other heavy-load vehicles, high power charging criteria are necessary in advance. IS-17017-Part-25, which offers low DC power (less than 7 kW) for light-load EVs, has recently been rectified by the BIS. Owing to this circumstance, digital communication affects DC EVSE and the standards for data transmission for EVs, which are specified in IS-17017-Part 24. The shared charging system (SCS) standard, which will be part of the IS-15118 series, is set up to provide AC and DC charging communications. The following list of charging standard characteristics is provided.

10.2.3.1 EV Charging Infrastructure

The focus of E-vehicle charging stations is on locations where cars are more likely to park. By doing this, operators of EVs may charge them while they are parked, saving time and avoiding the need to travel a distance to a public charging station.

10.2.3.2 Power Plug Points for Charging EV

Because they are the most expensive and may be detrimental to the battery health of an E-vehicle, a strong network of normal power charging points removes the need

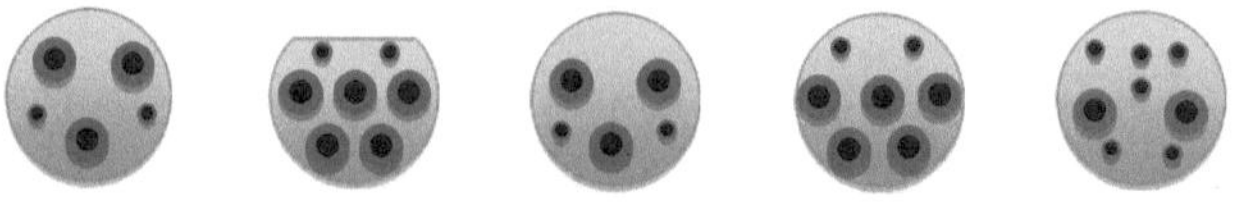

FIGURE 10.3 Various power plugs for charging EV various the country: (a) J1772-type 1, (b) Mennekes—type 2, (c) CCS1, (d) CCS2 and (e) GB/T.

reflection for high and ultra-high power charging points (if over-used). Figure 10.3 shows the typical power plugs for EV charging. These are KIA EV6, Tata Tiago EV, Tata Tigor EV, Hyundai KONO Electric, Mahindra XUV400, Tata Nexon EV, and AUDI E Tron EV are all the latest EVs in developed and developing countries vehicles have this type of charging plugs.

10.2.3.3 Cost Effectiveness

Continuous power charging stations save capital expenses since they are inexpensive, take up less space, and require less electricity. Low-voltage, single-phase, and three-phase distribution networks, which are generally present in buildings and public areas, can connect it.

10.2.3.4 Financial Stability of Electric Car Charging

The upfront expenditures of charging foundations minimize the necessity for government endowments and work on the reasonability of private area support in charging tasks [10]. A skillful rollout of the E-vehicle charging framework of a youthful E-vehicle market is essential for escalating the quantity of accessible charging focus. The assumed organization of several typical power charging stations focuses on high-power charging stations, guaranteeing to meet the E-vehicle charging requirements productively.

10.3 TYPES OF CHARGING INFRASTRUCTURE

Generally, E-vehicle charging infrastructure administration depends on its proprietorship and habit. E-vehicle charging frameworks are broadly categorized (Figure 10.4) as public, semi-public and private.

10.3.1 Private Charging

Utilization: Enthusiastic charging for specific E-vehicle holders appealed on one element.

Areas: Separable homes, committed to stopping land approachability.

Possession: Separable E-vehicle proprietors, E-vehicle holder proprietors/administrators.

Activity: Self-worked or charge point operators made due (for E-vehicle holder charging)

Domestic Supply Rating	Residential AC Rating	Turbocharging
Voltage 120V AC, Single Phase 250V AC, Single Phase 480V AC, Three Phase	Voltage 208V - 240 V AC, Single Phase 250V AC, Single Phase 480V AC, Three Phase	Voltage 380V - 600 V, Three Phase
Current Rating 12 A – 16 A	Current Rating 12 A – 80 A	Current Rating 400 A
Time of Charging 8-12 Hrs	Time of Charging 4-6 Hrs	Time of Charging 15 – 30 mins

FIGURE 10.4 Types of charging infrastructure.

10.3.2 Semi-Public Charging

Utilization: Mutual charging for a restrained organization of E-vehicle consumers.

Areas: Apartment buildings, office grounds, gated networks, shopping centres, medical clinics, colleges, government structures, etc.

Possession: Host properties, original equipment producers (OEMs) and charge point operators (CPOs)

Activity: Charge point operators accomplished

10.3.3 Public Charging

Utilization: Exposed for all E-vehicle handlers

Areas: Public parking areas, on-road stopping, charging courts, petroleum siphons, expressways, metro stations

Possession: Municipal specialists, public supply unit, charge point operators, have chattels

Activity: Charge point operators accomplished.

These are not protected classes; particular charging offices might show cross-type attributes. For manifestation, charging frameworks claimed by E-vehicle user proprietors/administrators for hostage habit is regarded as private; be that as it may, it very well may be opened to general society as a paid charging administration when users are obtainable for habit. Depending on whether they are accessible to people travelling alone or to all E-vehicle habits, EV charging stations at transportation hubs or metro stations may be semipublic or public.

10.4 E-VEHICLE CHARGING INFRASTRUCTURE FOR PUBLIC HABIT

Numerous state and local government authorities are in charge of the effective execution of the infrastructure for public charging. Nevertheless, now there is no system of collaboration between the numerous establishments. A functional gathering of E-vehicle charging establishment maintenance is vital in coordinating various government administrations, consisting of all fitting nodal and executing offices, counting state nodal agencies, distribution companies (DISCOMs), metropolitan companies, and metropolitan advancement specialists. Similarly, the state emergency response commission (SERC) and transport authorities are also addressed. The occupied gathering should incorporate high-positioning authorities from the energy or metropolitan advancement divisions. It might be going by a Chief Secretary to guarantee fundamental departmental coordination. For massive urban communities and metropolitan locales, embracing city boards for E-vehicle charging is cheered, grouped by the metropolitan partnership justices or heads of the portion DISCOMs. Nodal officials for E-vehicle charging bought to be relegated inside urban local bodies (ULBs) and DISCOMs, to prime the execution processes. Edge building among neighbourhood authorities will be fundamental for growth and evidence on prerequisites for E-vehicle charging infrastructure.

10.4.1 Operators of Charge Stations and EV Service Providers

The everyday activities of the EV charging foundation for semipublic and public charging offices are supervised and facilitated by charge point administrators (CPOs) and E-versatility administration suppliers (E-MSPs). In order to enable the uniform management of charging offices and their communication with the DISCOMs and to ensure efficient customer access to EV charging services, CPOs and E-MSPs are also responsible for setting up the long-term infrastructure, conventions, and cycles. Administrators of EV charging stations for mid or public use set, maintain, and collaborate with an organization of EV charging stations. They could own the EV chargers or operate them for the charge point owners' benefit. CPOs can manage a variety of strategies and charge point organizations that are both client-owned and self-claimed at the same time.

10.4.1.1 Ministry of Power Targets for Public Charging

The following minimal specifications for the location of public charging stations are provided by the Ministry of Power (MoP) in its *Charging facilities Guidelines and Specifications* [11].

- In a 3 km by 3 km grid, there should be one or more charging stations available.
- On both sides of highways and streets, one charging station will be installed every 25 km. Public charging stations may have any combination of connector types, as allowed by MoP. As long as they adhere to the specialized and security requirements laid down by the Central Electricity Authority, charging stations for E-2Ws and E-3Ws may use any charger (CEA).

10.4.1.2 MoHUA Has Prioritized Semipublic Charging

The configuration of EV charging in structures was included in the Model Building Byelaws (MBBL) of the Ministry of Housing and Urban Affairs (MoHUA) in 2016.

Corrections are made to MBBL-2016, as mentioned, to facilitate the infrastructure for charging EVs.

- A charging framework will be established for EVs at 20% of the premises' total vehicle holding and waiting limits.
- The structure premises should contain an extra power load, identical to the power expected for all charging focuses on being worked simultaneously, with a security element of 1.25. The revisions are pertinent to all structures but autonomous homes. Further arrangement standards for slow chargers (SC) are given in light of the number of EVs to be overhauled by portion and Fast charger (FC) standards are not required.

10.5 PLANNING AND PERMISSIONS

Study each site's space and power required for EV charging in connection to accessibility to build the optimum EV charging installations. To get the necessary connections, authorizations, and licences, cooperate with the nodal and executing authorities.

10.5.1 Establishment and Commissioning

Obtaining EVSE sticks to imperative particulars, contingent upon charging request, charging designs and required charging functionalities [12]. Introduce a unified framework board programming for the backend network of the board, including client enlistment and consent of the board, EV charger arrangement (by area and type of charger), and remote checking.

10.5.2 Activity and Billing

Control functional aspects including booking charging accessibility, revenue collection, live charger use monitoring, load regulating, execution diagnostics, and so on (Figure 10.5). Establish assistance fees for EV charging in accordance with

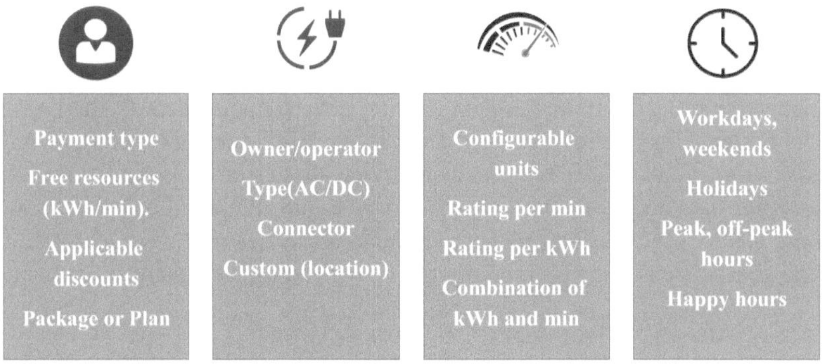

FIGURE 10.5 Activity and billing.

government guidelines or market norms. As deemed legally essential, provide the DISCOMs and other governmental bodies with the specified information.

The CPO may be a confidential substance or a public utility. Because of their advantages in managing the energy system, public energy utilities are playing a significant role in the execution of daylight charging foundations and projects in various parts of the world. In India, though, private enterprises make up the majority of CPOs, creating a market for public charging. E-portability specialist companies are offering charging services to EV consumers by linking them to various CPO-run EV charging networks. EV clients can find and manage exchanges at EV charging stations with the help of E-MSPs. Additionally, they permit EV members of one CPO to "wander" when required to the charging stations of other CPOs. CPOs will occasionally serve as E-MSPs as well.

10.5.3 Target Setting for Charging Infrastructure

The degree of vehicle ownership and predicted transport jolt patterns are given by the emphasis for EV charging arrangement shifting from one location to another. Long-term changes will also occur when EV infiltration increases. The installation of EV charging in structures and public charging arrangements has been given special attention by the MoHUA. These suggested goals may be adopted by local and state planning organizations, or they may order more forceful focuses for their regions.

10.5.4 Location Planning and Land Allocation

EV charging needs a place to put up the necessary equipment and to halt the EV for the charging duration. The allocation of land and choice of location are shown in Figure 10.6 [13]. For private and semipublic charging, this area is set aside in the stopping zones of autonomous residences, skyscrapers, or commercial and institutional buildings. Nonetheless, it is crucial to plan ahead for an arrangement of chargers to be distributed and placed strategically around a city or neighbourhood for public charging [14]. This section provides a methodology for the public charging system's area planning that combines top-down spatial analysis with base-up site selection. Also, it has institutional strategies and strategy adjustments that are anticipated to rise public charging.

10.5.4.1 Location Plan for Public EV Charging

The public charging framework's area-making arrangements identify the best locations for establishing public charging offices. It frequently finds acceptance on a range of scales, from a city-level activity to a small one. To ensure a planned public charging network, state nodal agencies (SNAs) or ULBs may direct or commission an area planning centre. CPOs setting up charge point enterprises may also complete areas with the goal of identifying optimal placement with significant charging demands.

10.5.4.1.1 Maximize Accessibility

The ease with which one may seek and go to public charging stations from any area is one measure of availability. They include calls for minimal assessed charges,

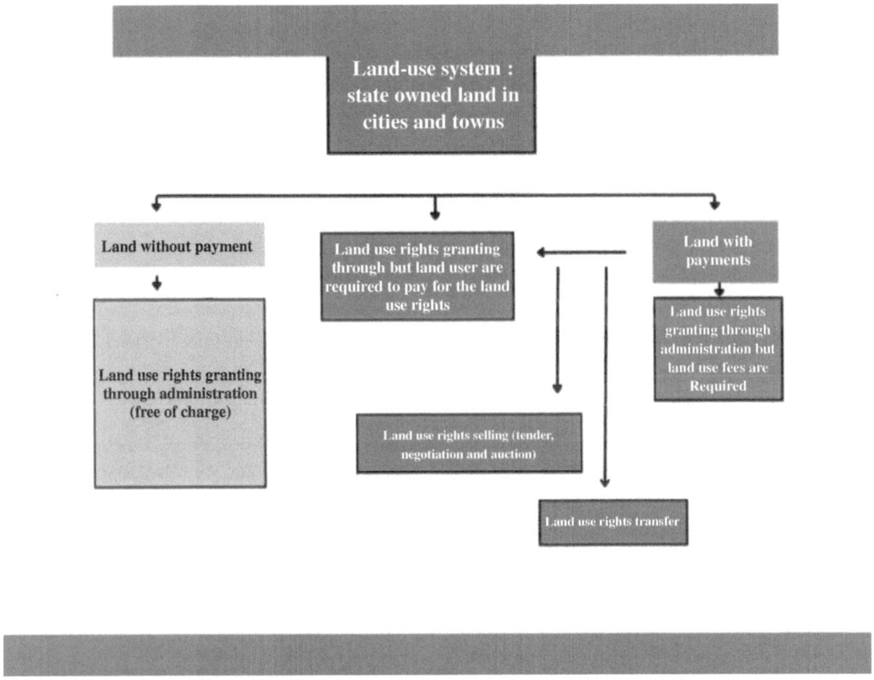

FIGURE 10.6 A choice selection of location planning and land allocation.

which demand a simple pricing system. To increase the accessibility of EV charging, network planning and site selection are both important. More distributed charging stations in the neighbourhood minimize the average distance that EV customers must travel to a public charging station. Moreover, the openness of charging stations can be impacted by their accessibility, ease of entry and exit, and closeness to major thoroughfares.

10.5.4.1.2 Maximize Utilization

To ensure high employment, the public charging infrastructures should be located in charge-request regions. Public charging requests in a particular location will depend on a variety of factors, such as the local population, business density, stopping accessibility, traffic volume, the presence of focus places like business facilities and travel stations or, on the other side, traveller complaints, etc. It also depends on whether there are any nearby private or semi-public charging stations.

10.5.4.1.3 Minimize Cost

The cost of the public charging architecture is primarily determined by the following variables: the equipment being used to supply E-vehicles, the cost of the land, and the cost of the workforce. By adopting a cyclic charging system of standard electrical charging points that are less affordable, take up less room, and use less energy at a random location, each can be fundamentally reduced.

10.5.4.2 Selection of EV Charging Infrastructure

The choice selection for EV Charging Infrastructure is shown in Figure 10.7. A computerized geospatial analysis or an on-the-ground exercise can be used to design an area for public charging infrastructure, depending on the breadth of the planning and the type of geographical information available. At a metropolitan or territorial scale, a hybrid approach to area planning is recommended. Future charging demands can be divided from the resulting public charging needs at a unit region level using a large-scale geospatial analysis. Local landowners and elected officials may be consulted when choosing a regional location for the installation of public charges.

10.5.4.3 Geospatial Analysis for EV Charging Demand Distribution

Depending on the size of the enterprise and the idea of the geospatial data available, region-making game plans for public charging infrastructure can be managed through a digital geospatial assessment or as an on-basis out. For managing region organizing at the metropolitan or regional scale, a mixed strategy is proposed. At the unit-district level, a large-scale geospatial study can pinpoint anticipated charging demand and the ensuing public charging required. Site assurance for the district-level public charging foundation should be feasible on-site, after consulting with local government members and landowners.

10.5.4.4 Site Selection and Planning

The location of the public recharging stations should enhance the transparency, openness, and ease of access for charging offices. A communicated arranging strategy may be used to choose the various charging locations for a particular charging request in a place, with varying arrangements of the number of chargers and power

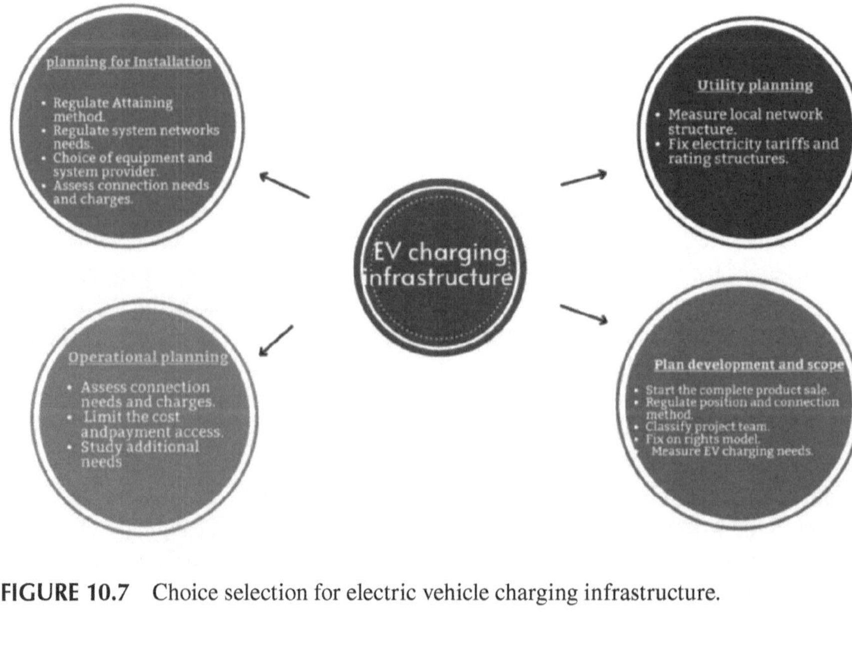

FIGURE 10.7 Choice selection for electric vehicle charging infrastructure.

levels as needed. This can diminish each site's space and power load prerequisites and empower more effective organization execution. Destinations for public charging might remember for road parking spaces, off-road public stopping, travel station stopping regions, or some other area with sufficient space and access for all EV proprietors. Responsibility for locales might shift and require numerous arrangements for held charging use. Site arranging is a significant part of incorporating EV charging in stopping regions. It relies upon the sort of stopping region as well as the number and sort of EV charge directs that need to be obliged at the site. For example, ordinary power chargers can be wall mounted, which is a more affordable and space-efficient choice. Charging establishments in private or, on the other hand, business structures might pick this choice. Platform-mounted charging gear grants better mobility for powerful chargers openly off-road stopping parcels and can have various charging focuses in a solitary EVSE. While making arrangements for EV charging coordination at the guaranteed site, the accompanying arranging rules ought to be kept as a main priority:

- Distribute space that is effectively open and apparent from the entrance of the site.
- Select the charging facility area to limit everyday work; what's more, wiring necessities, where conceivable.
- Follow all security arrangements for EV charging arranging as characterized by the Central Electricity Authority (CEA) (Measures relating to Safety and Electric Supply) (Amendment) Guidelines, 2019.
- Outline the parking spots saved for EVs accusing of suitable signage and markings.
- Give adequate space to vehicle flow, such as entering and leaving the charging sounds.
- Guarantee that the charging region is gotten against robbery and defacement. CPOs ought to occupy with place proprietors to bond to the arranging rules.

10.6 EVALUATION OF EV CHARGING DEMAND

Numerous elements of the charging foundation layout can be handled by an EV charging requirement appraisal. It may be used as information to set goals for the many of public EV chargers, as stated in this section. As we will see in later parts, it may also be used to analyze network constraints and the need for upgrades as well as for area planning for public charging infrastructure. According to per capita car ownership rates, EV infiltration rates, and vehicle usage patterns, EV charging demand depends on a metropolitan or provincial level [15].Such an evaluation needs to be led or charged by government organizing experts responsible for charging infrastructure since it's regularly used for public arranging operations. An EV charging request appraisal was obtained to concentrate on the ongoing interest in public charging for different automobile segments in order to examine the requirements of the public charging framework. This can be used to figure out how many public charging stations are required, and from there, annual targets for the public charging infrastructure can be established. The projects for evaluating EV charging requests and charging stations are listed below.

- *Step 1*: Because of target EV entrance rates, gauge EV deals for various vehicle fragments for skyline 2025 to 2030. Fragments can be separated into two wheelers, traveller and freight three wheelers, individual and business vehicles, and other Locomotive vehicles (LCVs).
- *Step 2*: Show up at the daily kilometres driven by every vehicle portion because of transport arranging information or information from city advancement plans.
- *Step 3*: In light of the standard battery limit and driving scope of every vehicle portion, compute the daily energy prerequisite for EV charging.
- *Step 4*: Distribute the charging offer to be satisfied at public charging infrastructure for various vehicle sections in light of existing research or through studies with existing EV customers. Individual two wheelers and vehicles, for instance, might be able to fulfil most of their charging requirements at home or the office and might only need to use public charging for 10% of their charging needs.
- *Step 5*: From Steps 3 and 4, work out the daily EV charging request at the public charging framework for various vehicle portions.
- *Step 6*: In light of the kinds of chargers accessible on the lookout, sorted by voltage level and power rating, indicate the charger types that will support the different EV fragments.
- *Step 7*: Determine the number of chargers needed for the infrastructure of general society using an estimated charger utilization, like 25%.

10.7 EFFECTIVE EV-GRID INTEGRATION AND CHARGING LOAD MANAGEMENT

At a 33% EV penetration rate by 2030, the total electricity mandate for EVs is anticipated to be 37 TWh (2018 Brookings India report). By 2030, this will organize less than two percentage of the nation's electricity demand. So, it is not anticipated that India will put the inclusive energy mandate for EVs to the test. However, EVs' large charging capacities and spatial concentration could cause significant instability in their power demand. This might provide challenges for the efficient deployment of EV charging connections and can affect grid stability for all power users when coupled with delays in local delivery capacity. To incorporate EV charging loads connected on electrical network provision and extension methods to improve power connections for EV charging at the position level, this chapter describes events to advance the operation of the standard grid infrastructure. Energy management methods and scattered charging network planning at the feeder and network level can improve grid utilization and ominously lessen the immediate need for expensive upgrades [16].

Growing EV adoption poses risks at two levels, from the DISCOM's service region to the feeder level, as a result of increased charging loads. A combined charging petition may, on the one hand, raise the maximum demand in a DISCOM's service area or create a new lobby (secondary) maximum, according to pragmatism in modelling the EV charging load in California, USA. On the other hand, in areas where the holding capacity of the power feeders is inadequate, irregular spots in the EV charging loads may significantly impact the distribution network. Unmanaged EV charging, often known as simple or dumb charging, can cause voltage fluctuations, voice distortions, power losses, and a buildup of consistency guides, which can disrupt the smooth operation of the energy distribution system. When EV charging stations consume power from an existing connection, it might create voltage fluctuations in the host formation's electrical circuit, which is known as dumb charging.

Interestingly, intelligent charging utilizes aloof and dynamic energy the executives measures to adjust charging request equally and limit the adverse effects of EV charging loads on the conveyance framework. Pilot programmes across the globe have demonstrated the efficacy of intelligent charging in moving a sizable amount of the EV charging load to off-busy hours while still meeting clients' charging demands. Intelligent charging works in conjunction with disconnected administrative techniques. Moreover, managed EV charging can provide greater sustainable power uptake by coordinating the best vehicle charging seasons with the maximum sustainable power age periods. Smart charging is also beneficial when an existing power association is used to provide electricity for EV charging. Intelligent chargers with the necessary skills can control the charging power levels to limit the charging load while still adhering to the permitted load limits. At the feeder level or inside a DISCOM administrative region, intelligent charging devices respond to signals to control the charging rate and provide recurrence reaction and load-adjusting services.

10.8 CONNECTING EVS TO THE ELECTRICITY GRID

Figure 10.8 shows the typical electric grid layout for EVs [17]. Manageable, dependable, and reasonable power is essential for a satisfactory charging framework arrangement. For a quick, versatile EV charging grid, the universal low tension (LT) power

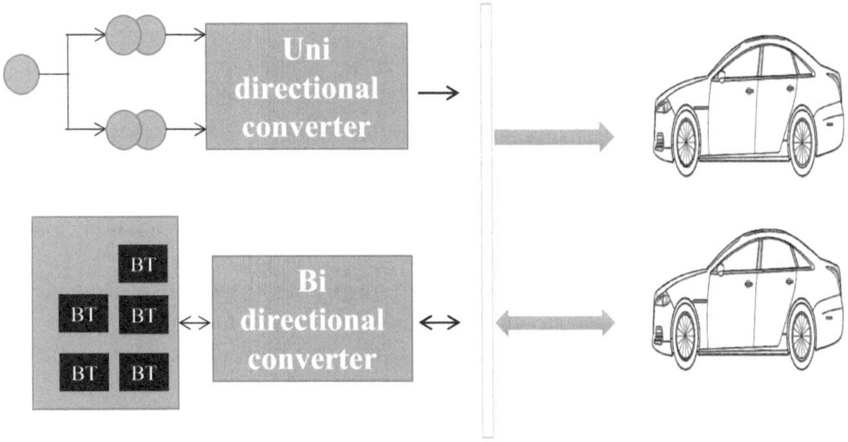

FIGURE 10.8 EVs to the electricity grid.

appropriation foundation should be utilized anywhere to give EV charging power associations. A dispersed way to deal with the charging framework, including typical power charging focuses, guarantees that most charging focuses can be associated with the LT power organization. This section investigates the administrative and administration arrangements that influence EV charging associations and discusses three techniques for organizing power supply to reserve or civic charging offices. India's power supply is heavily regulated, with guidelines at local and regional levels. Arrangement for power associations in EV charging goes beneath many guidelines and rules; some are general, whereas others have been planned explicitly for charging offices. In India, an electricity supply is in a very controlled fair, with protocol at state and central levels. Providing electricity connections for EV charging derives from protocols and strategies; few are general, and others have been articulated precisely for charging services.

10.8.1 Type of DISCOMs in the Supplied Power Connections

DISCOMs are liable to giving power associations with EV charging framework. They implement as well as accomplish the power supply guidelines and guidelines on-ground and communicate with various sessions of power buyers DISCOMs, and power association prerequisites that are particular from other customer classes. To take special care of these prerequisites, DISCOMs should carry out administrative measures, for example, EV levy classes, layout standard working strategies, and become comfortable with arranging and giving power associations with EV charging foundations. As the point of interaction between the power organization and CPOs or EV clients, DISCOMs can smooth out the most common way of giving power associations with charging foundations. SERCs ought to order that all DISCOMs attempt the accompanying processes:

- Give perfect open rules on the application cycle designed for metered associations for EV charging and create a solitary window framework for handling the applications.
- Endorse specialized pre-possibility checks for public charging associations; CPOs evaluate the practicality and assess the cost of obtaining the necessary authorized load for a planned charging office at a given area.
- Set the most extreme timetables have assisted with the investigation, affirmation of charging offices and grant of EV charging associations.
- Openly, it shares measures and prerequisites for various associations and related charges in an improved CPO design.
- Spread out perfect rules for proprietors of private charging (for example, homes and workplaces) on the prerequisites and cycles to apply for metered EV associations, exploit any practical advantages similar to EV-explicit duties, and tweak EV charging programmes.
- Make a commitment inside the group, similar to an E-versatility cell, to answer inquiries, coordinate with intrigued candidates and do site visits concerning EV charging Connections.

10.8.2 Charging on Power Demand

The alliance for an energy-efficient economy's review, "EV—A New Entrant To India's Power Consumer-Basket," evaluates the sporadic effects on charging requirements for 10,100 EVs (including 7,100 E-2Ws, 1,550 E-3Ws, 1,350 E-4Ws, and 100 E-transports) on the peak power interest of each of the four DISCOMs in Delhi. Because of the supporting data, the review encouraged the development of a model based on success to evaluate how EV charging alters the load profile of a DISCOM

- The state load dispatch focus provides load information for DISCOM over a variety of time periods (SLDC)
- EV populace situation because of freely accessible projected information.
- EV classes for various types of vehicles and chargers.
- For different EV components, charging requirements call for time-sensitive charging solutions.

The analysis reveals that facilitating speedy EV charging requests is only a small portion of the overall interest in the DISCOM administration zone. Nevertheless, given the current state of EV charging technology, EV charging may aggravate peak loads or end the off-peak hours for power demand. Additionally, with more than 23,000 EVs enrolled in Delhi in 2019, the evaluation took into account about 10,000 EVs in the DISCOM management region of Delhi. By 2030, the number is anticipated to increase to over 200,000 EV enlistments. It is typical for the predicted rate of Electricity charge requests to increase quickly within all-out power requests within a DISCOM administrative region. The influence evaluations of constrained charging loads on the lattice foundation at the feeder level can be used to predict how EV charging would affect the dispersion transformer (DT) level.

10.8.3 Models of EV Charging Execution

India's EV charging framework fair is youthful, with below 2,000 charging stations laid out nationwide as of March 2021. Notwithstanding, the fair is predicted to increase quickly in the following couple of ages, with organizations from different areas arriving at different focus in the worth chain [18]. Various partners are investigating action plans; also, execution associations arrange EV charging, focused on benefit thought processes or administrative necessities. This part recognizes the ordinary partner jobs in the execution of the charging framework and characterizes the regular models of execution grasped in India [19]. A regular charging infrastructure execution model includes different jobs, which one partner might take up or convey through associations between various partners. Aside from the setup of the charging foundation, different jobs incorporate land arrangement, power supply, EVSE supply, charging programming arrangements, and client administrations.

10.9 FUTURE SCOPE

The optimized erection of EV charging stations at regular intervals give rise to the increased usage of EVs in future [5]. EV charging stations in India are generally mounted in metropolises like Chennai, Bangalore, Hyderabad, Surat, Delhi, and Mumbai. Nevertheless, considering the want for similar EVs in the country, the authorities have plans to establish charging stations throughout the United States. The authorities of India plan to expand the charging station content in India in a phased way. In keeping with the consolidated recommendations for EVs issued by the Ministry of electricity on 14 January 2022, in Section 1, all huge metropolises with a populace of four million and over can be targeted to set up charging stations. In segment 2, the state capital and union houses will admit the charging stations. The government of India has been creating a 360-diploma problem to enhance public EV charging station shape by involving non-public and public enterprises. Numerous private gamers have come forward to put in charging stations for the motors and bring a large community in each part of the United States.

10.10 CONCLUSION

The implementation of advanced techniques to erect the charging station between intra-city charging requests may be helpful in the effective utilization of EVs. The vital discussions of the chapters are as follows. The effective planning in the erection of EV charging stations will be helpful in the mobility of EV in the urban and rural place. The fundamental need for EV charging and the different modes of charging have been explored. Upon the selection of charging infrastructure, the EVs can be charged at a faster mode which enables the utilization of EVs. The effective integration of grids will give outcomes of effective charging with the increased vehicle performance and battery life. As per populace thickness and stopping conduct, the corner stores are not reasonable to serve the intra-city charging request. The primary issue is the absence of administrations at corner stores. Improvement of data administration helps the driver choice and activity considering continuous information from charging organization and traffic signal framework. Traffic volume, distance to the closest quick charging station and the scope of accessible administrations are the fundamental factors that impact the use of a charging station along public streets. The enhancements on the EVs Charging module with the intelligent matrix is need to be focused from the administrator and driver perspective. Improvement of the charging request model decides the vital number of charging focuses at charging stations. The current refuelling organization and rest places along the roadways are appropriate for the charging stations to serve city-to-city charging requests.

REFERENCES

[1] H. Morais, New approach for electric vehicles charging management in parking lots considering fairness rules, *Electr. Power Syst. Res.* 217 (2023) 109107. https://doi.org/10.1016/j.epsr.2022.109107.

[2] S. Singh, M.K. Verma, Sustainable energy, grids and networks smart charging schedule of plug-in electric vehicles for voltage support : A prosumer-centric approach, *Sustain. Energy Grids Netw.* 33 (2023) 100972.

[3] M.E. El-hawary, M.E. El-hawary, Electric power components and systems the smart grid - state-of-the-art and future trends the smart grid: State-of-the-art and future trends, *Electr. Power Comp. Syst.* 42 (2014) 239–250. https://doi.org/10.1080/15325008.2013.868558.

[4] A. Ahmad, Z.A. Khan, M.S. Alam, A review of the electric vehicle charging techniques, standards, progression and evolution of EV technologies in germany a review of the electric vehicle charging techniques, standards, progression and evolution of EV technologies in Germany, *Smart Sci.* 0477 (2018) 1–18. https://doi.org/10.1080/23080477.2017.1420132.
[5] S. Pradhan, D. Ghose, Planning and design of suitable sites for electric vehicle charging station: A case study, *Int. J. Sustain. Eng.* 14 (2021) 404–418. https://doi.org/10.1080/19397038.2020.1862347.
[6] P. Van Den Bossche, *Electric Vehicle Charging Infrastructure.* Elsevier (2010). https://doi.org/10.1016/B978-0-444-53565-8.00020-8.
[7] G. Ala, G. Di Filippo, F. Viola, G. Giglia, A. Imburgia, Different scenarios of electric mobility : Current situation and possible future different scenarios of electric mobility : Current situation and possible future developments of fuel cell vehicles in Italy, *Sustainability* 12 (2020) 22. https://doi.org/10.3390/su12020564.
[8] D. Kettles, R. Raustad, *Electric Vehicle Charging Technologies Analysis and Standards: Final Research Project Report* (2017). https://rosap.ntl.bts.gov/view/dot/32266.
[9] M.N. Ahmad, *Electric Vehicle Charging Stations in India (The challenges and barriers being faced by charge point operators and a look at possible solutions).* Master's Programme. Lund University (2020) 1–40.
[10] Vatsala, R.H. Khan, Y. Varshney, A. Ahmad, M.S. Alam, R.C. Chaban, *Compendium of Technical Papers: ISGW.* Springer (2017). https://doi.org/10.1007/978-981-10-8249-8.
[11] S. Nair, N. Rao, S. Mishra, A. Patil, India's charging infrastructure: Biggest single point impediment in EV adaptation in India, In: *2017 IEEE Transp. Electrif. Conf. ITEC-India* 2017 (2018) 1–6. https://doi.org/10.1109/ITEC-India.2017.8333884.
[12] Y. Hu, Y. Pi, J. Cui, J. Sun, Research on electric vehicle charging station modeling, Dianli Xitong Baohu Yu Kongzhi, *Power Syst. Prot. Control.* 45 (2017) 107–112. https://doi.org/10.7667/PSPC160621.
[13] C. Ding, Land policy reform in China : Assessment and prospects, *Land Use Policy.* 20 (2003) 109–120. https://doi.org/10.1016/S0264-8377(02)00073-X.
[14] D. Efthymiou, K. Chrysostomou, M. Morfoulaki, G. Aifantopoulou, Electric vehicles charging infrastructure location: A genetic algorithm approach, *Eur. Transp. Res. Rev.* 9 (2017). https://doi.org/10.1007/s12544-017-0239-7.
[15] S. Mishra, S. Verma, S. Chowdhury, A. Gaur, S. Mohapatra, G. Dwivedi, P. Verma, A comprehensive review on developments in electric vehicle charging station infrastructure and present scenario of India, *Sustain.* 13 (2021) 1–20. https://doi.org/10.3390/su13042396.
[16] S. Powell, G.V. Cezar, L. Min, I.M.L. Azevedo, R. Rajagopal, Charging infrastructure access and operation to reduce the grid impacts of deep electric vehicle adoption, *Nat. Energy.* (2022). https://doi.org/10.1038/s41560-022-01105-7.
[17] Y. Dan, S. Liu, Y. Zhu, H. Xie, Tertiary control for energy management of EV charging station integrated with PV and energy storage, *Front. Energy Res.* 9 (2022) 1–16. https://doi.org/10.3389/fenrg.2021.793553.
[18] J.C. Ferreira, V. Monteiro, J.L. Afonso, A. Silva, Smart electric vehicle charging system, *IEEE Intell. Veh. Symp. Proc.* (2011) 758–763. https://doi.org/10.1109/IVS.2011.5940579.
[19] R. Brazil, Recharging the future, *Educ. Chem.* 54 (2017) 24–27.

11 Advanced Driver Assistance Systems

Bollepelly Manichandra and Yaojung Shiao
National Taipei University of Technology

11.1 INTRODUCTION

Advanced driver assistance systems (ADAS) simplify the driving process and are capable to enhance vehicle safety. ADAS is an innovative technology gaining widespread momentum in the automobile field to assist the driver. A combination of safety-critical standards and autonomous features possess immense automation by providing greater comfort. As per data compiled, on average around the world, numerous lives are lost on the roads every minute. Major crashes cause injury or death due to the driver's distraction or inattention [1]. Drivers who keep an eye away for less than 2–3 seconds from the roadway increase the risk of collision. Safety primarily depends on the different factors of roads contributing to accidents.

Major risk factors due to driving are traveling at excessive speed, reckless driving, violation of the rules, being tired, and consuming alcohol or drug increases the risk of a crash. Distractions that have the potential to cause impaired driving such as the usage of mobile phones and slow reaction times make it difficult to control the vehicle. Collisions occurring due to vehicle conditions and the severity of their consequences are directly linked to an increase in average speed, failure of brakes, struck steering wheel, tire flatness, insufficient headlights, and overloading [2].

Functional commands related to throttling, steering, and braking as shown in Figure 11.1 play a vital role in the advanced driver assistance system. These functions help to avoid accidents by giving the drivers early advice on how to save and

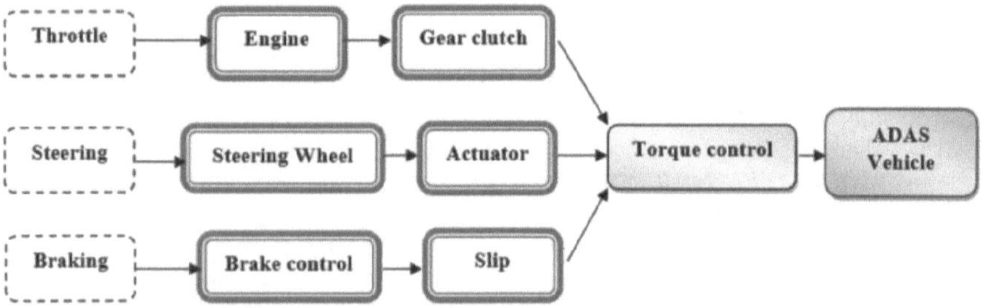

FIGURE 11.1 Functional commands of an ADAS vehicle.

DOI: 10.1201/9781003354901-11

correct the vehicle, reducing the impact of dangerous situations [3]. Therefore the importance of ADAS is to avoid accidents by giving early advice to the driver to save lives and to correct the vehicle, reducing the impact of dangerous situations.

11.2 ADAS TECHNOLOGY

Advanced driver assistance systems (ADAS) are a newly developed technology in modern automobiles function which actively assists human drivers in avoiding collisions as well it is a group of subsystems collectively used for autonomous driving. The major goals are to alert drivers, help prevent mistakes from happening, and support them as they do their driving responsibilities. The two primary factors that lead to accidents are loss of vehicle control and failure to avoid another vehicle. By taking over driving duties and easing the pressure on the driver, the majority of these systems also provide comfort benefits for the driver in addition to greater safety [4–6].

Perception of the driver's maneuver is the key role of the advanced driver assistant system's cooperation with the driver. Otherwise, the driver's command won't be honored by the assistant system and it will just override it. The assistance will take place on a variety of levels, with the operative level being the only place where the virtual driver will actually interact with the vehicle command. At first, the driver's level of allocation must be understood. Second, the virtual driver's capacity for assistance must be defined. As a result, the requirements for ADAS with human cooperation (taking control when needed) can be verified based on different levels of autonomous.

In most cases, cooperation from the assistant system can range from simple advice to full operation of the maneuver. Furthermore, the cooperation algorithm can reduce the intensity of the cooperation to avoid risks associated with driver drowsiness or complacency. Last but not least, when the system must reduce the level of cooperation, the driver's alertness or capacity must be known. Indeed, the decision control must determine whether or not the driver can regain vehicle control. Driver monitoring is not only useful for detecting tiredness but also for validating road scenes. Many false alarms can be avoided by monitoring where the driver is looking. This is based on the cooperation's overarching goal of assisting the driver by informing of events that the driver has not detected.

Technically, ADAS-related systems can be divided into four major categories which are discussed in Figure 11.2. (i) Longitudinal support, (ii) lateral support, (iii) driving vigilance system, and (iv) parking assistance.

11.2.1 Longitudinal Control

The ego vehicle's longitudinal motion based on its longitudinal dynamics such as velocity, acceleration, and jerk in the longitudinal direction is controlled by the throttle and brake inputs. Longitudinal support systems were the first advanced driver-assist systems available in the 1990s, measuring the distance between two vehicles on the same lane using embedded sensors inside the car [7]. Adaptive cruise control and a collision avoidance system are standard features on mid-range vehicles today.

FIGURE 11.2 Classification of Advanced Driving assistance systems.

11.2.2 Lateral Control

Lateral support systems are the first lane-keeping assistance featured system which is advanced to the longitudinal systems. In terms of lateral dynamics, this component helps to control the lateral motion of the ego vehicle. In this case, the controlled variable is the steering input, which governs the steering angle. The heading is concerned with the orientation of the ego vehicle, whereas the steering angle describes the orientation of the steerable wheels and thus the direction of motion [8].

Finally, driver vigilance monitoring is intended at identifying scenarios such as sleepiness alerts that use already existing information from various sensors such as steering wheel positions, maintained speed, total driving time, day time, or daylight to assess the driver's attention. When a complicated algorithm identifies driving impairment, an alarm is generated. To detect alcohol misuse, a breathalyzer-based device linked to the car's ignition is designed to prohibit intoxicated drivers from starting their vehicle when a critical alcohol level is surpassed. Distraction while driving is another prominent cause of accidents, particularly among young drivers. Driver vigilance monitoring systems may assess driver stress, weariness, and rage levels using cameras and sensors placed in the cockpit, which monitor the driver's face and expression in real time [9]. The method may be used to identify possible safety hazards caused by human drivers' unhealthy emotions.

The most advanced ADAS is currently the parking assistance system, which can able to detect the parking spot without the driver's presence. The perception or input layer for the car is provided by cameras in the front and back of the vehicle, as well as other lateral and longitudinal sensors.

11.3 ADVANTAGES OF ADAS TECHNOLOGY

ADAS technology helps in the reduction of road accidents by providing assistance in key driving circumstances. In addition to the existing features of safety, and increased driving comfort, these technologies may deliver greener and more prominent transportation in the near future. Currently, researchers are integrating current systems with web information and global positioning system (GPS) signals to determine the most effective driving route and speed. Such technologies may, for example, detect a red traffic light long before the driver notices it and lower speed appropriately.

An autonomous vehicle (AV) can drive itself and execute sophisticated activities without human intervention by recognizing its surroundings through sensors and actuators. AVs have enormous potential to improve safety while also reducing congestion, energy consumption, and pollution [10]. ADAS has played an important role in lowering the number of vehicle collisions on the road throughout time. It is primarily intended to improve driving safety and make journeys more comfortable.

Considering its numerous advantages ADAS has played an important role in reducing the number of vehicle collisions on the road throughout time. It is primarily intended to improve driving safety and make journeys more comfortable. Improvements to safety measures to encourage safe driving by incorporating key driving assistance functions such as: maintaining a safe distance and controlling speeding [11]. The ADAS system primarily functions as a co-pilot, alerting and notifying the driver of any possible road hazards. ADAS is intended to help drivers avoid crashes on the road by deploying technologies that alert them to possible hazards. ADAS alters the driving style of the car so that it can reduce fuel consumption. The ADAS system also lowers vehicle maintenance expenses.

11.4 VEHICLE AUTOMATION LEVELS

Every development at the automation level necessitates significant hardware and software technological breakthroughs, as well as efficient supervision of all safety-critical operations. An AV when can be driven on its own without any human intervention and is classed depending on automation levels ranging from entirely manual to fully automate.

Sensors work together to let the car perceive its surroundings more accurately, leading to automated driving. Adaptive cruise control is one of the vehicle automation stages employed in AVs (ACC). Furthermore, given information about the vehicle's position and localization, autonomous navigation, tracking, and obstacle avoidance may be accomplished [12]. Vision-based odometry, which is a cost-effective technology, a voice-controlled system similar to that developed for unmanned aerial aircraft can also be included in AV for progress.

There are six levels of automation as discussed in Figure 11.3 in terms of diminishing reliance on manual driving. At level 0, the vehicle is entirely manual (people monitor the driving environment), and at level 5, the ADAS transforms the vehicle into fully autonomous. No driver intervention is a scenario in a full AV where the car takes over lateral and longitudinal control while the driver remains Eyes-off [13].

LEVEL-01 No automation: Vehicles that do not have automation are referred to as manual vehicles. All activities including steering control, braking, and speeding cannot be accomplished without human interaction in these vehicles. The sensor system in such cars is non-existent.

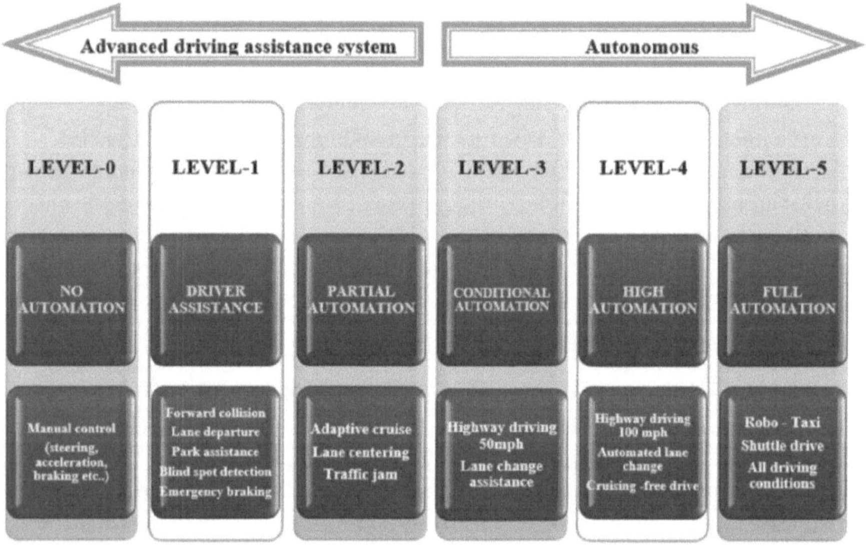

FIGURE 11.3 Six levels of automation process in vehicles that can be implemented.

LEVEL-02 Driver assistance: Vehicles with driver assistance are referred to as single-automated system vehicles. A vehicle classified as having this level of automation has at least one advanced driver assistance (ADA) feature. Although the mobility of such cars is still monitored by a person, the vehicle includes a function that maintains velocity in certain scenarios for the benefit of comfort and ease.

LEVEL-03 Partial automation: These cars are equipped with at least two sophisticated driver assistance technologies (ADASs). Although cars with this level of automation are highly pleasant in terms of driving mechanics, they still need a human to overrule them and be ready to intervene at any time.

LEVEL-04 Conditional automation: This vehicle is classified as level 3, and conditional automation can take over complete control from the vehicle's operating system during preferred routes when operating conditions are encountered. During the chosen routes, the driver must be alert and focused on the road, even while the vehicle is in unmanned driving mode.

LEVEL-05 High automation: This vehicle is classified as level 4, and it may finish an excursion without the assistance of a driver; however, the vehicle does have certain limitations, such as being to the region and being prohibited from operating at a guaranteed speed.

LEVEL-06 Full automation: The car is well-equipped for hands-free driving in all situations. Fully automated cars have self-driving technologies and no human interaction while driving (Figure 11.4).

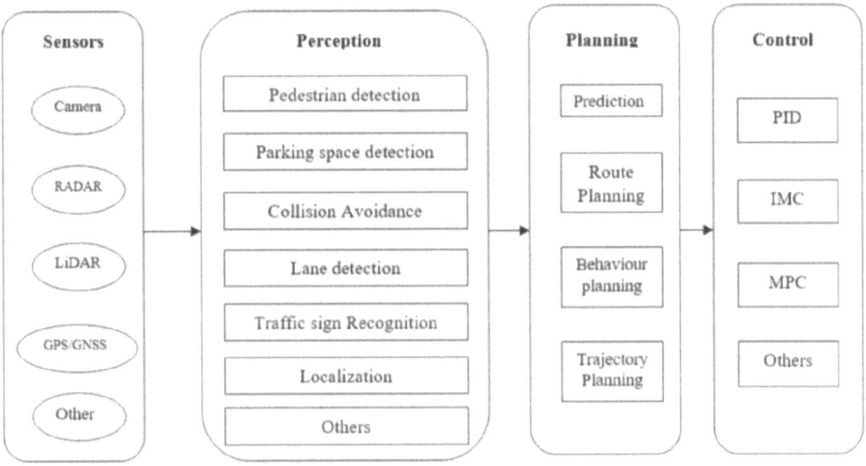

FIGURE 11.4 Shows the software and hardware architecture.

11.5 COMPONENTS OF ADAS

11.5.1 HARDWARE ARCHITECTURE

11.5.1.1 Sensors

Currently, drivers make the majority of judgments behind the wheel; however, with ADAS, certain decisions may be made by the systems. Sensors are employed to guarantee that proper safety precautions are performed and that the necessary data about the surrounding environment is collected [14]. As part of ADAS, many orthogonal and complementary sensors are deployed, which collectively give information and redundancy to increase driver safety and performance. GPS sensors, IMU sensors, distance measurement unit (DMU) sensors, cameras, Lidar sensors, and radar sensors are examples of sensors that may be found in practically every self-driving vehicle. By assessing data redundancy, the intelligent system integrates these observations in order to increase accuracy.

11.5.1.2 Mapping Systems

Vehicle-to-network communication, vehicle-to-vehicle, vehicle-to-infrastructure: Localization of autonomous cars in the outer by the usage of cameras, mobile networks, communication towers, and other technologies. GPS is a very effective technique for vehicle localization that may be used to supplement the localization of such methods or to locate autonomous cars when such methods fail. This has implications for outdoor navigation as well as intelligent transportation systems for car autonomy.

V2X technology enables the system to interact with its components as well as with the outside world, calculating the distance to passing automobiles and estimating absolute and relative speed, which may aid in accident prevention. Furthermore, it makes the system "alert" to unexpected events such as the sudden appearance of people or stray animals, the approach of an adjacent vehicle from the incorrect direction, and the incidence of potholes, road breaks, and so on. Vehicle-to-network communication links the vehicle to the cellular infrastructure and the cloud, allowing drivers to access in-vehicle services such as traffic updates and video streaming.

Sensors collect, store, and update geographical and infrastructural information to govern a vehicle's precise location. Even if GPS coverage fails, this information is kept and relayed to the control system. Wireless networks play a critical role in both internal and external connections. Wireless solutions, in addition to crucial systems such as tire pressure monitoring and anti-lock brake systems, give additional flexibility for regular automobile communications. The Internet of Things (IoT) and automated predictive maintenance systems (APM) promise to connect smart devices within automobiles with cloud-based apps to deliver increasingly more complex services [15–16].

11.5.1.3 Actuators

Actuators are in charge of controlling mechanical movement based on the energy supply. Motoring, braking, and steering systems can be powered by pneumatic, hydraulic, thermal, or rotatory actuators. The electrification of vehicle actuation systems has been a key enabler of ADAS. This enables various ADAS systems to communicate with other electrical components in the vehicle without causing interference.

11.5.2 Software Architecture

Software is regarded as the foundation of hardware functions. The system becomes clever enough to make judgments about software architecture. With the introduction of digitalization technologies, ADAS systems are becoming more effective and increased adoption. This is accomplished by the architecture's following three subdivisions.

11.5.2.1 Perception or Processor

Processor subsystems are utilized in ADAS applications for anything from creating a real-time 3D spatial representation of a car's surroundings to calculating closeness and threat levels depending on the environment. It gives the AV the capacity to grasp sensory input acquired from various types of attached sensors, assess the surroundings, and monitor their own physical characteristics in order to make intelligent judgments like as categorizing an item in a recorded picture [17]. This subsystem gathers the data from various sensors surrounding the vehicle and acts as in charge.

11.5.2.2 Planning or Scheduling

The planning subsystem uses the data collected from the perception subsystem to determine the most optimum behavior to be taken by forecasting the states of the other vehicle. It gives the AV decision-making authority [18]. The planning system operates by integrating perceptive data and processing it to take appropriate action, such as detecting objects, forecasting an accident, monitoring blind areas, and deciding the vehicle's velocity. Complex AI algorithms are used to prepare and execute such decisions.

11.5.2.3 Control

The control system is the last component of the system, also a valuable part of the overall design of an AV. This subsystem is also in charge of precisely following the trajectory supplied by the planning subsystem and it engages the relevant actuators to generate the required movement. The control system finally establishes the connection and interacts with the surrounding car's environment to change the final control components (throttle, brake, and steering).

11.6 ADAS SENSORS IN THE MODERN AV

Sensors are necessary to pick up everything the driver cannot see or hasn't detected in addition to what the driver can see. ADAS uses a variety of automobile sensors to give drivers information about their surroundings, enhancing driving safety [19]. It is feasible to improve the ADAS functionalities by integrating multiple technologies because each type of sensor has recognized advantages and disadvantages. As shown in Figure 11.5, cameras, radar, Lidar, global positioning systems/global navigation satellite systems, and sonar/ultrasound sensors are just a few of the many distinct kinds of ADAS sensors (Table 11.1).

11.6.1 Camera Sensors

In light of their comparatively inexpensive cost and robust usage, cameras are the most often utilized detectors for an AV. Cameras in ADAS/AD systems can be characterized in a number of ways, including by placement (external cameras and in-cabin cameras), application-oriented such as human vision cameras as well as technology-based like stereo cameras and infrared cameras. Human vision cameras are frequently used as backup cameras, surround view cameras, and cameras to replace rear-view mirrors.

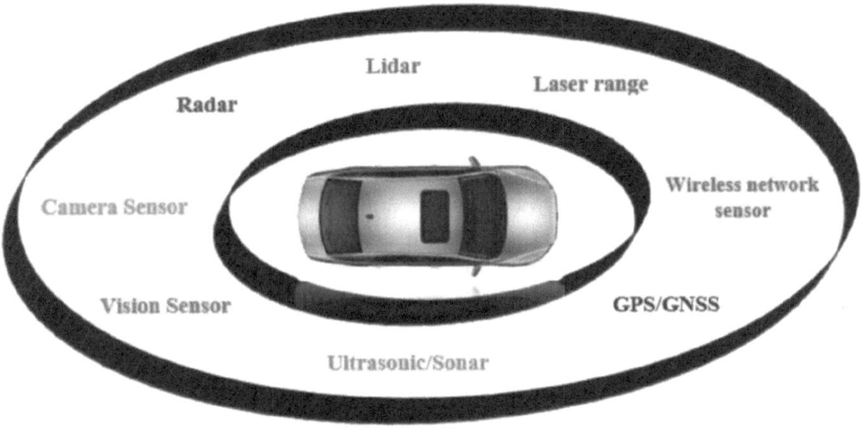

FIGURE 11.5 Different types of sensors used in ADAS.

TABLE 11.1
Comparison of Various Types of Sensors

	Features	Range of Detection (m)
Camera	Pedestrian recognition Obstacle detection Lane tracking Road information tracking Auxiliary positioning	0–100
Radar	Long/Medium distance obstacle measuring	Approximately to 300
Lidar	Obstacle detection 3D modeling of the surrounding environment for distance and speed calculation	Upto 200
GPS/GNSS	Localization Mapping	0–10
Ultrasonic	Small-range distance detection Blind spot monitoring Parking and reverse assistance	0–3

The camera's field of vision may range from a few millimeters to around 100m. Additionally, they frequently consume little electricity and are tiny and light. The camera picture is helpful for jobs since it offers a lot of information at fast frame rates [20]. An AV camera generally consists of a lens, an image sensor, a camera module, and image processing components that control the hardware and digitally edit recorded images. Various cameras may be used with different lenses, each with a different focal length. The image sensor, which serves to detect and transmit the information required to make a picture, converts the fluctuating attenuation of light waves into electrical signals.

11.6.2 Lidar Sensors

Light detection and ranging (Lidar) provides 3D depth information by emitting a laser pulse and detecting the pulse that is reflected back from objects. Lidars are equipped with higher autonomous driving technologies, allowing them to collect accurate 3D data from the environment. For autonomous systems, there are Lidars to cover various ranges of distances up to 250m. Up to 50m or less is the range covered by short-range lidar. Typically, it uses a near-field camera to monitor and detect things [21]. It offers additional safety detection in situations when cameras find it challenging to gauge distance. Between fifty meters and a few hundred meters are covered by long range Lidar. Sometimes referred to as the middle range occasionally has a range between 50 and 100m. In order for an AV to operate at high speeds in motorway conditions, the Lidar in this range is essential.

For sharper readings of the desired surroundings, the Lidar sensor is often installed on the vehicle's roof. To protect eyes from lasers, it also has a spinning unit for power transfer to each pulse, while other factors like pulse length, divergence angle, and exposure direction all affect safety ratings. The core of object detection is LiDAR, which can transmit millions of light pulses per second to an item's surface in a well-defined pattern and measure the time it takes for the reflection to return to the receiver [22]. It is capable of mapping the surroundings dynamically and three-dimensionally using its spinning axis. LiDAR produces a 3-d cloud image of things as opposed to a two-dimensional image as the camera does. The LiDAR's point cloud output gives autonomous computing the information it needs to locate items and calculates the position of the vehicle with respect to other objects by processing and computing the larger amount of data collected.

11.6.3 Radar Sensors

Same like the Lidar the radar also produces a point cloud of detections that the velocity immediately measured. In addition to recognizing materials and objects that test the camera and lidar, radar is also capable of detecting glossy, dark, and translucent ones. New potential for safety systems and autonomous cars is anticipated to be unlocked by radar as it continues to evolve. Indeed, developments in integrated circuit architecture have made it possible to create radars operating at very high frequencies and use advanced signal processing and machine learning techniques [23]. Radar functions by projecting electromagnetic waves into space and gathering their reflection to calculate a target's distance from the radar, its velocities, and the angle of arrival. As a radar employs these waves to determine the target's position, it is frequently referred to as a "4-dimensional sensor." and velocity in 3-dimensional space. The radar is categorized as follows by the range of sensing an object (i) short range, (ii) medium range, and (iii) long range.

Furthermore, modern radars have the ability to categorize things that are far away. The majority of modern radars use a totally solid-state technology that keeps prices down and provides higher dependability. In the automotive industry, radar has been widely used for ADAS applications, demonstrating its viability as a useful sensor modality.

11.6.4 GPS/GNSS Sensors

Instantaneous data concerning position and time are sent from a collection of artificial satellites in orbit because of their high orbits. The GPS is a component of what is known as the GNSS, or global navigation satellite system. For most of the equipment, the range of location accuracy is between 100 and 10 m. A minimum of four satellites may always be seen from a certain location on Earth's surface according to the way the satellites are launched. By comparing the time difference between a radio signal that the GPS receiver picks up and a signal that is transmitted from each satellite, it is possible to determine the user's location. To compute the distance to the satellite, it employs the speed of propagation. To estimate the GPS user's location, 3-dimensional directions of a signal from a single satellite must be received for each dimension of the user's position [24].

For each dimension of the GPS user's position, a separate satellite signal must be received in order to estimate. Also utilized to improve GPS location accuracy is the environment's geometry. A 3D laser range scanner produces a 3D depiction of the region where the GPS receiver receives GPS signals. Given that the accuracy of the GPS location depends on the scattering and reflection of radio signals from such structures, the 3D model is useful in identifying the signals that are reflected by the structures in the environment.

11.6.5 Wireless Sensor Networks

Utilizing data from wireless access points, wireless sensor networks help autonomous cars locate their current location. When GPS signals are not received or are degraded by multipath effects, wireless sensor network-based localization is useful. The fingerprinting approach is one of the most used methods for locating using wireless sensor networks. The service set identifier and media access control are parameters that are employed in this approach [25]. For localization, a database of wireless sensor networks that are accessible in the area where autonomous cars move is created.

11.6.6 Ultrasonic/Sonar Sensors

The ultrasonic sensor is commonly employed in automobiles. The reverse assist and parking assist features, it is often mounted on the bumper at the back, front, and sides of the automobile. Low-range sensors that produce brief, high-frequency sound pulses are known as ultrasonic sensors. The most common components of ultrasonic sensors are a transmitter, a receiver, and a temperature sensor. High-frequency sound waves are sent out to determine the distance between neighboring objects [26]. The microprocessor converts these sensors' time signals into distance signals. They are perfect for low-speed, close-range tasks like parking, backing up, stopping at stop signs, and keeping a safe distance from other people and cars and also perform very well in poor visibility and uncommon weather conditions.

11.6.7 Laser Range Sensors

The depth data obtained by laser range sensors are used to supplement the GPS position data. Sensors that can measure the depth or focused to long distance (range) laser beams are used as laser range sensors to detect the distance of the object [27]. The TOF laser range sensor's basic operation involves timing how long it takes the laser beam to go to the object, reflect, and return to the sensor. The duration of traversal is multiplied by the speed of light to determine the depth information.

11.6.8 Vision Sensors

Given their ability to provide real-time navigation and localization, vision sensors are the most common sensors used in autonomous driving applications. They are popular because they guarantee great accuracy with minimal processing. They are used in many different autonomous driving activities, including lane marking, object identification, curb recognition, and traffic light and traffic sign detection. To comprehend the camera data from different vision sensors, deep learning or computer vision algorithms must be computationally expensive [28]. However, after the data has been understood, it offers a rich semantic representation of the environment. In the dark, infrared cameras with more advanced sensing capabilities work better. The different kinds of vision cameras are:

- *Stereo cameras*: Produce distance estimation through 3D construction
- *Thermal cameras*: Improves detection accuracy under challenging weather conditions and environment time.
- *High frame rate cameras*: Simplifies detection and tracking of moving objects.
- *Event-triggered cameras*: Reduced bandwidth for data transmission and low latency reactive path for collision avoidance

11.7 ADAS FEATURES

The goal of ADAS development is to automate, upgrade, and improve vehicle technology for better driving and safety. By providing technologies that warn the driver of issues, putting in place precautions, and taking control of the vehicle if required, safety features are intended to prevent collisions and crashes. Adaptive features can automate lighting, blind spot detection, provide adaptive cruise control and help drivers avoid collisions, automate parking assistance, incorporate traffic warnings from satellite navigation, alert drivers to potential obstacles, recognize traffic signs, detect pedestrians, help with lane departure, and provide many other features as shown in Figure 11.6. There are four categories of ADAS, which are assistance, monitoring, controlling, and warning.

- *Assistance*: Assistance systems assist cars in making minor modifications to improve driving safety based on information from the surroundings.
- *Monitoring*: Monitoring system uses cameras/sensors in order to enhance visibility into safety-critical data for emergency braking or to avoid the collision.

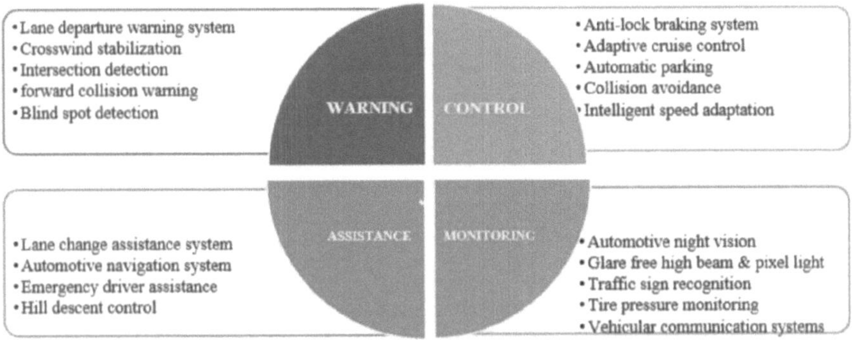

FIGURE 11.6 Illustrates the types of ADAS features.

- *Control*: In the event of an approaching accident, control systems can take over and steer the car.
- *Warning*: Drivers can anticipate potential safety issues in real time with the use of automatic in-cab notifications provided by warning systems.

11.7.1 Adaptive Cruise Control

The adaptive cruise control (ACC) technology was developed for driving on highways where traffic congestion is lower than in urban areas. The device assists the driver in maintaining a safe driving speed and longitudinal distance relative to other vehicles in the surroundings. A range sensor is required for the system to identify cars traveling ahead and quantify their distance and speed. ACC features two working modes: cruise mode, which maintains a constant speed, and follows mode, which keeps a safe distance from other vehicles. The general block diagram of ACC is depicted in Figure 11.7, which will automatically change the rate of speed to maintain the right distance between any impediment and the vehicle [29]. This may be accomplished by using a laser or radar beam to compute the distance between two objects.

To operate the vehicle, onboard sensors' data is employed that helps to detect another vehicle ahead and then accelerate when permits, such systems for traffic controlling may use a radar, laser, or camera arrangement. Future iterations of intelligent autos are expected to require the use of ACC technology. By preserving appropriate vehicle space and removing driving mistakes, the system improves passenger safety and convenience.

The following are the primary types of adaptive cruise control:

- Vehicles with full speed range can bring the automobile to a halt at 0 mph (0 km/h) and must be re-activated.

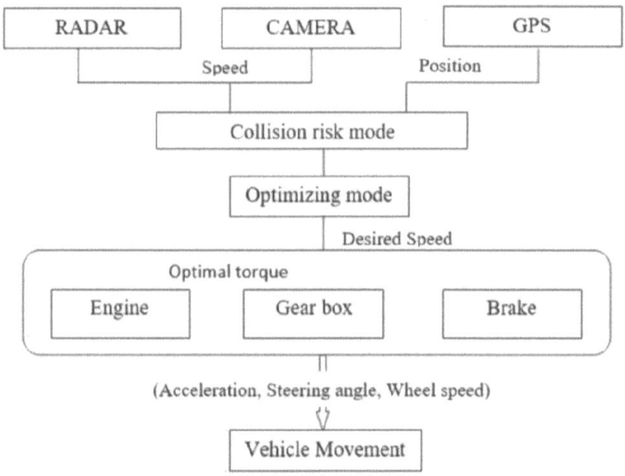

FIGURE 11.7 General block diagram of ACC.

- Vehicles equipped with partial cruise control need the driver's attention when it cut off below a minimum speed.
- Fully equipped adaptive cruise control vehicles can control the speed automatically and respond to speed non-vehicular on-road activity.

11.7.2 Collision Avoidance System

A collision avoidance system is a mechanism meant to assist prevent collisions with stopped or traveling in the same direction vehicles. It is classified as an active safety system since it operates based on signals and information acquired, and it often either alerts the driver to a risky situation or assists in key tasks like steering when braking. Radar, laser, and cameras (with image recognition) are among the technologies and sensors used to detect an impending accident [30]. CAS assemble with small radar detectors at the front of the vehicle, continually emitting short bursts of high-frequency radar waves. With this information, a control algorithm unit may nearly instantaneously calculate another car's position, distance, speed, and relative velocity. A collision avoidance system (CAS)/collision mitigation system, is designed to reduce the severity of crashes.

11.7.3 Anti-Lock Braking System

In advanced automobiles, an anti-lock braking system (ABS) is employed to avoid wheel slide and locking following braking. It is also known as a safety anti-skid braking system because it maintains contact with the road surface to control over the vehicle [31]. ABS acts at a considerably faster and more effective pace than most drivers can. The controller is given to adjust the necessary torque to ensure optimum slip response time, allowing the driver to steer easily.

As illustrated in Figure 11.8, the ABS incorporates four wheel speed sensors, and at least two brakes hydraulic valves and a central electronic control unit (ECU). The controller, an ECU, is configured to ignore variances in wheel rotation speed below a key threshold and gets data from each individual wheel speed sensor. It triggers the valves to lower hydraulic pressure to the brake at the affected wheel, reducing the braking force on that wheel, if it notices a wheel rotating noticeably slower than the vehicle's speed, a sign of impending wheel lock. Every brake has a brake line valve that the ABS controls. There are three locations for the valve in certain

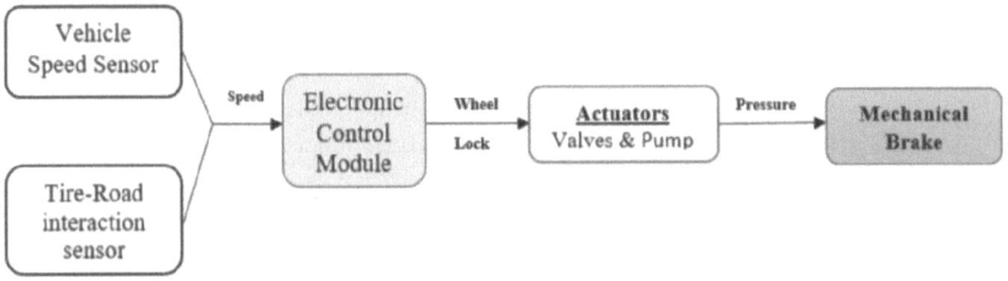

FIGURE 11.8 Illustrates the block diagram of the anti-lock braking system.

systems: When the valve is in the open position, master cylinder pressure may flow directly through to the brake. The valve isolates that brake from the master cylinder by closing the line in position two. This prevents the pressure from increasing further if the driver applies more force to the brake pedal. Once the valves have released pressure from the hydraulic brakes, the ABS pump returns it. The valve is signaled to open when the controller notices the wheel sliding. To provide the necessary pressure while minimizing slippage, the controller will change the pump's state. On wet, snowy, or slippery roads, ABS decreases total crash involvement and causes deadly collisions with other cars.

11.7.4 Intelligent Speed Adaptation

Intelligent speed adaptation (ISA) is an in-vehicle device that aids drivers in adhering to speed limits. It informs, cautions, and discourages drivers from exceeding the legal local speed limit. Intelligent speed assistance is a technology that adopts the condition of the vehicle and alerts the mandated speed to the vehicle speed. If the driver is suspected of speeding, the speed can be decreased automatically or the driver can be informed.

The vehicle speed restriction is automatically detected based on the speed limit signs across the road. ISA technology uses GPS and speed limit maps and updates the vehicle speed to adjust as per restrictions to the road speed limit. ISAs are classified into three types: (i) Informing or advising ISA provides feedback to the driver in the form of a visual or audible indication [32]. It can alert the motorist to current speed restrictions and speeding. (ii) ISA enhances the upward pressure on the accelerator pedal whether it is supportive or warning. By pressing the accelerator harder, the supporting system can be disabled. (iii) Any speeding is prevented by intervening or requiring ISA. Local speed limits can be detected based on four types of vehicle speed such as:

11.7.4.1 Position-Based Systems

GPS is based on a network of satellites that are constantly sending radio signals. To pinpoint the receiver's location within a few meters, GPS receivers gather up these transmissions and compare the signals from several satellites to establish the receiver's location. This is done by measuring the time between the signal's delivery by the satellite and its reception by the receiver.

11.7.4.2 Radio Beacons

Bollards, which are radio beacons placed along the road, work by transmitting information to a receiver within the car. The fitted receiver on the car receives the data that the beacons are continuously broadcasting as the car drives by each beacon. This information may include containing local speed limits, school zones, variable speed limits, and traffic alerts. To allow quick modifications, beacons may be connected to a primary computer.

11.7.4.3 Optical Recognition Systems

Optical recognition technology has mostly been employed to identify speed signs, road markings, and items along the side of the road. In order for a scanner or camera system to record information on a speed sign or other similar signal, the vehicle must pass it. The speed limit information is gathered as soon as the system detects a sign and is contrasted with the vehicle's speed. Until it notices a speed sign with a different restriction, the system applies that speed limit.

11.7.4.4 Dead Reckoning

In dead reckoning (DR), the path traveled by the vehicle is predicted using a mechanical system connected to the drive assembly. The DR system determines the vehicle's approximate location, the applicable speed limit, and the vehicle's speed by superimposing this path onto a computerized map. The digital map's data may then be used by the system to warn of impending dangers or sites of interest as well as to issue alerts if the speed limit is being violated. Some of the most advanced GPS-based navigation devices now available on the market fall back on DR if the GPS signal is lost.

11.7.5 Automatic Parking

The automated parking system improves the ease and security of driving in confined spots where careful attention and skill are needed to guide the vehicle. In this, the vehicle can be moved from the driving lane to an empty parking space by a maneuvering system to execute parallel, perpendicular, or angle parking. To ensure collision-free movement within the available space, the parking maneuver is accomplished by synchronized steering angle and speed control that takes into consideration the real situation in the surroundings.

In the automotive parking system, the vehicles are kept or stacked in a particular arrangement that helps the vehicle drive its entry path to the particular parking spot without any driver's assistance. As shown in Figure 11.9 the block diagram of the automated parking system explains the working criteria from receiving data from a sensor to parking in the slot. Angled parking spots make it simpler for automobiles to enter and exit the space since they don't force them to make a sharp bend [33].

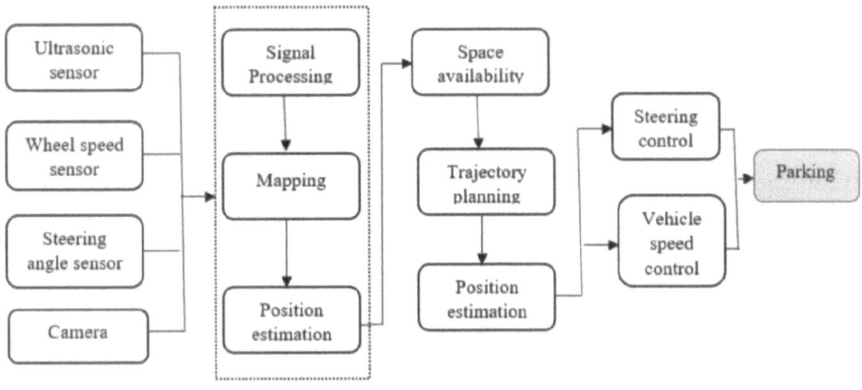

FIGURE 11.9 Block diagram of automatic parking system.

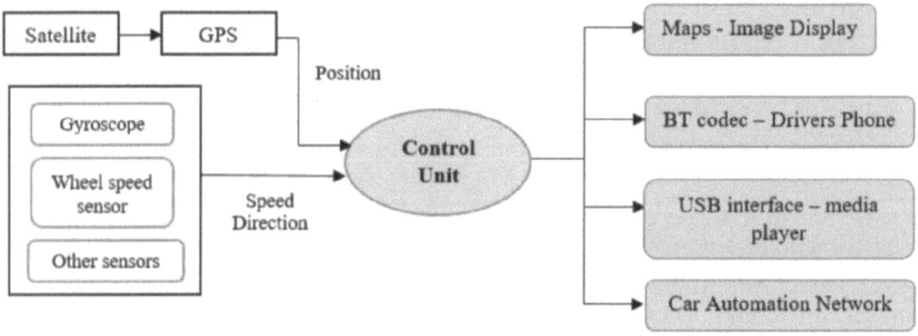

FIGURE 11.10 Block diagram of automatic navigation system.

As long as cars keep inside the lines of their parking places, this also lowers the risk of an accident. The advantages of automated parking systems include the ability to put more automobiles in a small space, a reduction in the time and cost of installation, improved security for vehicles and personal property, improved pedestrian safety, and a decrease in hazardous emissions (Figure 11.10).

11.7.6 Automotive Navigation System

The majority of the time, satellite navigation is used by automotive navigation systems as part of the car's controls for exact position, then data of position is connected to a location on the road. It contains offline maps (often 3D and HD) that can be updated through OTA services, location search, real-time traffic monitoring, voice help, and more. Additionally, it frequently incorporates voice assistants, POI (Points of Interest), address books, prior destinations, and address-based searches. Most combine a satellite link to the GPS with a computerized road map to show the vehicle's location in real-time [34]. Drivers may learn which route may be the quickest to take to arrive at their destination by keeping an eye on the real-time traffic, which is a really helpful feature. An accurate and current map is essential for providing navigational advice in vehicle guidance for positioning from a starting point to a destination.

11.7.7 Hill Descent Control

An automotive safety feature called hill descent control (HDC) enables the safe descent of inclines. Although the feature is primarily designed to be used in rocky terrain, you may use it to securely and gradually down a steep slope. When the driver activates HDC, the car uses its traction control and anti-lock braking systems to reduce tire slide and make it easier for you to down the slope. The driver may focus on navigating possible risks up ahead without having to press the throttle or brake pedals. The ABS pulses cause the tires to momentarily slide on a particular surface. The driver may switch off HDC and accelerate once the road levels out of the vehicle. By adjusting the braking pressure applied to each tire, which results in each being braked to its maximum traction, HDC lowers the amount of wasted braking potential. The computers handle the remainder after the driver selects a reasonable pace. The average speed of the car's four wheels is used to determine its speed.

11.7.8 Lane Centering

To maintain a vehicle centered in its driving lane, a method called lane centering was developed. As illustrated in Figure 11.11 this ADA system continuously assesses the location of the car with respect to the left and right lane markers and makes modifications to keep the car centered between them. ACC and lane centering are two assist systems used by combination to allow almost hands-free driving in some scenarios, lane centering is a crucial technology for autonomous vehicles [35].

To locate lane markings, the majority of lane-centering systems employ one or more cameras to transmit real-time visual data that is processed using a variety of methodologies. Laser, radar, infrared, and other sensors are also used by certain systems. In general, lane-centering systems function by continuously checking the position of lane markers in relation to the location of a vehicle, and if necessary, automatically correcting the vehicle's position if it veers off the center of the lane which can be accomplished by the car automatically steering to maintain a parallel orientation.

11.7.9 Emergency Driver Assistance

A new assistance system called emergency assist combines the features of lane assist and ACC. The recently created dynamic steering has the ability to change both the steering forces and the steering angle that the driver sets. Driving faults are eliminated or made up for in a way that prevents the driver from being shocked by the vehicle's response. Various escalation phases are activated by the system to awaken or alert the driver when it notices no action of steering, braking condition, or accelerating the pedal. This helps the driver for an emergency stop to avoid the dangerous situation, and then the vehicle itself makes a few small steering adjustments [36]. Finally, the system constantly brakes the car until it comes to a stop. Greater comfort and convenience when traveling at high speeds since the driver no longer has to worry about unintentionally exerting too much steering power and losing control of the vehicle. In reaction to steering input, it enables more agile activation.

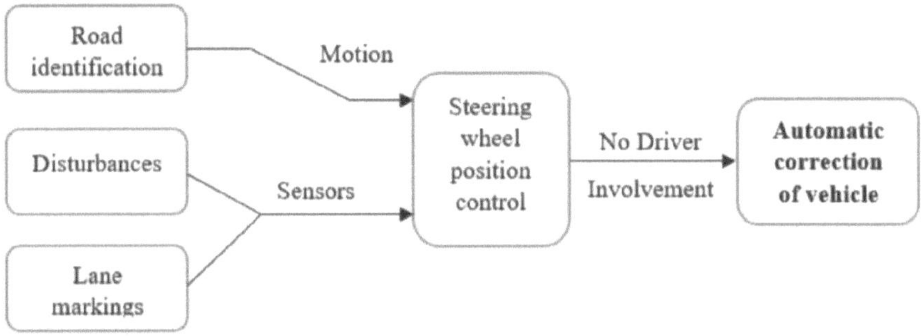

FIGURE 11.11 Block diagram of Lane centering system.

11.7.10 Lane Change Assistance System

The technology called Lane Change Assist (LCA) helps vehicles change lanes safely. It makes use of sensors to identify and alert drivers to oncoming vehicles in nearby lanes. It also aids in preventing lane change accidents on the road by alerting you to cars approaching your blind zones. As shown in Figure 11.12 to assist you in making a non-violent lane change, LCA will alert you with an audio or display warning when it detects that you are changing lanes in a particular manner. The lane change assistance system consists of two control units with identical structural layouts, one of which actively supports driving and the other of which issues warnings. One of the two main types of LCA still requires the driver to steer and brake, while the other already automates steering and speed for keeping up with the driver's distance.

Rear radar sensors, which are found in the car's corners in the back, are specifically used for lane change assistance. The control units combine with a built-in radar sensor to create a single independent component. A plastic cover is installed on the control unit over the radar sensor. Radar beams may flow through this radome without being obstructed since it is made of a unique substance [37]. Millimeter-wave radars produce high-frequency electromagnetic waves that bounce off objects and return to the sensor to provide information to LCA systems. The system can also issue audio and/or haptic warnings to draw the driver's attention to the possible risk when the driver takes the initiative to turn the vehicle to other lanes. The lane change assistance, an active safety device, assists in preventing accidents by avoiding potentially dangerous circumstances.

11.7.11 Glare-Free High-Beam and Pixel Light

A camera-driven dynamic lighting management method known as a glare-free high beam selectively shades areas and slices out of the high beam pattern to shield other road users from glare while continually giving the driver the widest possible field of vision. Although headlights and headlamps are sometimes used interchangeably, in the most formal sense, a headlamp refers to the actual equipment that is connected to the front of a vehicle to illuminate the road ahead, while a headlight refers to the light beam that is generated and dispersed by the device. Modern headlights must provide both a low and a high beam and are electrically powered. To accomplish this, numerous pairs of single-beam lamps, a pair of dual-beam lamps, a combination of single-beam and dual-beam lamps, or a combination of all three, may be used.

As illustrated in Figure 11.13, the glare-free beam mostly depends on the road environment, and these light projections are divided into two categories: low beams

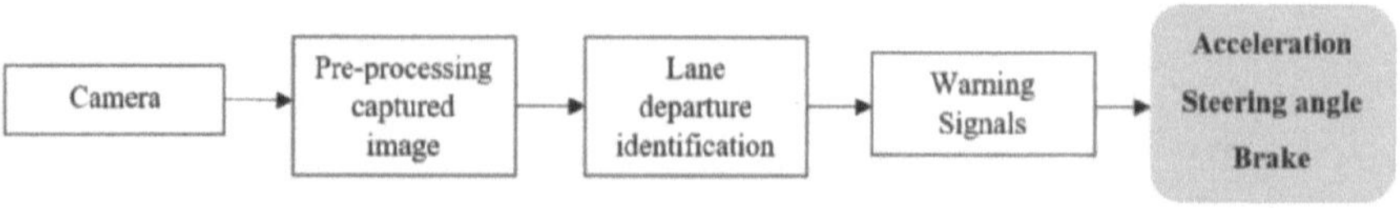

FIGURE 11.12 Block diagram of Lane change assistance system.

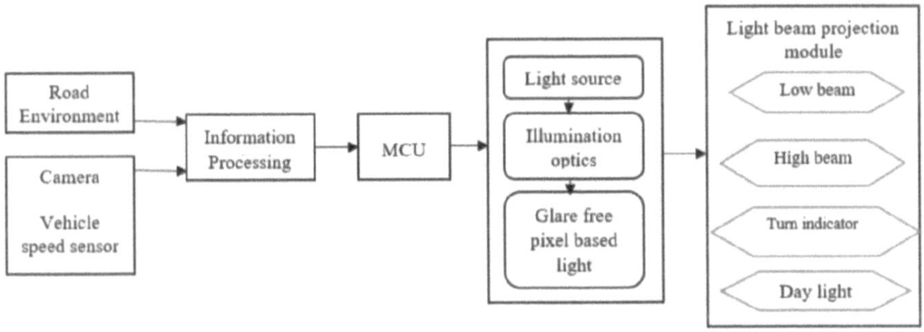

FIGURE 11.13 Block diagram of glare-free high-beam and pixel light.

and high beams. Low beam headlights (also known as dipping, passing, and meeting beams) provide a pattern of light that is intended to give forward and side lighting while limiting the light that is aimed to reduce glare for other road users. This beam is meant to be used if there are any other cars up ahead, whether they are approaching from behind or overtaking. High-beam headlights emit a strong amount of light that is aimed at the eyes of other road users. As a result, they should only be used when driving alone since the brightness they emit will blind other motorists. Due to the continuously moving vehicles that must be shadowed out of the beam, sophisticated sensors, microprocessors, and actuators are needed to change the beam patterns. Moveable shadow masks adjusted inside the headlamp's light stream can be used to create dynamic shadowing. Alternately, the effect may be produced using the pixel light approach, which involves selectively dimming addressable LED emitters or reflector components.

11.7.12 Vehicular Communication System

Typically, the development of vehicular communications is a component of intelligent transportation systems (ITS). Safety and the elimination of the disproportionately high cost of traffic collisions are the primary drivers behind vehicle communication systems. Vehicles and roadside equipment communicate with one another through computer networks known as vehicular communication systems, sharing information like traffic updates and safety alerts. They may be successful in preventing collisions and traffic jams. Emerging vehicular technologies include vehicular communication (VC) systems, active road signs, and toll collecting [38]. They are made up of network nodes, including cars and roadside infrastructure units (RSUs), which have wireless communication, processing, and sensing modules built right into them. There are a number of applications that may be made possible by vehicle-to-vehicle and vehicle-to-infrastructure communication to improve transportation safety.

Vehicle-to-vehicle (V2V) communication allows for the electronic transmission of information regarding a vehicle's position, speed, and direction. Because of the technology enabling V2V communication, cars may transmit and receive Omni-directional signals up to 10 times per second, giving them a 360-degree "knowledge" of other adjacent vehicles. Vehicles equipped with the necessary software may use communications from nearby automobiles to see potential crash hazards as they appear [39]. Then, to alert drivers, the gadget may employ a combination of aural, tactile, and visual alerts or any one of them by itself. To avoid crashes, drivers can respond to these indications. The present accident avoidance systems, which employ radars and cameras to identify potential collisions, are extended and improved through V2V communication.

11.7.13 Traffic Sign Recognition

Due to congestion caused by nearby cars, it is very likely that a driver would miss some traffic signs under the existing traffic management systems. A car can now understand the traffic signs placed on the road thanks to a technology called traffic-sign recognition (TSR) that the working block diagram illustrated in Figure 11.14. Convolutional neural networks are being used to create contemporary traffic sign recognition systems, primarily in response to the needs of autonomous and self-driving automobiles [40]. The detecting system must recognize a range of traffic signals in various settings, not only speed limitations. To find the traffic signs, image processing techniques are used. The three main categories of detection techniques are color-based, shape-based, and learning-based techniques. Many contemporary automobiles, trucks, and vehicles include forward-facing cameras that may be used to evaluate traffic signs.

11.7.14 Automotive Night Vision

Automobile night vision systems employ thermographic cameras that can visualize the forward environment beyond the car headlamp range. This technique is based on night vision optical devices that have been electronically upgraded and work in one of three modes: active lighting, thermal imaging, or picture augmentation. Automotive night vision is divided into two systems: 1) Active systems: Infrared light, which is invisible to humans, is used by active systems to illuminate the road in front of the vehicle. Active systems may be divided into two categories: gated and non-gated. These systems use extra visual data generated by relaying infrared light in the form of a pulse achieving a long range of up to 250 m. This system can provide a high-contrast image of inanimate objects, persons, animals, and the images of the surrounding vehicle. Smaller sensors can able to detect rain and snow through a camera mounted on the rear-view mirrors. 2) Passive systems: These are the system that works on detecting the heat of the object or substance that comes across the path. In passive infrared systems, it uses thermographic cameras to collect the thermal radiation that objects have previously given off.

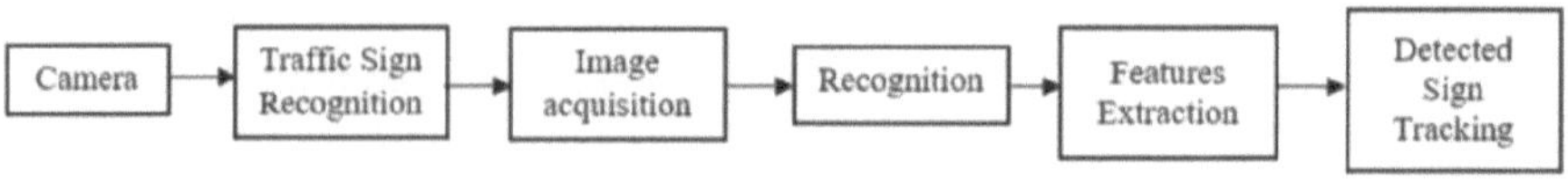

FIGURE 11.14 Block diagram of traffic sign recognition.

11.7.15 Tire Pressure Monitoring

An electronic system that uses sensors to detect the air pressure in a vehicle's pneumatic tires. This system works by using a gauge, a display or a low-pressure warning light when its gets low-pressure or flattened tires. There are two distinct forms of TPMS, (i) direct monitoring systems—direct tire pressure monitoring devices take measurements, detect low pressure, and alert the driver. Direct systems can provide precise alerts and can instantaneously notify the driver if the pressure in any one of the vehicle tires is rapidly decreased or brought on by a puncture since they include sensors in each of the wheels. Direct tire pressure monitoring devices can also spot a slow loss of air over time. (ii) Indirect monitoring systems—these systems compare the rotational speed of one tire to that of a tire located at a different location on the vehicle using the sensors attached to detect wheel speed which helps in the ABS of the vehicle. One tire will roll at a slightly different number of rotations per kilometer than the other three tires if it has low air pressure because its circumference will fluctuate [41]. The vehicle's onboard computer is configured to read the same signal that supports ABS systems and may now also alert the driver when one tire is operating with a lower inflation pressure than the rest.

11.7.16 Blind Spot Detection System

Blind spot monitoring is a sensor system that locates blind spots that are nearby to the vehicle and alerts the driver. Warnings might be tactile, visual, aural, or vibrating. Blind spot monitors could look at more than just the car's sides and back. This system can also warn the driver when oncoming traffic is coming from the sides. The blind spot monitoring system (BSD) uses an IR sensor to find cars in neighboring lanes that the driver might not be able to see in his or her side mirrors. The BSM system warns the motorist that there is a car in the adjacent lane and advises them to proceed with caution while changing lanes. Therefore, these systems have characteristics that improve safety and drivability. BSMs utilize sensors to find cars in neighboring lanes that the driver might not be able to see right away.

11.7.17 Crosswind Stabilization

A very recent and sophisticated driver assistance technology for driving a vehicle in windy situations is called crosswind stabilization (CWS). To calculate how much help to provide the driver in a specific situation, whether it is at various speeds or during turning the CWS employs lateral acceleration, steering angle, and velocity sensors. The following two forms of unequal torque distribution may also contribute significantly to CWS: (i) the car's one-side brakes unevenly. By doing this, the automobile would turn in the direction of the brake wheels. The car will proceed in a straight path as a result of the modest turn's torque distribution. (ii) Applying additional force on a single pair of axles. A wheel will revolve at a greater rpm when a set of axles is given additional power. Another name for this is torque vectoring. Compared to a wheel with a lower rpm, one with a higher rpm will cover a greater distance. The fundamental idea behind CWS is that an unfavorable force (a crosswind) is acted upon by a force acting in the opposite direction of an equal force. CWS may use the force sensor data to correctly aid the driver in a particular scenario.

11.7.18 Lane Departure Warning System

The lane departure warning system (LDWS) is a device that helps to avoid crashes and also to alert the driver when the car starts to depart its lane. This system detects the path of lane to reduce accidents or causes of the collision. As shown in Figure 11.15 the block diagram explains that the camera placed close to the rear-view mirror is used by the lane departure system to identify lane markings. To function effectively, this device keeps an eye on the paint stripes on the left and right sides. If the car changes lanes, the device flashes a warning light or makes a sound or vibration to notify drivers. Only when lane markers are clearly visible on the road will the system work. If the system warns you of a potential lane departure and you don't respond in time, Lane Assist will gently guide you back into the lane.

11.7.19 Forward Collision Warning

The road ahead is scanned by the forward collision warning while you are driving. If you're going to collide with a car that is traveling more slowly or that is stopped, it will alert you. Your car's front sensors are able to gauge how close you are to vehicles in front of you. These are generally radar- or camera-based. If the vehicle has to slow down because of an unexpected change with the cars in front of it, the system alerts the driver with warnings that may include audio, images, vibrations, a rapid brake pulse, or a combination of cautions.

11.7.20 Intersection Assistant

A cutting-edge driving assistance technology called Intersection Assistant keeps an eye on cross traffic at intersections and road junctions. In today's car traffic, intersections are relatively complicated locations. Numerous traffic incidents involving cars interacting in these regions happen as a result of this intricacy. Each driver must evaluate factors such junction geometry, right-of-way laws, and the actions of other drivers; failure to do any one of these responsibilities by the driver might result in an accident. Therefore, in these intricate situations, an intersection aid system needs to take all forms of probable driver errors into account. This makes avoiding crashes at intersections very difficult. The program is an addition to the Rear Traffic Alert

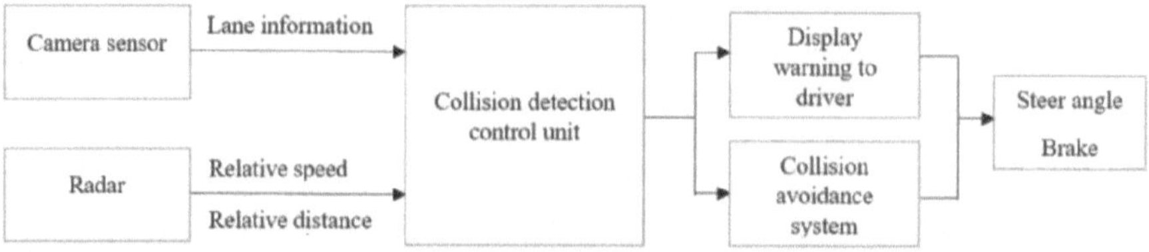

FIGURE 11.15 Block diagram of lane departure warning system.

program. At crossroads, highway exits, and parking lots, Intersection Assistant now alerts drivers when cars are coming from the sides. Some intersection turn assistance systems will use the brakes to stop a collision if these warnings are insufficient to notify a collision.

11.8 ADAS FEATURES IMPROVING DRIVER SAFETY

Reduced accident risk and decreased driver effort are the two goals of a vehicle's driver safety systems. The weather and the circumstances of the vehicle have the most influence on increasing driving safety. For a safer driving experience, data on the road may be continually captured and drivers can get pertinent information through info-cars or intelligent traffic signals. Through driving monitoring, driver conditioning, and driver fitness sense, safety must still be enhanced. The aid capability will only be helpful to the driver for correct and safer driving if the driver is aware that he or she is sleepy. The driver safety system is only a comfort feature that always assumes full responsibility for the acts and thoughts of the driver.

11.8.1 Driver Drowsiness Detection

The purpose of drowsiness detection systems in ADAS is to stop accidents caused by driver inattention, exhaustion, and lack of awareness. To decrease human errors and accidents, a variety of programs and technologies have been developed. These can include automatic emergency braking and alerting drivers to objects in their blind spots. Driver drowsiness detection systems described in Figure 11.16 can employ cameras, eye-tracking sensors, and other hardware to monitor visuals. Drowsiness can be identified by the frequency of yawning, eye blinking, eye-gaze movement, head movement, and facial expressions. The driver's fatigue may be identified with

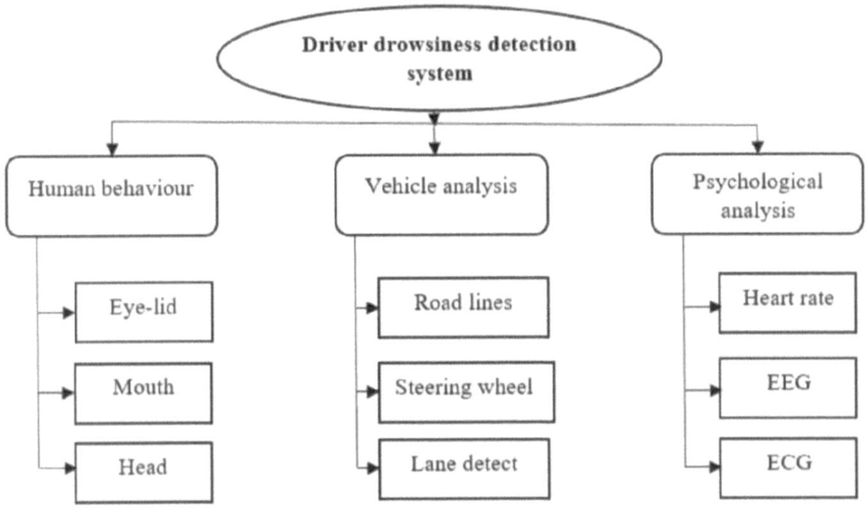

FIGURE 11.16 Block diagram of driver drowsiness detection.

a decent level of accuracy using both direct and indirect monitoring approaches. Psychological performances serve as the foundation for the bulk of direct measurements. More attention is paid to changes in steering and lateral position in the indirect measures. Additionally, the devices may watch for irregular steering, pedal use, and lane departure and alert the driver when necessary. It has been demonstrated that sleepy drivers steer with far higher variability than alert ones.

11.8.2 Driver Monitoring System

By reducing distractions, which are common during night-time and long distance driving, a driver monitoring system helps to reduce the risks. The technology attempts to keep drivers vigilant and conscious of their surroundings, particularly at high speeds when a fleeting mistake might have disastrous consequences. The main cause of traffic deaths is driver distraction; despite the increased technologies that are accessible in a car, the driver needs to be careful about road conditions and traffic. For better comprehension of the driver maneuver, consider their posture, head position, iris and gaze detection, and eye-lid position, as well as their seat belt detection, operational condition (eating or drinking), and driving status (fatigue/tension, drowsiness/arousal, distraction, drinking). Driver monitoring systems aid in making drivers more attentive and aware when operating a vehicle. This technology makes driving safer for everyone on the road by decreasing driver distraction and boosting focus on the road.

11.8.2.1 Future Scope of ADAS

In the previous decade, the automotive and transportation sectors have greatly benefited from ADAS technology, which is largely focused on the driving experience. In the coming years, ADAS technology for autonomous driving will advance to the point where they are almost universal. The large road infrastructure with lane markings, expressways, and well-paved motorways provide a number of difficulties for ADAS when considering the driving circumstances. Future ADAS work will be based on five major trends: embedded vision, sensors, networking, infrastructure for automotive systems, and the development of more adaptable automotive human-machine interfaces (HMIs). The ability of sensor technologies to properly and reliably execute complicated tasks must be improved for assistance systems and fully autonomous driving. The automotive sector will need to modernize its approach to developing new automotive technology that significantly improves the infrastructure of autonomous cars.

11.8.2.2 Conclusion

ADAS is a collective term for the various technological comforts and improvements in modern vehicles. Fully autonomous cars must include ADAS systems because they have the ability to significantly improve the safety and comfort of drivers, pedestrians, and property as well as the efficiency of our vehicles and transportation systems. In this chapter, we learned about the ADAS systems with all the enhanced features and various sensors that are useful in ADAS features of self-driving cars. Even when a car has many improved features, the distraction of the driver leads to accidents, so for a safer driving experience driver needs to be conscious of the sensor technology alerts. To increase dependability and eventually transition to fully autonomous driving, today's vehicle architecture must undergo major modifications of automotive electronics.

REFERENCES

[1] Brunet, J., Da Silva Dias, P., Yahiaoui, G., Real-time driving risk assessment for onboard accident prevention: Application to vocal driving risk assistant, ADAS, and autonomous driving, In: *SIACESA 2018—Electric Components and Systems for Automotive Applications, Components and Systems for Automotive Applications*, pp. 269–274, Springer, New York, 2019.

[2] Reinholtz, C., Hong, D., Wicks, A., Bacha, A., Bauman, C., Faruque, R., Fleming, M., Terwelp, C., Alberi, T., Anderson D., et al., Odin: Team victor tango's entry in the DARPA urban challenge. In: Buehler, M., Iagnemma, K., Singh, S. (eds). *The DARPA Urban Challenge*, pp. 125–162, Springer, Berlin, 2009.

[3] Altendorf, E., Weßel, G., Baltzer, M., Canpolat, Y., Flemisch, F., Joint decision making and cooperative driver-vehicle interaction during critical driving situations. *I-com*, 15(3), 265–281, 2016.

[4] Kukkala, V. K., Tunnell, J., Pasricha, S., Bradley, T., Advanced driver-assistance systems: A path toward autonomous vehicles, *IEEE Consumer Electronics Magazine*, 7(5), 2018.

[5] Voinea, G. D., Postelnicu, C. C., Duguleana, M., Mogan, G. L., Socianu, R., Driving performance and technology acceptance evaluation in real traffic of a smartphone based driver assistance system, *International Journal of Environmental Research and Public Health*, 17(19), 2020.

[6] Winner, H., Hakuli, S., Lotz, F., Singer, C., *Handbook of Driver Assistance Systems Basic Information, Components and Systems for Active Safety and Comfort*, Springer, New York, 2016.

[7] Rahman, M. M., Strawderman, L., Lesch, M. F., Horrey, W. J., Babski-Reeves, K., Garrison, T., Modelling driver acceptance of driver support systems. *Accident Analysis & Prevention*, 121, 134–147, 2018.

[8] Perez, J., Milanes, V., Onieva, E., Combined longitudinal and lateral control for automated vehicle guidance. *Vehicle System Dynamics*, 52(2), 261–279, 2014.

[9] Yang, H., Ciftci, U., Yin, L., Facial expression recognition by de-expression residue learning. In: *Proceedings of the IEEE Conference on Computer Vision and Pattern Recognition (CVPR)*, Salt Lake City, UT, USA, 18–22 June 2018, pp. 2168–2177.

[10] Kukkala, V. K., Tunnell, J., Pasricha, S., Bradley. T., A survey of advanced driver assistance systems and current challenges. *IEEE Consumer Electronics*, 7(5), 18–25, 2018.

[11] Takahama, T., Akasaka, D., Model predictive control approach to design practical adaptive cruise control for traffic jam. *International Journal of Automotive Engineering*, 9, 99–104, 2018.

[12] Inagaki, T., Sheridan, T. B., A critique of the SAE conditional driving automation definition, and analyses of options for improvement, *Cognition, Technology and Work*, 21, pp. 569–578, 2019.

[13] Greenwood, P. M., Lenneman, J. K., Baldwin, C. L., Advanced driver assistance systems (ADAS): Demographics, preferred sources of information, and accuracy of ADAS knowledge. *Transportation Research Part F: Traffic Psychology and Behaviour*, 86, 131–150, 2022. doi:10.1016/j.trf.2021.08.006

[14] Cicchino, J. B., Effectiveness of forward collision warning and autonomous emergency braking systems in reducing front-to-rear crash rates. *Accident Analysis & Prevention*, 99, 142–152, 2017.

[15] Hata, A. Y., Osorio, F. S., Wolf, D. F., Robust curb detection and vehicle localization in urban environments. In: *Proceeding of 2014 IEEE Intelligent Vehicles Symposium Proceedings*, Ypsilanti, MI, USA, 8–11 June 2014, pp. 1257–1262.
[16] Schreiber, M., Knöppel, C., Franke, U., Laneloc: Lane marking based localization using highly accurate maps. In: *Proceedings of the 2013 IEEE Intelligent Vehicles Symposium (IV)*, Gold Coast, Australia, 26–28 June 2013, pp. 449–454.
[17] Abraham, H., Reimer, B., Mehler, B., Learning to use in-vehicle technologies: Consumer preferences and effects on understanding. *Proceedings of the Human Factors and Ergonomics Society Annual Meeting*, 62(1), 1589–1593, 2018.
[18] Benson, A., Teft, B. C., Svancara, A. M. and Horrey, W. J., *Potential Reduction in Crashes, Injuries and Deaths from Large-Scale Deployment of Advanced Driver Assistance Systems*, AAA Foundation for Traffic Safety, Washington, DC, 2018.
[19] Barakat, B., Crump, C., Cades, D., Rauschenberger, R., Schwark, J., Hildebrand, E., Young, D., Eye tracking evaluation of driver visual behavior with a forward collision warning and mitigation system. *Proceedings of the Human Factors and Ergonomics Society Annual Meeting*, 59, 1321–1325, 2015.
[20] Maurer, M., Gerdes, J. C., Lenz, B., Winner, H., *Autonomous Driving: Technical, Legal and Social Aspects*, pp. 439–442, Springer, New York, 2016, doi:10.1007/978-3-662-48847-8.
[21] Niclass, C., Inoue, D., Matsubara, H., Ichikawa, T., Soga, M., Development of automotive LIDAR. *Electronics and Communications in Japan*, 98, 28–33, 2015, doi:10.1002/ecj.11672
[22] Sivaraman, S., Trivedi, M. M., Looking at vehicles on the road: Survey of vision-based vehicle detection, tracking, and behaviour analysis. *IEEE Transactions on Intelligent Transportation Systems*, 14(4), 1773–1795, 2013.
[23] Engelhardt, M., Pfeiffer, F., Biebl, E., A high bandwidth radar target simulator for automotive radar sensors. In: *Proceedings of the 13th European Radar Conference (EuRAD), IEEE*, 245–248, 2016.
[24] Awathe, A., Varunjikar, T., Ganguli, S., *System Identification Techniques to Improve ADAS Feature Performance* (SAE Technical Paper 2022-01-0091) 2022.
[25] Prauzek, M., Konecny, J., Borova, M., Janosova, K., Hlavica, J., Musilek, P., Energy harvesting sources, storage devices and system topologies for environmental wireless sensor networks: A Review. *Sensors*, 18, 2446, 2018.
[26] Thong-un, N., Hirata, S., Orino, Y., Kurosaswa, M. K., A linearization-based method of simultaneous position and velocity measurement using ultrasonic waves. *Sensor and Actuators A*, 233, 480–489, 2015.
[27] Gao, H., Cheng, B., Wang, J., Li, K., Zhao, J., Li, D., Object classification using cnn-based fusion of vision and lidar in autonomous vehicle environment. *IEEE Transactions on Industrial Informatics*, 14(9), 4224–4231, 2018.
[28] Hasirlioglu, S., Kamann, A., Doric, I., Brandmeier, T., Test methodology for rain influence on automotive surround sensors. In: *2016IEEE 19th International Conference on Intelligent Transportation Systems (ITSC)*, 2242–2247, 2016.
[29] Botzer, A., Musicant, O., Perry, A., Driver behaviour with a smartphone collision warning application-A field study. *Safety Science*, 91, 361–372, 2017.
[30] Raffone, E., Fossanetti, M., Rei cEng, C., *A Humanized Vehicle Speed Control to Improve the Acceptance of Automated Longitudinal Control*, *SAE International Journal of Advances and Current Practices in Mobility*, 5(2), 610–621, 2023.
[31] Kuttolamadom, M., Mehrabi, M., *Analysis of Stability of Motion of Intelligent Vehicle Systems* (SAE Technical Paper 2022-01-0096), 2022.
[32] Liu, K., Jia, J., Zuo, Z., Ando, R., Heterogeneity in the effectiveness of cooperative crossing collision prevention systems. *Transportation Research Part C: Emerging Technologies*, 87, 2018, 1–10.

[33] Wang, Z., *Motion Estimation of Connected and Automated Vehicles under Communication Delay and Packet Loss of V2X Communications* (SAE Technical Paper 2021-01-0107), 2021.
[34] Min, K.W., Choi, J.D., Design and implementation of autonomous vehicle valet parking system, In: *Intelligent Transportation Systems (ITSC), 2013 16th International Conference on IEEE*, pp. 2082–2087, IEEE, New York, 2013.
[35] Ziegler, J., Bender, P., Schreiber, M., Making bertha drive-an autonomous journey on a historic route. *IEEE Intelligent Transportation Systems Magazine*, 6(2), 8–20, 2014.
[36] Becker, J., Colas, M. A., Nord, B. S., Bosch's vision and roadmap toward fully autonomous driving. In: Meyer, G., Beiker, S. (eds). *Road Vehicle Automation, Part II*, pp. 49–59, Springer, Switzerland, 2014.
[37] Marino, R., Scalzi, S., Netto, M., Nested PID steering control for lane keeping in autonomous vehicles. *Control Engineering Practice*, 19(12), 1459–1467, 2011.
[38] Varunjikar, T., Awathe, A., Kushwaha, P., Chen, C., *Simulation Study to Evaluate Robustness of the Lane Centering Feature* (SAE Technical Paper 2021-01-0109), 2021.
[39] Abboud, K., Omar, H. A., Zhuang, W., Interworking of DSRC and cellular network technologies for V2X communications: A survey, *IEEE Transactions on Vehicular Technology*, 65(12), pp. 9457–9470, 2016.
[40] Ellahyani, A., El Ansari, M., El Jaafari, I., Traffic sign detection and recognition based on random forests. *Appl Soft Computer*, 46, 805–815, 2016.
[41] Liu, Y., Ball, J., Abdelwahed, S., He, G., *An Automatic Emergency Braking System for Collision Avoidance Assist of Multi-Trailer Vehicle Based on Model Prediction Control* (SAE Technical Paper 2021-01-0117), 2021.

12 An Insight Review on Cognitive Bias for Advanced Driver Assistance System

R. Vidya
SASTRA Deemed to be University

S. Yamini
National Institute of Technology Tiruchirappalli

A. Cyril Thomas
SASTRA Deemed to be University

Sandeep Singh
National Institute of Technology, Puducherry

S. Moses Santhakumar
National Institute of Technology Tiruchirappalli

12.1 INTRODUCTION

Automated vehicles (AV) and advanced driver assistance systems (ADAS) let the drivers to free up the time usually spent monitoring the roads, allowing them to spend their time more efficiently in other relaxing activities during the time usually spent driving. Numerous nations are actively promoting the development of automated driving in terms of laws and regulations, technological advancement, and the building of infrastructure [1]. Driving assistance and driving automation are therefore an important concern in the contemporary world. In recent years, engineering and media's concern has been shifted towards the fast emerging area of advanced driver assistance systems (ADAS). Self-parking, Adaptive Cruise Control (ACC)—maintaining safe distances from vehicles ahead. safety challenges and adaptive techniques are also a significant part in these systems [2]. Driving will be replaced by automation. The mental workload of driver is a vital parameter for the safety of AVs. While ADAS can relieve the driver from continuous responsibilities of driving, it can increase their mental workload giving rise to reactions that are slow and inattentive [3].

DOI: 10.1201/9781003354901-12

This in turn has raised apprehensions concerning the capability of driver to monitor the automation and the performance of driver while intervening in critical situations [4]. Similarly, automated vehicles possess the utmost worth only if they can do activities that are impossible for humans to do. Nevertheless, even in the context of fully automated driving (AD), drivers are considered to have a role if they opt to turn off the automation and assume manual control. Despite the advantages such as environmental friendliness, time-saving, increased mobility and comfort, individuals are often hesitant to embrace new technology Autonomous vehicles are a comparatively new phenomenon, with only limited research being conducted in recent years. The major focus is on the technical aspects, practicability, and the impact of AVs on safety and congestion. However, there is very little research carried out to understand the potential behavioural shifts due to introduction of the autonomous vehicles. The tri-level study conducted by Indiana University to identify the accident causes concluded that human error was the likely or definite cause of over 90% of all crashes. A lot of analogous studies conducted in various parts of the world also arrived at similar conclusions [5]. Supporters of AD also conclude that substituting the driver with state-of-the-art sensing, computing, communication, and robotic-control technologies, would eliminate the influence of human error while driving. However, the human error could also be a result of misperception, misjudgement or biased behaviour of the driver. Even though the technology is assumed error-free, research on driver behaviour and recent experiences with ADAS does not support these assumptions. The data for the complete effect of AD is inadequate since the frequency and severity of crashes are unknown. With the present technology, one cannot evidentially say that a fully automated vehicle will be harmless than an experienced middle-aged driver [6]. The application of AV faces hurdles in many perspectives of which human cognitive bias of driver is of significant importance since the driver is an integrated component of the vehicle and road system. Deeper insight into the literature reveals that bias, outcome and individual reactional behaviour in acceptance of these technologies is different and that the success of the technologies and the former are intertwined and needs exploration. ADAS uses camera, sensors for obstacle detection and response will take place based on that. The improvement of ADAS technology based on the developer is only concentrated on a major scale; driver's safety and behaviour aspects need to be studied for its successful implementation.

This chapter aims to present a comprehensive review of how cognitive biases are linked to these technologies and serve as a relative performance measure for the adoption of these technologies. The importance of human component and supervision and the vitality of their contribution were discussed by numerous researchers [7]. However, driving differs from person to person and at the same for the same driver under varied circumstances thereby the efficiency and adherence need optimisation. Driver error close to 90% had been a major cause of road accidents [8–11]. Improved driving behaviour leads to safety improvement [12–14]. Several human factors explored through numerous researchers and found to cause road accidents. Speed is the major factor contributing to fatal accidents constituting the foremost reason hindering the safety of road. Previous researchers had discussed distraction factors [5, 9, 14], driver behaviours [15–22], bias like trust, confidence [23–25], lane departures vehicles [26].

Non-adaptive systems for collison avoidance with function that are fixed will have issues about trust and performance acceptance [27]. Autonomous cars are a great breakthrough [28], yet the interaction to be established well, has unanswered questions [14], and can have mistrust issues affecting the system functionality [29].

The remainder of this chapter is organised as follows. The Segment 2 of the paper deals with technologies pertaining to ADAS on a broader perspective, Segment 3 presents classification into longitudinal, lateral and combination of lateral and longitudinal driving assistance with discussion of their behavioural driving component, Segment 4 deals with the discussion on cognitive bias and Segment 5 is dedicated to discussions of their interlinking and recommendations thereon.

12.2 ADVANCED DRIVER ASSISTANCE SYSTEMS

ADAS would aid drivers in driving parking operations thereby increasing safety. ADAS integrated with technological advancements either in vehicle or out vehicle mechanisms are carefully coined to aid the driver enabling greater safety thereby aiming to reduce crashes [30]. The primary motive being minimising accidents enhancing safety [4, 31–34]. [35] discussed the importance of Adaptive Cruise Control (ACC) as a milestone in the history of driving assistance. Many objectives are framed to study the car-following system like safety, reaction of driver [36].

There are six levels of automation in driving as per the Society of Automotive Engineers (SAE) scaling from 0 to 5 [37] as presented in Table 12.1.

12.2.1 Classification of Driver Assistance System (DAS)

1. *Passive*: Only warning by DAS
2. *Active*: Operation control (pedal, gear, steering) by driver and/or DAS

TABLE 12.1
Six Levels of Automation in Driving as Per SAE Standards

Level 0	No automation • Lane departure warning (LDW) • Forward collision warning (FCW)
Level 1	Assisted • Lane keeping assistance (LKA) • Adaptive cruise control (ACC)
Level 2	Partial automation • Parking assistance
Level 3	Conditional automation • Traffic jam Chauffeur
Level 4	High automation • Parking garage pilot
Level 5	Full automation • Robot taxi

12.2.2 Classification of DAS Based on Manoeuvre

1. Longitudinal control/warning
2. Lateral control/warning
3. Fusion of longitudinal and lateral control/warning

The various features of ADAS are depicted in (Figure 12.1). The typical view of various features involved in ADAS can be sourced at [38]. The description of features is presented in Table 12.2. They are employed for aiding under varied and risky traffic conditions. The main motive of assistance systems is to give the drivers prior warning with ample time for their reaction to the situation encountered.

12.3 AWARENESS AND ACCEPTANCE OF ADAS IN INDIA

ADAS technology is in an infant stage with low levels 1 and 2 of SAE standards is yet to be implemented in India on a massive scale and is a potential application that can aid in decreasing congestion and collision, thereby enhancing safety. Acceptance and adaptability of drivers to these modern systems in India needs to be studied. Presently, ACC, lane assistance and braking during emergency are only offered in ADAS features due to the limitations in infra sector and less awareness. To project in on a bigger scale and implement on a mass scale is the biggest challenge in Indian

FIGURE 12.1 Features of ADAS.

TABLE 12.2
Features of ADAS

Features	Description
Adaptive cruise control (ACC)	ACC aims for keeping a safe distance from front automobiles by using Radar thereby preventing accidents.
Lane departure warning assistance system (LDA)	It traces the vehicle position probably within the lane and thereby alert the driver of any variation from the lane by taking into account like lateral position, velocity width of lane and road geometry giving a warning in terms of sound or vibration.
Intelligent speed assistance system (ISA)	This figures out vehicle speed pertaining to speed restrictions prescribed and control it in case if it exceeds either by providing warning signals or through automated system bringing back to desired speed.
Braking assistance system (BAS)	Pertaining to the situation, braking assistance system applies brake thereby avoiding collision of vehicles.
Electronic stability control (ESC)	Helps in maintaining the vehicle control during unexpected road geometry or difficult unpredicted situations. This involves continuous monitoring of tyre and steering movement without causing slip.
Adaptive head lights (AHL)	This enables in providing more light on turnings and hills indicating the risks with respect to movement and direction of the steering.
Pedestrian detection system (PDS)	This configures to reduce hit on pedestrians and colliding with them by providing automatic braking.
Blind spot detection system (BSD)	This assist in location of vehicles in blind spot regions and provides visual sign and sounds.
Forward collision warning/mitigation (FCM)	Alerting of driver in terms of flashing light, beep or vibration is done by the system when there would be probable cases of collision of vehicles by maintain safe distance required. In case of Mitigation system, the vehicles moving ahead are found and tends to apply the brake in case of no response from the driver.
Driver drowsiness detection (DDD)	This helps in detecting the fatigue condition of driver or his drowsiness based on visual features of drivers and prevents accidents.
Traffic sign recognition (TSR)	This will identify different road traffic signs and aid the driver for smooth driving operations.
Driver monitoring system (DMS)	This makes uses of sensors to find the alertness of driver mainly through tracking the eye of driver with the help of detectors, giving indication to them in terms of lights or sounds when difficult conditions are met. In case the driver is not responding brakes are applied automatically.

industry. Indian players need to study and optimise the technical learning capabilities owing to Indian Conditions. This can be helpful for global players and suppliers to explore the possibilities of the Indian market and integrate

12.3.1 ADAS Indian Players

The Indian market pertaining to ADAS will experience significant growth in another 10 years. The increasing demand for vehicles coupled with higher features can be a positive factor for its market. Further owing to the increasing fatalities on road, Government of India had planned to improve the safety by making the ADAS Integration mandatory in commercial vehicles both light and heavy that will be sold after 2022. Figure 12.2 shows the Indian companies working in ADAS.

Robert Bosch Engineering Company
- Driver assistance systems
- Automatic emergency braking
- Predictive pedestrian protection system with built-in emergency stop

Denso India Pvt. Ltd.
- Automotive Technologies primarily focused on Environment
- Safety and Convenience
- To support OE Customers especially for Indian terrain & weather conditions

Wabco India
- Collision Mitigation System
- Active braking on moving and stopping Vehicles

Continental Automotive Components (India) Pvt. Ltd.
- Long Range and Short-Range Radars
- Front, Driver Monitoring and Satellite Cameras
- Automated Driving Control Units

Mando Automotive India Pvt. Ltd.
- To provide AEB radar and camera to Mahindra
- Developing devices to detect large animals

Delphi Automotive
- ADAS Equipment
- DS diagnostic software with integrated OEM

Infineon Technologies
End-to-end automotive data connectivity and sensor

Hella India
- Body Electronics and Energy Management
- Steering systems and Sensors
- Actuators to the passenger and commercial vehicle segments

FIGURE 12.2 List of Indian companies working in ADAS.

12.3.2 Driver Operational and Behaviour Adaptation for DAS

Controls on lateral and longitudinal behaviour are closely linked while the former is closely linked with full automation of driverl whereas the latter is associated with normal category [39]. Entrusting vehicle control of one or both will have an impact on driving behaviours for resuming back manually. Longitudinal control is better for dealing automatic deceleration failures in comparison with combined control [40]. The studies on longitudinal, lateral and combined longitudinal and lateral Manoeuvre are discussed in Tables 12.3–12.5 respectively [41–54].

12.4 COGNITIVE BEHAVIOUR ADAPTATION

Human behaviour often inhibits to adapt new technology even if that technology provides demonstrable benefits. The automotive industry is in the early stages of a potentially transformative evolution where personally owned, driver-driven vehicles likely will travel alongside shared and self-driving cars. Cognitive bias of driver plays a dominant part in driving automated systems. The efficiency of driving might vary from person to person and also for the same driver it varies depending on the situations, which implies that the productivity and obedience can be improved. The studies of traffic safety and human behaviour have emphasised that speed is the first contributor to fatal crashes [55].

Even though the DAS is developed along with driving behaviour adaptation, it should be monitored or the driver usage of the assistance system should be measured.

TABLE 12.3
Longitudinal Manoeuvre

Parameter Studied	Author	Summary
Individual drivers' usual speeding tendencies (intelligent speed adaptation)	[41]	The impact of intelligent speed adaptation (ISA) was examined using a hierarchical Bayesian model wherein random effects was correlated and found that speeding tendency of driver and characteristic conditions of road as confounding factors
Varying time headway (rear-end collision avoidance system)	[42]	Assessment index developed to avoid rear end collision for commercial heavy vehicle in mixed traffic conditions.
Collision time and headway (time) for collision avoidance system (CAS)	[43]	Optimisation done by algorithms that are personalised for the behaviour of road axis longitudinally.
Car-following style—individual (ACC)	[44]	Algorithm for car-following style learning developed to predict how close will the vehicle be.
Pedal deflection timing during approach (ACC/FCC)	[45]	Forward driving assistance system pertaining to dynamic behaviour of driver

TABLE 12.4
Lateral Manoeuvre

Parameter Studied	Author	Summary
Time to lane crossing, future offset distance and steering behaviour (lane departure warning system)	[46]	Algorithm for multimode TLC and FOD based on driver behaviour characteristics for lane change behaviour and how the driver makes steering movement.
Lane change assistance	[47]	Response during lane changes to capture longitudinal position and speed before manoeuvre, gap acceptance.
Driver's ability and Time to collision (lane change assistance)	[25]	Collision avoidance systems for hazardous lane changes
Preview time and assistance gain (steering assistance system)	[48]	Driver's steering behaviour, developed capable of varying the preview time depending on situation.

TABLE 12.5
Combined Longitudinal and Lateral Manoeuvre

Parameter Studied	Author	Summary
Acceleration, RPM, gravity and speed of throttle (driver monitoring system)	[49]	Classified into Normal, drowsy, drunk, distracted, aggressive and for monitoring driver and reducing accident risk. 2D convolutional neural network (CNN) developed
Passing collision warning system	[50]	Passing manoeuvres studied PCWS developed for drivers to select proper passing gaps.
Individual safe inter-vehicle distances (lane change assistance with ACC)	[51]	Identification of driver behaviour characteristics in real-time individualities for both lateral and longitudinal maneuverers.
Pedal operation (personalised driver assistance system)	[52]	Recordings—audio and video
Manual parameters to be fine-tuned by the driver (autopilot)	[53]	References of driver for a particular trip
Age-related changes (semi-autonomous vehicle	[54]	Older ageing drivers with cognitive challenges and impact on driving capacity.

According to the usage, the system automatically needs to adjust the parameters. The future vision will impact regulatory and technological advances that in turn will affect consumer expectations and cause behaviour shift. Because, even if the benefits of a world of shared self-driving cars seem self-evident, companies should not assume that consumers will reach a similar conclusion. Peoples' cognitive biases suggest many individuals may be reluctant to relinquish their personally owned and driver-driven vehicles. The entire process requires not only an accurate and complex AD system but also the cooperation and adaptability of human drivers [56]. Improved driving behaviour leads to safety improvement [57]. Deeper insight into the literature reveals that bias, outcome and individual reactional behaviour in acceptance of these technologies are different and that the success of the technologies used in ADAS needs exploration. This section aims to present a comprehensive review of how cognitive biases impact these technologies and serve as a relative performance measure for the adoption of these technologies.

Autonomous vehicles are yet to reach human-level perception [58]. For this, we need to know about cognitive bias of the driver for the acceptance of the technology development, legal challenges [59], the human factors and ergonomics [60–63]. AD systems pose many challenges to the driver in the forms of technical, moral, and legal fronts which may impose personal and economic risks to the users [64]. Behavioural economics and psychology (social) research shows profound and consistent biases that may not lead to optimal choices. It also led to the discovery of possible ways to overcome these mental constraints, developing new options of mobility.

The success of the artificial intelligent systems will also depend on the decision biases and user preferences in using ADAS, why people trust ADAS, importance of trust and moral dilemmas in AV [65] for keeping the drivers engaged. Trust is a versatile idea that can specify to believe that others will show the same behaviour with compassion, honesty, certainty, and skill [66]. [67] have postulated in their reviews that trust is a mediating aspect between the assumption of automated features and the drivers will depend on the automated technology. Any form of error will cause distrust.

12.4.1 Integration of Cognitive Bias and ADAS

The insight of people about the autonomous driving system and their benefits and risks influences their acceptance of the model [68–69]. Cognition is an imperative factor that influences the attitude and decision-making towards evolving technologies. In psychology and cognitive science, cognition denotes the process of an individual's sensory input being transformed, managed and used for different stimuli. The mental perspective of the driver is a significant parameter to be considered while driving autonomous vehicles [70]. To be specific, how impressions, emotions, information processing, repetition of task, perception, intuition, change of environment, current mind-set, opinions, ease of actions, nearest availability of access, memory, time availability, illusions, loss aversion and risk aversion would impact the driving behaviour is often being overlooked.

Understanding these challenges would help in better and safe designing of ADAS systems and should be incorporated in system design itself (Figure 12.3). Most of

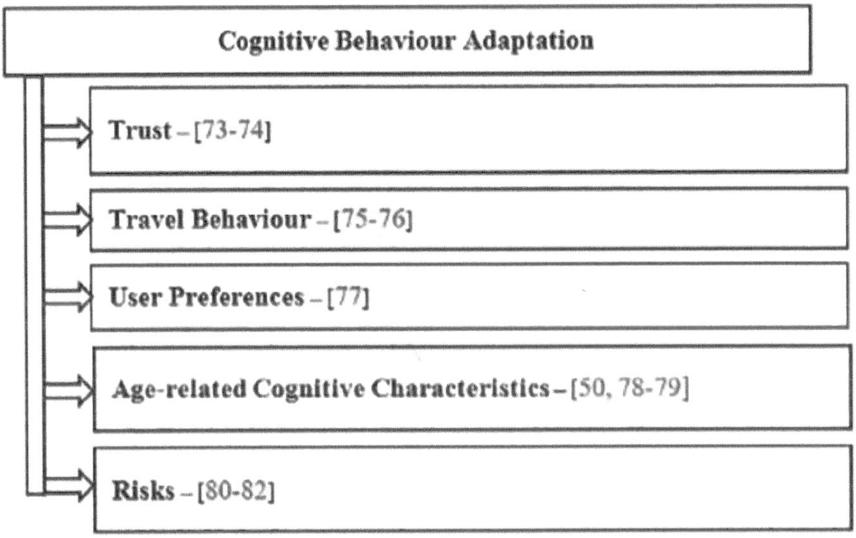

FIGURE 12.3 Typical view of the integration of cognitive bias and ADAS technology.

the times, driver error is quoted as a cause. AD systems suggest void of human will abbreviate the crash occurrence by 90%. They assume that sensors, computation, and control technologies substituting the driver would reduce the human error. A careful examination of this presumption clarifies that people think that technology is error-free and error of human are due to not properly understanding or unsuitable driver behaviour, Unfortunately, research on driver behaviour is not backing these assumptions.

Although the common reason for driver error quoted in the literature are fatigue, speeding and distraction; it does not talk more about the cognitive biases of human influencing their driving behaviour [71]. We should also note that although automation can be helpful in detecting and controlling the automobiles, we could not neglect driver's violations. The reason for exhibiting these types of actions may be due to conformity bias where one driver just follows the action of other drivers in order to fit with the group and conform their driving with others [72]. Most of the time we can notice that people blindly follow the actions of others in traffic signals that might also lead to accidents.

Drivers' decisions in traffic are mostly governed by their beliefs regarding the expected behaviour of fellow drivers. Typical directions of rules are followed by the drivers; These unsuitable beliefs give rise to negative results. A vital analysis of accidents or any disastrous situation may not only occur due to failures (software) in vehicles in moving traffic but also due to driver behaviour. The crash experience with automated vehicles are high in number compared to conventional vehicles. Even in fully AD systems, the assumptions made by the system about other drivers will be wrong and they cannot consider the behaviour of human drivers while making such assumptions. In summary, neither the technology nor human will be foreseeing the mentality of road users.

A cognitive load of driver needs to be tracked through many assessment techniques that include signals of physiological nature, and driving behavioural measures. Tracking physiological signals is very vital for assessing cognitive outcomes. Physiological monitoring can also be conducted. In automated vehicles, many features like self-parking, automatic lateral control device, cruise control adaption are available [2]. Further the ways in which drivers process the information also play a major role in understanding whether these ADAS will be error-free. For instance, if the driver decides by overweighing the information, which is easily available to them, or easily recallable to them from recent memory, then it would lead to a biased decision and might also lead to potential consequences such as accidents and crashes.

Implementation of automation in driving creates many ethical issues. An ADAS-featured vehicle's driving mainly depends upon the pre-installed AI and ML algorithms. The AI and ML algorithms do not take moral decisions in critical situations. A delayed decision during a critical situation in AD causes major accidents. In recent years, most of the automation industries are focusing on the manufacturing of ADAS-supported cars. The number of accidents caused by the ADAS programmed cars increased notably. According to the Reuters article, ADAS featured car (autopilot) involved 11 fatal deaths in 33 accidents since 2016. In 2022, United States Safety Regulators released the accident statistics of partially and fully automated vehicles. Out of 400 cars involved in the crashes, 273 cars have ADAS features. Recently, an ADAS-featured car erroneously detect horse chariot as a truck. This incident shows a major design flaw in the ADAS system [73–82].

When the automation is very progressive, the role of driver will also be challenging. Since the driver will be monitoring a system excluding the mental model as most of the system interfaces do not have enough information required by operators to effectively understand system operations [83]. While monitoring many tasks together, the response time of driver may increase as the cognitive workload is high. Some people can only focus on a particular task as their attention span is very limited. When the driver lacks practice or situational awareness during the disengagement of automation, it might also lead to accidents. Having an experience in automation driving can help the drivers to execute tasks at a low knowledge level, decrease the difficulty in understanding the traffic situation and better monitoring [84–86].

Cognitive load detection is vital for ADAS. Cognitive load if high will result in mental fatigue which reduces the driver's ability to drive with precaution. They can result in dangerous conditions for co-passengers on the road and the driver himself. The environment also will play a pivotal role in ADAS limitations that includes conditions of road and lane, lighting as well as weather. Raindrops detection and rain speed can be used to adjust the speed of wipers to get a good view for drivers when inclement weather conditions prevail [87–88].

12.5 CONCLUDING REMARKS AND FUTURE DIRECTIONS

Technology is a common substitute for humankind. The mental workload of driver is to be considered regarding safety aspects. Automated driver assistance systems are also growing at a faster rate as they are considered to minimise the work load of drivers and reduce the errors that are caused by them. The technicalities and adoption of ADAS will alone not be enough for the success of the intelligent systems, it also has been based on driver using these systems. The emotional co-efficient is to be taken for cognitive assessments. The individuals subjected to anger before driver are liable to more risk [89]. Anger is the major perspective that will lead to rash driving. This comes into picture when the comparison is made. Optimistic bias [90] and false idea of notion [91] were identified to be the significant factors in influencing the risk-taking behaviour of drivers. Drunk and drive will create risks [92–95]. We can observe a lot of risk-taking behaviour in violent situations. The unrealistic positivity of drivers will reduce the fear of negative consequences.

Along with the rise of automation, there is an extensive literature, which also widely talks about the human-automation interaction. While ADAS can relieve the driver from the continuous responsibilities of driving, it can increase their mental workload leading to dizzy and slow driving [3]. This in turn has raised apprehensions concerning the ability to track the automation and the behaviour of driver while intervening in critical situations [96].

12.6 FUTURE DIRECTIONS

The driving process can be simplified by having favourable and assuring features offering safety enhancement. Therefore, the distraction of driver and inadvertence can be reduced. This can lead to safer travel combatting fatal accidents state-of-the-art technologies with diverse and advanced qualities can be developed to understand the future key trends. Human-machine interface with touchscreen technology may help drivers and can offer better solutions for improved driving. Improved driver monitoring architecture to be incorporated in designing vehicle system by integrating with smart sensors. Real-time systems, development tools are to be designed owing to ISO safety standards. Vehicles can be equipped with many processors for different operations necessitating a wide range of sensors. The loss of friction can be predicted by the integration of data from acceleration sensors, tyre-pressure sensors, electronic stability control and anti-lock braking system. Identification and tracking of hazard is very vital and serve as input for alerting the drivers. Sensor-based improved vision systems can help in lane change, unidentified vehicles, traffic sign recognition and parking services. In addition to warning the driver of his mistake, these vision systems with sensors integration can aid in decision-making and help to act decisively to prevent accidents and further damage. They serve as an integral part of both in-vehicle and external connectivity. They provide compliance for normal automobile applications. They inform the drivers conditions of vehicle and road. IoT (Internet of Things) can integrate vehicles communication systems with smart vehicles and offer greater safety alerting the driver. These systems have to be adopted on a large scale and are still under research. Further, this also necessitates meaningful research in this regard. This can encounter challenges that require fast, reliable and correct judgement by machines. Further, this needs to be linked with drivers' expertise to machine cognition. This necessitates multidisciplinary research. Several physiological measurements should be measured for the detection of driver's engagement mentally. Assistance in action to be offered which has complexity at several levels and vehicle—driver interaction. simple level lateral steer is found in many vehicles [96–97]. These systems conciliate the steering control flaws wherein disturbances and inaccuracies associated with human driving behaviours can be reduced. Driver behaviour changes are also to be taken [98]. Rear-view cameras are an example of inside–vehicle screens. They deliver information that is invisible. The infrared cameras can be located in front of driver and can show the front scenario of vehicle and can alert the driver. The outside-vehicle lighting systems can sense the possible hindrances for the drivers. The latter systems will make driving simple [98–101]. Single graphical user interface will have concerns of usage and challenges in human driver ergonomics [102–104]. Further driving conditions can be different between driving aspects of city driving and highway [105]. Designing such an adaptive interface incorporating the diversities will be an open challenge.

The problems encountered due to the cognitive part of driver while adapting these autonomous vehicles, has to be given due importance since it is new to the driver or something which he/she is not used to. This chapter draws attention to review the impact of driver behaviour on adaptation of ADAS. It also signifies how the cognitive biases, heuristics and behavioural instincts of a driver would resist him/her to adopt these ADAS technologies. Acceptance of ADAS depends on technology and driver behaviour. To accept the system–cognitive behaviour adaptation is vital. Therefore, the former and latter needs to be integrated for successful implementation and further efforts in future to be designed based on driver's personalisation factors. Future studies also must accommodate the associated circumstances of cognitive biases and risky driving behaviours. Not only this from future perspective, the interactions of AVs with human-operated vehicles also need to be studied and what can be the effects on traffic safety. Future research must address how these interactions can be impacted at various levels of automation.

REFERENCES

[1] Z. Yu, K. Jiang, Z. Huang, P. Zhang, Z. Feng, Will I start an automated driving system? Report on the emotions, cognition, and intention of drivers who experienced real-world conditional automated driving. *Cognition, Technology and Work*, 24(4) (2022) 641–666. doi:10.1007/s10111-022-00706-2.

[2] P. Zmud, Towards an understanding of the travel behavior impact of autonomous vehicles. *Transportation Research Procedia*, 25 (2017) 2500–2519. doi:10.1016/j.trpro.2017.05.281.

[3] E.T. Greenlee, P.R. DeLucia, D.C. Newton, Driver vigilance in automated vehicles: Hazard detection failures are a matter of time. *Human Factors and Ergonomics Society*, (2018) 18720818761711. doi:10.1177/0018720818761711.

[4] M. Kyriakidis, C. Weijer, B. Arem, R. Happee, The deployment of advanced driver assistance systems in Europe. In: *22nd Intelligent Transportation Systems World Congress*, 2015. doi: http://dx.doi.org/10.2139/ssrn.2559034.

[5] M. Kyriakidis, J.C. de Winter, N. Stanton, T. Bellet, B. Van Arem, K. Brookhuis, R. Happee, A human factors perspective on automated driving. *Theoretical Issues in Ergonomics Science*, 20(3) (2019) 223–249. doi:10.1080/1463922X.2017.1293187.

[6] T.A. Dingus, F. Guo, S. Lee, J.F. Antin, M. Perez, M. Buchanan-King, J. Hankey, Driver crash risk factors and prevalence evaluation using naturalistic driving data. *Proceedings of the National Academy of Sciences*, 113(10) (2016) 2636–2641. doi:10.1073/pnas.1513271113.

[7] T.A. Dingus, S.G. Klauer, V.L. Neale, A. Petersen, S. E Lee, J.D. Sudweeks, M.A. Perez, J. Hankey, D.J. Ramsey, S. Gupta, S, C. Bucher, Z.R. Doerzaph, J. Jermeland, R.R. Knipling, *The 100-Car Naturalistic Driving Study, Phase II: Results of the 100-Car Field Experiment.* Virginia Polytechnic Institute and State University, Blacksburg; National Highway Traffic Safety Administration, Washington, USA, 2006.

[8] M. Sivak, B. Schoettle, *Road Safety With Self-Driving Vehicles: General Limitations and Road Sharing with Conventional Vehicles.* Transportation Research Institute, University of Michigan, Ann Arbor, MI, 2015.

[9] L. Bainbridge, Ironies of automation. *Automatica*, 19(6) 775–779 (1983) doi:10.1016/0005-1098(83)90046-8.

[10] National Center for Statistics and Analysis (NHTSA), *An Examination of Driver Distraction.* NHTSA, Washington, DC, 2009, https://perma.cc/LLH5-CECY (accessed on February 15, 2023).

[11] R.D. Robertson, S.R. Meister, W.G.M. Vanlaar, H.M. Mainegra, Automated vehicles and behavioural adaptation in Canada. *Transportation Research Part A: Policy and Practice*, 104 (2017) 50–57. doi:10.1016/j.tra.2017.08.005.

[12] F. Jiménez, F. Aparicio, Aportación de los ITS a la sostenibilidad y mejora del transporte por carretera. *Revista DYNA*, 83(7) (2008) 434–439.

[13] G. Silberg, *Self-Driving Cars: Are We Ready?* KPMG LLP, Amstelveen, Netherlands, 2013, https://perma.cc/526Y-DG9K (accessed on February 15, 2023).

[14] W. Knight, Driverless cars are further away than you think. *MIT Technology Review*, https://perma.cc/49L6-HU8E (accessed on February 15, 2023).

[15] J.D. Lee, K.A. See, Trust in automation: Designing for appropriate reliance. *Human Factors*, 46(1) (2004) 50–80. doi:10.1518/hfes.46.1.50_30392.

[16] M.S. Young, N.A. Stanton, What's skill got to do with it? Vehicle automation and driver mental workload. *Ergonomics*, 50(8) (2007) 1324–1339. doi:10.1080/00140130701318855.

[17] A.H. Jamson, F.C. Lai, O.M. Carsten, Potential benefits of an adaptive forward collision warning system. *Transportation Research Part C: Emerging, Technologies*, 16(4) (2008) 471–484. doi:10.1016/j.trc.2007.09.003.

[18] H. Farah, H.N. Koutsopoulos, Do cooperative systems make drivers' car-following behavior safer? *Transportation Research Part C: Emerging Technologies*, 41 (2014) 61–72. doi:10.1016/j.trc.2014.01.015.

[19] A. Várhelyi, C. Kaufmann, A. Persson, User-related assessment of a driver assistance system for continuous support: A field trial. *Transportation Research Part F: Traffic Psychology and Behaviour*, 30 (2015) 128–144. doi:10.1016/j.trf.2015.02.008.

[20] W. Rongben, G. Lie, T. Bingliang, J. Lisheng, Monitoring mouth movement for driver fatigue or distraction with one camera. In: *Proceedings of the 7th International IEEE Conference on Intelligent Transportation Systems (ITSC '04)*, 2004, pp. 314–319. doi:10.1109/ITSC.2004.1398917.

[21] L.M. Bergasa, J. Nuevo, M.A. Sotelo, R. Barea, M.E. Lopez, Real-time system for monitoring driver vigilance, *IEEE Transactions on Intelligent Transportation Systems*, 7(1) (2006) 63–77. doi:10.1109/TITS.2006.869598.

[22] Z. Tao, The study of fatigue detection system of drivers based on the eye status. Zhejiang Sci-Tech University, Hangzhou, China, Ph.D. thesis, 2012.

[23] J.D. Lee, N. Moray, Trust, self-confidence, and operators' adaptation to automation. *International Journal of Human-Computer Studies*, 40 (1994) 153–184. doi:10.1006/ijhc.1994.1007.

[24] R. Parasuraman, V. Riley, Humans and automation: use, misuse, disuse, abuse. *Human Factors: The Journal of the Human Factors and Ergonomics Society*, 39(2) (1997) 230–253. doi:10.1518/001872097778543886.

[25] H. H. Tan, C.S. Tan, Toward the differentiation of trust in supervisor and trust in organization. *Genetic, Social, and General Psychology Monographs*, 126(241) (2000) TM-97-C057.

[26] D. Shinar, E. Schechtman, Headway feedback improves inter vehicular distance: A field study. *Human Factors: The Journal of the Human Factors and Ergonomics Society*, 44(3) (2002) 474–481. doi:10.1518/0018720024497682.

[27] H. Muslim, M. Itoh, Trust and acceptance of adaptive and conventional collision avoidance systems. *IFAC-Papers on Line*, 52(19) (2019) 55–60. doi:10.1016/j.ifacol.2019.12.086.

[28] G. Silberg, R. Wallace, *Self-driving Cars: The Next Revolution*. KPMG LLP and Center for Automotive Research, Amstelveen, Netherlands, 2012, https://perma.cc/7GN8-DPYB (accessed on accessed on February 15, 2023).

[29] P. Leonardi, Why do people reject new technologies and stymie organizational changes of which they are in favor? Exploring misalignments between social interactions and materiality. *Human Communication Research*, 35 (2009) 407–441. doi:10.1111/j.1468-2958.2009.01357.x.

[30] K.A. Brookhuis, D. De Waard, W.H. Janssen, Behavioural impacts of advanced driver assistance systems-an overview. *European Journal of Transport and Infrastructure Research*, 1(3) (2001) 245–253. doi:10.18757/ejtir.2001.1.3.3667.

[31] P. Gao, R. Hensley, A. Zielke, A roadmap to the future for the auto industry. *McKinsey Quarterly*, https://perma.cc/2HWT-WHHG

[32] V. Prathamesh, S.P. Wadkar, Need of advanced driver assistance systems. *International Journal of Current Engineering and Technology*, 7 (2017) 167–169.

[33] V. Rajaram, S.C., Subramanian, Heavy vehicle collision avoidance control in heterogeneous traffic using varying time headway. *Mechatronics*, 50 (2018) 328–340. doi:10.1016/j.mechatronics.2017.11.010.

[34] V. Lucie, M. Šucha, L. Viktorová, Driver's acceptance of advanced driver assistance systems: What to consider? *International Journal for Traffic and Transport Engineering*, 8(3) (2018) 320–333. doi:10.7708/ijtte.2018.8(3).06.

[35] K. Bengler, K. Dietmayer, B. Farber, M. Maurer, C. Stiller, H. Winner, Three decades of driver assistance. *IEEE Intelligent Transportation Systems Magazine*, 6(4) (2014) 6–22. doi:10.1109/MITS.2014.2336271.

[36] S.B. Li, K.Q. Li, Rajamani, R.J.Q. Wang, Model predictive multi-objective vehicular adaptive cruise control. *IEEE Transactions on Control Systems Technology*, 19(3) (2010) 556–566. doi:10.1109/TCST.2010.2049203.

[37] SAE, *Surface Vehicle Recommended Practice: Taxonomy and Definitions for Terms Related to Driving Automation Systems for On-road Motor Vehicles* (Standard J3016). On-Road Automated Vehicles Standards Committee, SAE International, Warrendale, PA, https://www.sae.org/standards/content/j3016_201806/ (accessed accessed on February 15, 2023).

[38] Synopsys, *What is ADAS?* Synopsys Solutions, Sunnyvale, CA, 2023, https://perma.cc/7PYA-SEZP (accessed on February 21, 2023).

[39] O. Carsten, F.C.H. Lai, Y. Barnard, A.H. Jamson, N. Merat, Control task substitution in semi-automated driving: Does it matter what aspects are automated? *Human Factors: The Journal of the Human Factors and Ergonomics Society*, 54 (2012) 747–761. doi:10.1177/0018720812460246.

[40] N. Strand, J. Nilsson, I.C.M. Karlsson, L. Nilsson, Semi-automated versus highly automated driving in critical situations caused by automation failures. *Transportation Research Part F: Traffic Psychology and Behaviour*, 27 (2014) 218–228. doi:10.1016/j.trf.2014.04.005.

[41] K. Matsuo, M. Sugihara, M. Yamazaki, Y. Mimura, J. Yang, K. Kanno, N. Sugiki, Hierarchical Bayesian modeling to evaluate the impacts of intelligent speed adaptation considering individuals' usual speeding tendencies: A correlated random parameters approach. *Analytic Methods in Accident Research*, 27 (2020) 100125 doi:10.1016/j.amar.2020.100125.
[42] V. Rajaram, S.C. Subramanian, Heavy vehicle collision avoidance control in heterogeneous traffic using varying time headway. *Mechatronics*, 50 (2018) 328–340. doi:10.1016/j.mechatronics.2017.11.010.
[43] M.C. Panou. Intelligent personalized ADAS warnings. *European Transport Research Review*, 10(2) (2018) 59. doi:10.1186/s12544-018-0324-6.
[44] H. Chu, L. Guo, Y. Yan, B. Gao, B, H. Chen, Self-learning optimal cruise control based on individual car-following style. *IEEE Transactions on Intelligent Transportation Systems*, 22(10) (2021) 6622–6633. doi:10.1109/TITS.2020.2981493.
[45] J. Wang, L. Zhang, D. Zhang, K. Li, An adaptive longitudinal driving assistance system based on driver characteristics. *IEEE Transactions on Intelligent Transportation Systems*, 14(1) (2012) 1–12. doi:10.1109/TITS.2012.2205143.
[46] H. Zheng, M. Zhao, An investigation on coordination of lane departure 1warning based on driver behaviour characteristics. *International Journal of Vehicle Autonomous Systems*, 15(1) (2020) 77–99. doi:10.1504/IJVAS.2020.104823.
[47] V.A. Butakov, P. Ioannou, Personalized driver/vehicle lane change models for ADAS. *IEEE Transactions on Vehicular Technology*, 64(10) (2014) 4422–4431. doi:10.1109/TVT.2014.2369522.
[48] K. Yang, Y. Liu, X. Na, X. He, Y.J. Wu, X. Ji, Preview-scheduled steering assistance control for co-piloting vehicle: a human-like methodology. *Vehicle System Dynamics*, 58(4) (2020) 518–544. doi:10.1080/00423114.2019.1590607.
[49] M. Shahverdy, M. Fathy, R. Berangi, M. Sabokrou, Driver behavior detection and classification using deep convolutional neural networks. *Expert Systems with Applications*, 149 (2020) 113240. doi:10.1016/j.eswa.2020.113240.
[50] U. Hassein, M. Diachuk, S. Easa, Development of a passing collision warning system prototype for overtaking trucks on two-lane highways. In: *Conference: 2018 Canadian Society for Civil Engineering Annual Conference at* New Brunswick, Fredericton. Building Tomorrow's Society, Fredericton, 2018, TR18-1–TR18-11.
[51] B. Zhu, Z. Liu, J. Zhao, W. Deng, *Driver Behavior Characteristics Identification Strategy for Adaptive Cruise Control System With Lane Change Assistance* (SAE Technical Paper, No. 2017-01-0432), 2017. doi:10.4271/2017-01-0432.
[52] I. Del Campo, V. Martínez, J. Echanobe, E. Asua, R. Finker, K.A. Basterretxea, Versatile hardware/software platform for personalized driver assistance based on online sequential extreme learning machines. *Neural Computing and Applications*, 31(12) (2019) 8871–8886. doi:10.1007/s00521-019-04386-4.
[53] V. Butakov, P. Ioannou, Driving autopilot with personalization feature for improved safety and comfort. In: *IEEE 18th International Conference on Intelligent Transportation Systems*, 2015, pp. 387–393. doi:10.1109/ITSC.2015.72.
[54] F. Knoefel, B. Wallace, R. Goubran, I. Sabra, S. Marshall, Semi-autonomous vehicles as a cognitive assistive device for older adults. *Geriatrics*, 4 (2019) 1–13. doi:10.3390/geriatrics4040063.
[55] A.C.E. Spek, P.A. Wieringa, W.H. Janssen, Intersection approach speed and accident probability. *Transportation Research Part F: Traffic Psychology and Behaviour*, 9(2) (2006) 155–171. doi:10.1016/j.trf.2005.10.001.
[56] C. Gold, M. Körber, C. Hohenberger, D. Lechner, K. Bengler, Trust in automation-before and after the experience of take-over scenarios in a highly automated vehicle. *Procedia Manufacturing*, 3 (2015) 3025–3032. doi:10.1016/j.promfg.2015.07.847.

[57] F. Jimenez, F. Aparicio, J. Paez, Evaluation of in-vehicle dynamic speed assistance in Spain: algorithm and driver behaviour. *IET Intelligent Transport Systems*, 2(2) (2008) 132–142. doi:10.1049/iet-its:20070019.

[58] W. Schwarting, M.J. Alonso, R. Daniela, Planning and decision-making for autonomous vehicles. *Annual Review of Control, Robotics, and Autonomous Systems*, 1 (2018) 187–210. doi:10.1146/annurev-control-060117-105157.

[59] M. Maurer, J.C. Gerdes, B. Lenz, H. Winner, *Autonomous Driving: Technical, Legal and Social Aspects*. Springer, New York, 2016.

[60] J.M. Lockhart, M.H. Strub, J.K. Hawley, L.A. Tapia, Automation and supervisory control: a perspective on human performance, training, and performance aiding. *Proceedings of the 37th Annual Meeting of the Human Factors and Ergonomics Society*, 37(18) (1993) 1211–1215. doi:10.1177/154193129303701802.

[61] P. Carayon, Human factors of complex sociotechnical systems. *Applied Ergonomics*, 37(4) (2006) 525–535. doi:10.1016/j.apergo.2006.04.011.

[62] G.H. Walker, N.A. Stanton, P.M. Salmon, D.P. Jenkins, A review of sociotechnical systems theory: A classic concept for new command and control paradigms. *Theoretical Issues in Ergonomics Science*, 9 (2008) 479–499. doi:10.1080/14639220701635470.

[63] D.D.C. Christopher, E. Alexander, D. Felix, D.W. Joost, How to keep drivers engaged while supervising driving automation? A literature survey and categorisation of six solution areas. *Theoretical Issues in Ergonomics Science*, 20(3) (2019) 332–365. doi:10.1080/1463922X.2018.1528484.

[64] T. Zhang, W. Zeng, Y. Zhang, D. Tao, G. Li, X. Qu, What drives people to use automated vehicles? A meta-analytic review. *Accident Analysis & Prevention*, 159 (2021) 106270. doi:10.1016/j.aap.2021.106270.

[65] D.A. Frank, P. Chrysochou, P. Mitkidis, A. Dan, Human decision-making biases in the moral dilemmas of autonomous vehicles. *Scientific Reports*, 9 (2019) 13080. doi:10.1038/s41598-019-49411-7.

[66] D.H. McKnight, N.L. Chervany, Trust and distrust definitions: One bite at a time. In: R. Falcone, M. Singh, & Y. H. Tan (eds) *Trust in Cyber-societies*. Springer, Berlin, Heidelberg, pp. 27–54, 2001.

[67] J.D, Lee, K.A. See, Trust in automation: Designing for appropriate reliance. *Human Factors*, 46(1) (2004) 50–80. doi:10.1518/hfes.46.1.50_30392.

[68] H. Liu, R. Yang, L. Wang, P. Liu, Evaluating initial public acceptance of highly and fully autonomous vehicles. *International Journal of Human-Computer Interaction*, 35(11) (2019) 919–931. doi:10.1080/10447318.2018.1561791.

[69] M. Chikaraishi, D. Khan, B. Yasuda, A. Fujiwara, Risk perception and social acceptability of autonomous vehicles: A case study in Hiroshima, Japan. *Transp Policy*, 98 (2020) 105–115. doi:10.1016/j.tranpol.2020.05.014.

[70] J. Stapel, F.A. Mullakkal-Babu, R. Happee, Automated driving reduces perceived workload, but monitoring causes higher cognitive load than manual driving. *Transportation research part F: Traffic Psychology and Behaviour*, 60 (2019) 590–605. doi:10.1016/j.trf.2018.11.006.

[71] I.Y. Noy, D. Shinar, W.J. Horrey, Automated driving: Safety blind spots. *Safety Science*, 102 (2018) 68–78. doi:10.1016/j.ssci.2017.07.018.

[72] P. Delhomme, Comparing one's driving with others': Assessment of abilities and frequency of offences. Evidence for a superior conformity of self-bias? *Accident Analysis & Prevention*, 23(6) 493–508. doi:10.1016/0001-4575(91)90015-W.

[73] W.J.H. Adam, E. Nicholas, The mind in the machine: Anthropomorphism increases trust in an autonomous vehicle. *Journal of Experimental Social Psychology*, 52 (2014) 113–117. doi:10.1016/j.jesp.2014.01.005.

[74] J.S. Metcalfe, A.R. Marathe, B. Haynes, V.J. Paul, G.M. Gremillion, K. Drnec, C. Atwater, J.R. Estepp, J.R. Lukos, E.C. Carter, W.D. Nothwang, building a framework to manage trust in automation. *Proceedings of SPIE 10194, Micro- and Nanotechnology Sensors, Systems, and Applications*, IX (2017) 101941. doi:10.1117/12.2264245.

[75] S. Aggelos, B. Martin, C. Francesco, Impacts of automated vehicles on travel behaviour and land use: An international review of modelling studies. *Transport Reviews*, 39(1) (2019) 29–49. doi:10.1080/01441647.2018.1523253.

[76] D.J. Fagnant, K.M. Kockelman, The travel and environmental implications of shared autonomous vehicles, using agent-based model scenarios. *Transportation Research Part C: Emerging Technologies*, 40 (2014) 1–13. doi:10.1016/j.trc.2013.12.001.

[77] J.H. Chana, I. Robert, S. Yoram, User preferences regarding autonomous vehicles. *Transportation Research Part C: Emerging Technologies*, 78 (2017) 37–49. doi:10.1016/j.trc.2017.01.010.

[78] J. Yang, J.F. Coughlin, In-vehicle technology for self-driving cars: Advantages and challenges for aging drivers. *International Journal of Automotive Technology*, 15 (2014) 333–340. doi:10.1007/s12239-014-0034-6

[79] D. Souders, N. Charness, Challenges of older drivers' adoption of advanced driver assistance systems and autonomous vehicles. In: J. Zhou, G. Salvendy (eds) *Human Aspects of IT for the Aged Population Healthy and Active Aging; Lecture Notes in Computer Science*. Springer, New York, 2016, pp. 428–440. doi:10.1007/978-3-319-39949-2_41.

[80] P.A. Pavlou, Consumer acceptance of electronic commerce: Integrating trust and risk with the technology acceptance model. *International Journal of Electronic Commerce*, 7 (2003) 101–134. doi:10.1080/10864415.2003.11044275.

[81] E. Adell, Acceptance of driver support systems. In: J. Krems, T. Petzoldt, & M. Henning (eds) *Proceedings of the European Conference on Human Centred Design for Intelligent Transport Systems*, 2010, pp. 475–486.

[82] L. Carter, F. Bélanger, The utilization of e-government services: Citizen trust, innovation and acceptance factors. *Information Systems Journal*, 15(1) 2005 5-25. doi:10.1111/j.1365-2575.2005.00183.x

[83] N.B. Sarter, D.D. Woods, How in the world did we ever get into that mode? Mode error and awareness in supervisory control. *Human Factors*, 37(1) 1995 5-19. doi:10.1518/001872095779049516.

[84] J. Paxion, E. Galy, C. Berthelon, Mental workload and driving. *Frontiers in Psychology*, 5, 2014. doi:10.3389/fpsyg.2014.01344.

[85] M.S. Young, N.A. Stanton, What's skill got to do with it? Vehicle automation and driver mental workload. *Ergonomics*, 50(8) 2007 1324-1339. doi:10.1080/00140130701318855.

[86] A.F. Larsson, K. Kircher, J.A. Hultgren, Learning from experience: Familiarity with ACC and responding to a cut-in situation in automated driving. *Transportation Research Part F: Traffic Psychology and Behaviour*, 27 (2014) 229–237. doi:10.1016/j.trf.2014.05.008.

[87] S. Gormer, A. Kummert, S. Park, P. Egbert, Vision-based rain sensing with an in-vehicle camera. In: *Proceedings of the 2009 IEEE Intelligent Vehicles Symposium*, 2009, 279–284.

[88] J.C. Halimeh, M. Roser, Raindrop detection on car windshields using geometric-photometric environment construction and intensity-based correlation. In: *Proceedings of the IEEE Intelligent Vehicles Symposium*, Xi'an, China. Institute of Electrical and Electronics Engineers (IEEE), 2009, pp. 610–615. ISBN: 9781424435036

[89] J.S. Lerner, D. Keltner, Fear, anger, and risk. *Journal of Personality and Social Psychology*, 81(1) (2001) 146. doi:10.1037/0022-3514.81.1.146.

[90] N.D. Weinstein, Unrealistic optimism about future life events. *Journal of Personality and Social Psychology*, 39(5) 806. doi:10.1037/0022-3514.39.5.806.

[91] E.J. Langer, The illusion of control. *Journal of Personality and Social Psychology*, 32(2) (1975) 311. doi:10.1037/0022-3514.32.2.311.
[92] D.M. DeJoy, The optimism bias and traffic accident risk perception. *Accident Analysis and Prevention*, 21(4) (1989) 333–340. doi:10.1016/0001-4575(89)90024-9.
[93] T.B. Hammond, M.S. Horswill, The influence of desire for control on drivers' risk-taking behaviour. *Transportation Research Part F: Traffic Psychology and Behaviour*, 4(4) (2001) 271–277. doi:10.1016/S1369-8478(01)00028-6.
[94] N. Harre, C.G. Sibley, Explicit and implicit self-enhancement biases in drivers and their relationship to driving violations and crash-risk optimism. *Accident Analysis & Prevention*, 39(6) (2007) 1155–1161. doi:10.1016/j.aap.2007.03.001.
[95] M.S. Horswill, F.P. McKenna, The Effect of Perceived Control on Risk Taking 1. *Journal of Applied Social Psychology*, 29(2) (1999) 377–391. doi:10.1111/j.1559-1816.1999.tb01392.x.
[96] M. Kyriakidis, J.C. de Winter, N. Stanton, T. Bellet, B. Van Arem, K. Brookhuis, R. Happee, A human factors perspective on automated driving. *Theoretical Issues in Ergonomics Science*, 20 (2017) 223–249. doi:10.1080/1463922X.2017.1293187.
[97] R. Hayama, S. Kawahara, S. Nakano, H. Kumamoto, Resistance torque control for steer-by-wire system to improve human-machine interface. *Vehicle System Dynamics*, 48 (2010) 1065–1075. doi:10.1080/00423110903267405.
[98] P. Yih, J. Ryu, J.C. Gerdes, Modification of vehicle handling characteristics via steer-by-wire. *Proceedings of the 2003 American Control Conference*, 2573 (2003) 2578–2583.
[99] M. Mulder, M. Mulder, M.M. van Paassen, D.A. Abbink, Haptic gas pedal feedback. *Ergonomics*, 51 (2008) 1710–1720. doi:10.1080/00140130802331583.
[100] J.H. Lim, O. Tsimhoni, Y. Liu, Investigation of driver performance with night vision and pedestrian detection systems-Part I: Empirical study on visual clutter and glance behavior. *IEEE Transactions on Intelligent Transportation Systems*, 11 (2010) 670–677. doi:10.1109/TITS.2010.2049843.
[101] W. Hahn, Vision enhancement-concepts for the future? In: National Highway Traffic Safety Administration, *Proceedings of the International Technical Conference on the Enhanced Safety of Vehicles*, 1995, pp. 1490–1496.
[102] D.B. Kaber, J.M. Riley, K.W. Tan, M.R. Endsley, On the design of adaptive automation for complex systems. *International Journal of Cognitive Ergonomics*, 5 (2001) 37–57. doi:10.1207/S15327566IJCE0501_3.
[103] K.J. Vicente, J. Rasmussen, Ecological interface design: Theoretical foundations. *IEEE Transactions on Systems, Man, and Cybernetics: Systems*, 22 (1992) 589–606. doi:10.1109/21.156574.
[104] S. Damiani, E. Deregibus, L. Andreone, Driver-vehicle interfaces and interaction: Where are they going? *European Transport Research Review*, 1 (2009) 87–96. doi:10.1007/s12544-009-0009-2.
[105] L. Rothrock, L.R. Koubek, F. Fuchs, M. Haas, G. Salvendy, Review and reappraisal of adaptive interfaces: Toward biologically inspired paradigms. *Theoretical Issues in Ergonomics Science*, 3 (2002) 47–84. doi:10.1080/14639220110110342.

13 The Sensors in Future Vehicles

Sundara Subramanian Karuppasamy
National Taipei University of Technology

N. Jeyaprakash
China University of Mining and Technology

Che-Hua Yang
National Taipei University of Technology

13.1 INTRODUCTION

The industrial revolution has uplifted the technological advancements in humans. During this revolution era, every sector has attained better growth in terms of advancements, economy, and so on. Starting in the 1800s, this revolution mechanized the basic sectors to an advanced level and this revolution has transformed the agricultural and handicrafts-based economy into the manufacturing (machine)-based economy. It also helps in building the nation's GDP and other economic strategies that paved the way for technological advancements in the entire nation [1–4]. For a predefined timeline (years), this revolution is focused on developing certain sectors until the outcomes were reached. Till now four industrial revolutions have been put forth by the committee. Industrial revolution 1.0 started in the 1840s and focused on transforming the handicrafts into machineries. Industries like textile, glass, mining, and agriculture got benefited from this revolution. Water and steam have been used as the primary energy source for developing these industries. To reach its outcomes, more labor force is needed which leads to overexploitation of the manpower [5,6]. In the 1870s, the second industrial revolution had started and aimed at the assembly line and mass production of products. Sectors like iron and steel got benefited and these industries used oil and electricity as the major energy source. Since most sectors got mechanized in this revolution, the exploitation of the human force had been controlled [7,8]. Nearly after 100 years (1970s), the third industrial revolution started and focused on the partial automation of the industrial sectors by means of computers using logical controls. The primary aim is to attain digitalization by implementing programmable logic controllers and other software [9,10]. Now, we are in the fourth industrial revolution era which is aimed at executing full automation through smart systems. Every progress is monitored by AI systems and their responses created a huge impact on the technological development and sustainability [11–13].

DOI: 10.1201/9781003354901-13

The automobile sector got much attention after the second industrial revolution by means of mass production and assembly line transformation, the usage of automobiles has increased in an abundant manner. Fossil fuels are quite commonly used as the primary source of energy for these automobiles [14]. Automobiles containing the two-stroke and four-stroke engines use petrol and diesel for the energy requirements [15]. Moreover, many technological advancements had been made to increase the fuel economy and efficiency of these automobiles. But the environmental impacts created by these traditional automobiles were more hazardous to the biodiversity. The major drawback of using these traditional automobiles is the pollution of all kinds followed by the ecological imbalance.

Since the traditional automobiles emit harmful gases to the environment, our earth has faced a lot of serious issues like the greenhouse effect, global warming, etc [16–18]. Due to these effects, the researchers are focusing on alternative sources of energy for transportation. According to the 48th Intergovernmental Panel on Climate Change reports, if we don't control the emission of toxic gases (CO_2) into the atmosphere, we have only 12 years left until the point of no return and we have to reverse the current trend in a huge way to avoid natural calamities. Rapid and significant changes have to be made to the automobiles used for transportation. The reports also suggested the usage of electric vehicles to overcome these emissions [19].

Electric vehicles (EVs) gained their importance after the environmental impacts made by the traditional internal combustion (IC) engine vehicles. It replaces the power generation and transmission systems used in traditional vehicles by means of batteries and electric motors [20,21]. The batteries/fuel cells act as the power source of these vehicles. The necessary trust to actuate the vehicle is provided by the power sources through electrochemical reactions [22]. Mostly the electric vehicles are reported to have zero emissions while operating and the by-product is water. Each and every system in the EVs is monitored by the various range of sensors that help in the proper functioning of the whole vehicle. These sensors aid in providing information and diagnosing the system's malfunctions periodically [23]. Figure 13.1 represents the schematic illustration of the eneralized electric and traditional combustion vehicles. This chapter is focused on addressing the need for electric vehicles along with their types. Furthermore, the sensors used in electric vehicles were elaborated along with their characteristics and presented.

13.2 TYPES OF ELECTRIC VEHICLES

This section deals with the types of electric vehicles with examples. Figure 13.2 lists the electric vehicle types.

13.2.1 Battery Electric Vehicles

Battery electric vehicle (Figure 13.3) belongs to the class of electric vehicles that operates solely on batteries. The entire vehicle gets power from the battery pack. This vehicle doesn't have any other power source. Based on the battery pack capacity, the vehicle's performance is decided. Larger battery packs will result in better performance and driving range. Mostly these vehicles have a driving range of 250 km

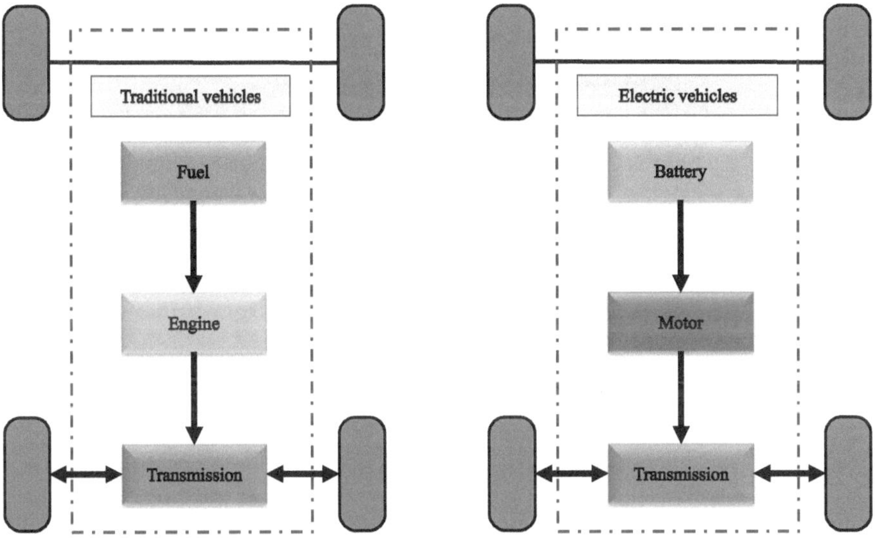

FIGURE 13.1 Basic design of traditional and electric vehicles.

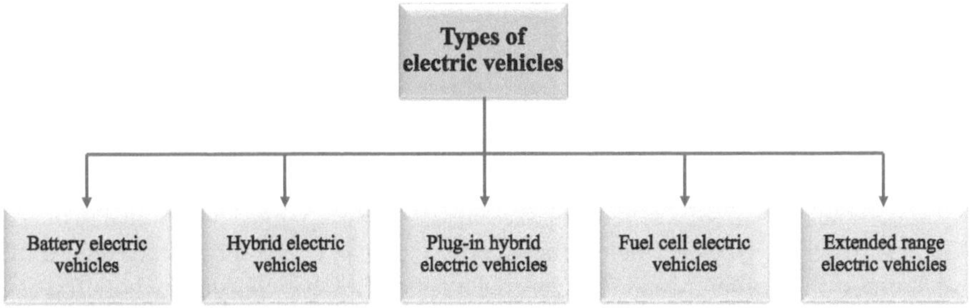

FIGURE 13.2 Types of electric vehicles.

per full charge. The battery pack also gets energy from the regenerative braking mechanism. Since this vehicle is solely dependent on the battery, it is good for urban areas and short-distance drives. Some of the major advantages of the battery electric vehicles are noise-free operation, pollution-free driving, and eco-friendly due to the absence of toxic gas emissions [24,25]. Nissan leaf model and all cars belonging to the EVGO series suit as best examples for battery electric vehicles.

13.2.2 Hybrid Electric Vehicles

This type (Figure 13.4) has two or more power source giving the necessary trust for the vehicle's motion. Mostly, the IC engine and battery pack were present in these vehicles. Modern hybrid vehicles have a regenerative braking mechanism that aids in charging the battery pack along with the IC engine [26,27]. This type of electric vehicle is further classified into series hybrid, parallel hybrid, and series-parallel hybrid

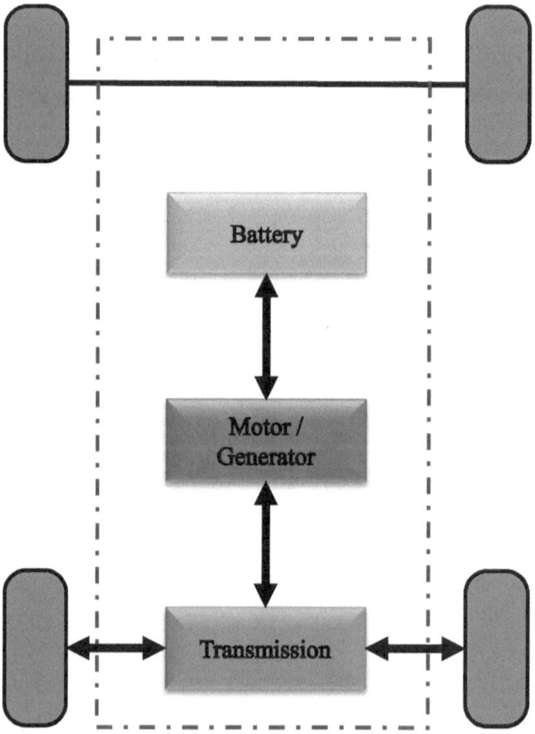

FIGURE 13.3 Schematic representation of a BEV.

vehicles. In series hybrid vehicles, the electric motor is charged by either one of the two sources. In parallel hybrid vehicles, the electric motor is charged by both the battery pack and IC engine by means of a regenerative braking system. The complications of series and parallel hybrid were overcome by implementing the series-parallel hybrid combination where both the IC engine and electric motors are connected to the wheels. As per the driver's convenience, the wheel can operate on both the IC engine and electric motor or one power source can drive the whole vehicle [28,29]. Some examples of hybrid electric vehicles include Toyota Prius fourth generation, Ford Escape, and Ford Fusion hybrid vehicles.

13.2.3 Plug-in Hybrid Electric Vehicles

As the name defines, this electric vehicle is the same as the hybrid electric vehicle but it requires an outside power source (Figure 13.5). An external power source (power grid) will energize the battery and this battery operates the electric motors for the vehicle's movement. The external power grid is used to charge the battery whereas the IC engine-battery pair propuls the vehicle [30–32]. First-generation Chevrolet volt and Mitsubishi Outlander PHEV serve as the best examples of this kind.

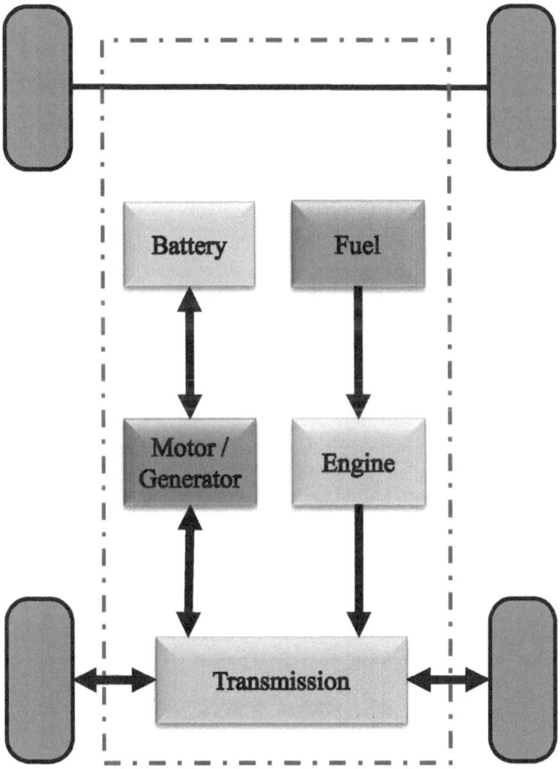

FIGURE 13.4 Schematic representation of a HEV.

13.2.4 Fuel Cell Electric Vehicles

These vehicles operate on the electrochemical reaction that takes place in the fuel cell. Mostly hydrogen ions were used as the anode material whereas the oxygen ions contribute to the cathode material. The electrolyte can be either solid or liquid or gas electrolyte. During the process, the anode (H_2) undergoes oxidation where it will lose the electrons. These electrons pass through the outer circuit thereby producing electric charges whereas the remaining protons in the anode pass through the electrolyte and reach the cathode. In the cathode, the O_2 ions were reduced and combine with the anode protons to form water. The outer circuit will produce the necessary electricity to power the electric motor of the vehicle. This fuel cell electric vehicle (Figure 13.6) is 100% pollution free since the by-product is the water that drains from the fuel cell chamber. The hydrogen ions reach the anode through the hydrogen storage tank whereas the atmospheric oxygen ions serve as the input for the cathode. The combined effect of oxidation and reduction reaction takes place at the anode and the cathode is responsible for the production of electric charges [33–35]. Both oxidation and reduction reactions are collectively called as the redox reaction. These cars are termed as the cars of the future due to their high performance and pollution-free

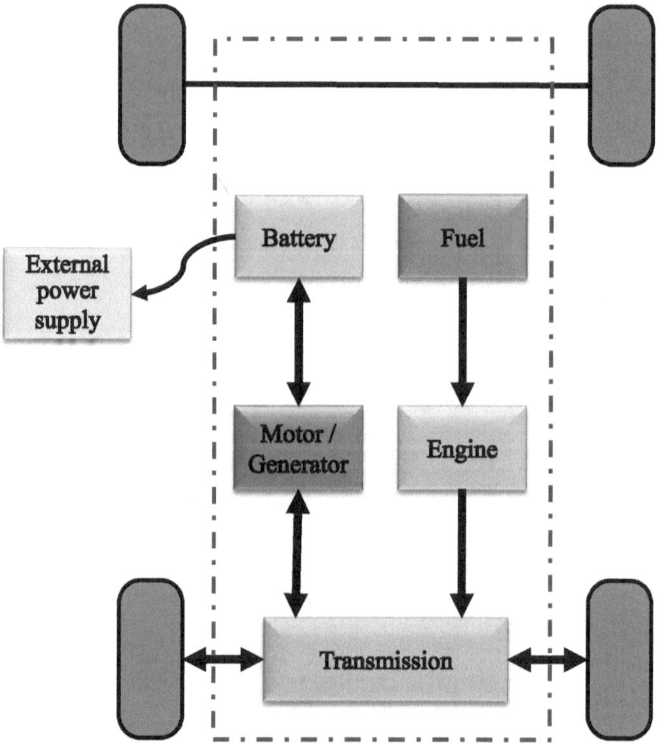

FIGURE 13.5 Schematic representation of a PHEV.

features [36]. Storage of hydrogen and its highly flammable nature are considered as the major drawbacks. Moreover, researchers are working on these electric vehicles to minimize the effect of their drawbacks. Toyota Mirai, Hyundai Nexo, and Foton BJ6123FCEVCH-1 models serve as examples of this kind.

13.2.5 Extended-Range Electric Vehicles

This type of electric vehicle is considered as the advanced version of battery electric vehicles. These vehicles have a substitute IC engine which is used to charge the battery pack (Figure 13.7). But this IC engine will not accelerate the wheels as in hybrid and PHEVs [37,38]. BMW i3 model uses the power produced by the IC engine only for charging the pack.

13.3 SENSORS USED IN ELECTRIC VEHICLES

Since the industrial revolution 4.0 is focused on fully automated systems, sensors play a significant role in building such systems. These sensors were implemented in electric vehicles for mainly two purposes: (i) To reduce the human work (efforts) and (ii) To gain information about each unit and diagnose its malfunction.

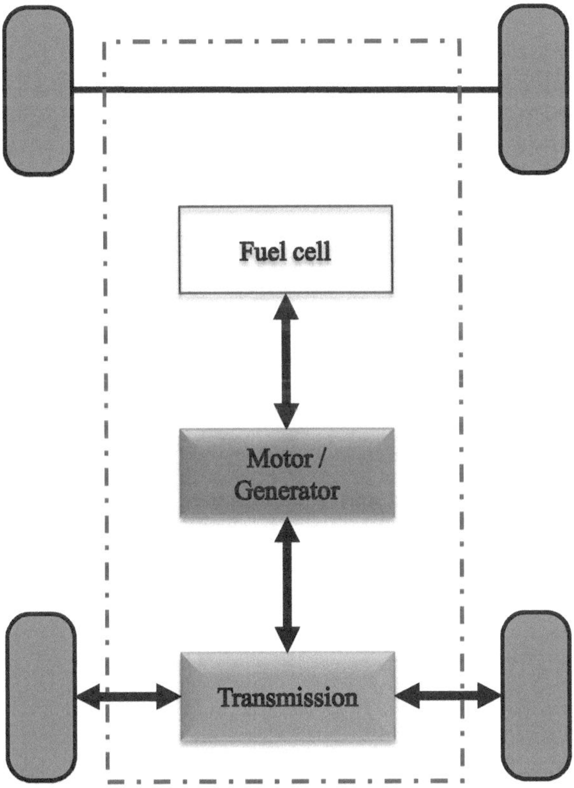

FIGURE 13.6 Schematic representation of a FCEV.

13.3.1 Sensors Based on Categories

Based on the sensor's application in EVs, the sensors are classified into five major categories as in Table 13.1 [39].

13.3.1.1 Traffic Management Sensors

In electric vehicles, various sensors were involved in the traffic management unit. This unit monitors the roadways with the aid of sensors that are installed on the roads and inside the vehicles. The main purpose of this unit is to monitor the traffic by integrating the information from the detection unit that is installed on roads and the sensors in the electric vehicles. A predictive or predefined algorithm uses this information and acts accordingly for effective traffic management. It also monitors the vehicle's position, vehicle's driving speed, driving direction and so on. These sensors aid in managing the traffic on roads, intelligent parking, avoids traffic jams and accidents [40,41].

Some of the sensors used in traffic management are listed as follows: Inductive loop detectors, laser sensors, RFID sensors, laser sensors, LIDAR sensors, magnetic and piezoelectric sensors. Magnetic, piezoelectric and inductive loop sensors are employed in collecting the data about the traffic flow, vehicle's speed, and its

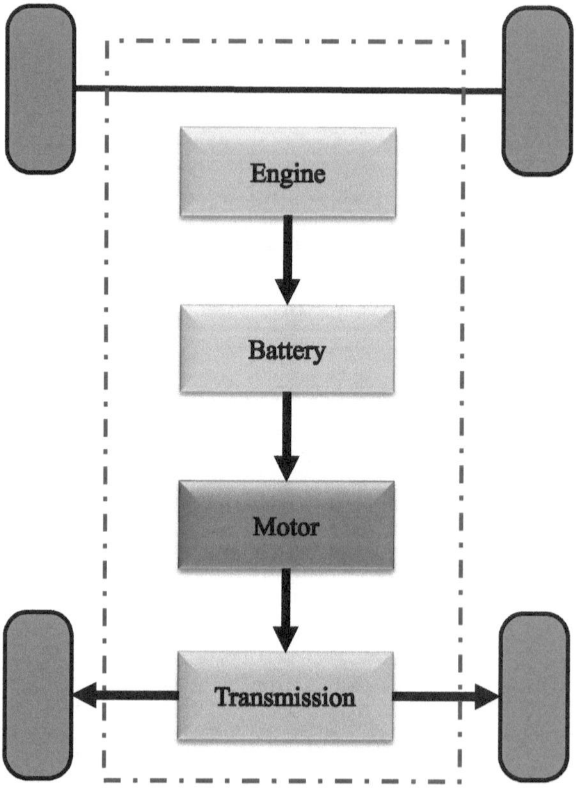

FIGURE 13.7 Schematic representation of a EREV.

TABLE 13.1
Sensor Classification Based on Application Category

Category	Sensor Types
Traffic management	Cameras, position sensors, laser sensors, LIDAR sensors, ultrasonic sensors
Vehicle diagnostic	Pressure sensors, temperature sensors, level sensors, gas sensors
Environment	GPS, IR sensors, moisture sensors, pressure sensors
User interface	Cameras, EEG sensors, temperature sensors, ECG sensors
Vehicle safety	Night vision cameras, speed sensors, ultrasonic sensors, ESC unit

occupancy. RFID sensors are implemented in intelligent parking platforms. Laser based sensors are used to collect the information from the roads about any obstruction. These sensors are coupled with a feedback system that helps the driver to change the speed or to stop the vehicle when it encounters any obstruction on the road. In the case of an autonomous vehicle, the LIDAR technology-based sensor can work in 360° and gives feedback to the system in case of any obstruction on roads. This feedback is fed into the predefined algorithm and aids in stopping the vehicle's movement. Thus, the sensors used for traffic management controls the vehicle's position, speed, traffic flow, driving direction, etc.

13.3.1.2 Vehicle Diagnostic Sensors

In case of any failure in electric vehicle, we can't avoid the money spent to fix these failures. Hence, there is a need in diagnosing these failures before it happens. Certain sensors are used to diagnose the fault in components used in electric vehicles. These sensors collect information that helps to identify the faults in an earlier manner and avoids fatal accidents [42,43].

Some of the diagnostic sensors are as follows:

i. *Temperature sensor*: This sensor monitors the temperature of the vehicle's fluid, battery pack, environmental temperature, engine temperature, and electric motor.
ii. *Pressure sensor*: Most accidents were caused by the imbalance in the vehicle's tire pressure, these sensors alert the driver by sensing the tire pressure of the electric vehicles.
iii. *Level sensor*: This sensor controls the fuel flow into the engines, measures the fuel level in the fuel storage tanks and tells the driver in case of fuel shortage.
iv. *Voltage sensor*: This sensor regulates the voltage to electric motors, avoids overcharging and deep discharging from battery packs, and monitors the voltage fluctuations while charging through an external grid (plug-in EVs).
v. *Speed and gas sensors*: The speed sensor plays a significant role in traction tracking system (TTS) and antilock brake system (ABS) whereas the gas sensor is used to monitor the exhaust gases that are released into the environment.

13.3.1.3 Environment Sensors

These sensors help in collecting the data about the external environment in which the electric vehicles operate. The prevailing environmental conditions are also considered as important criteria for the good working of the components in electric vehicles. The road temperature, ambient pressure and temperature were recorded by these sensors. These sensors help in real-time monitoring of the external environment and warn the driver if the values exceed the safety limits [44,45]. Some of the sensors used for this purpose are listed below:

i. *GPS, IR sensor*: Aids in continuous real-time monitoring of the prevailing environment.
ii. *Pressure and temperature sensors*: Help in sensing the road temperature, and atmospheric pressure which are the important criteria for charging the electric vehicle.
iii. *Moisture sensors*: Collects information about the humidity.
iv. *Cameras*: Images collected by the cameras are processed to know about the road condition like traffic, obstruction, etc.

13.3.1.4 User Interface Sensors

The lethargicness of the user (drivers) is responsible for accidents. The drowsiness (sleep) of the driver accounts for the emergencies. To overcome this, certain sensors are implemented to reduce the risk of accidents caused by the drivers. These sensors analyze the drowsiness level of drivers and with the help of AI-based algorithm, the vehicle's speed is controlled to avoid fatal accidents [46,47]. Some of the examples are listed as follows:

i. *Camera, head and eye sensors*: Detects the driver's behaviors and analyze the drowsiness thereby alerting the driver by means of steering vibrations and automatic braking.
ii. *Temperature sensors*: Analyse the health conditions of the drivers.
iii. *EEG sensors*: Electroencephalogram sensors are used in examining the driver's brain signal that acts as the input for AI-based algorithm which helps in avoiding fatal accidents [48].

13.3.1.5 Vehicle Safety Sensors

These types of sensors are implemented to ensure the well-being of the drivers and passengers through earlier prediction of obstruction and help in avoiding fatal accidents. These sensors not only help in avoiding fatal accidents but also ensure safe parking and fine navigation through traffics [49,50]. Some of the sensors used in vehicle safety are given below:

i. *Night vision cameras*: These cameras consist of infrared (IR) based unit that helps in providing better visibility on dark and challenging roads [51].
ii. *ESC system*: The electronic stability control system combines the information from the speed sensors mounted on wheels and steering angle sensors in order to maintain the vehicle's stability [52].

13.3.2 Sensor Types

Most of the electric vehicles use position, pressure, temperature, and voltage sensors in their systems. Thus, these sensors are discussed as follows:

13.3.2.1 Position Sensors

As the name suggests, the position sensor is a sensing device that stores the information about the position of the electric vehicles. In order to navigate to the desired location, the position sensor collect the data about wheel's position and maps with the GPS position. In most electric vehicles, resolvers and encoders have been used as position sensor [53].

13.3.2.1.1 Resolvers

Resolvers are analog devices that are used to calibrate the degree of rotation. It is often described as the rotary transformer which contains three windings (primary and secondary windings). The primary windings are mounted on the shaft whereas the two secondary windings are aligned at 90° to each other. It collects data about the changes in the shaft angle while rotation and these resolvers are excited by the AC supply [54].

13.3.2.1.2 Encoders

Encoders measure the rotary motion since it converts the mechanical motion into electrical signals and these encoders are categorized into three basic types. The absolute position of the vehicle is found by using the absolute encoder when the vehicle runs in the unusual stage whereas the incremental encoder shows the output in digital form in binary digits with a reference mark (angle zero). On the other hand, the incremental sinusoidal encoders display the output in a sinusoidal waveform rather than digital output. Usually, all the encoders are energized by the DC supply [55].

13.3.2.2 Battery Monitoring Sensors

The battery monitoring unit consists of voltage, pressure, and temperature sensors. The voltage sensor records the voltage fluctuation from the battery. It monitors the battery thereby regulating the voltage flow to the electric motor. This sensor also checks the fluctuations happening from the outside power source (onboard charger). Shunt resistors and hall sensors were used as the voltage sensors [56]. The pressure sensor ensures the environmental pressure thereby making the electric vehicle for efficient charging. Mostly thermistors were employed in measuring the temperature [57]. It stores the battery's temperature and notifies the driver if the limit exceeds the safety limit. Optimum temperature is required for charging the battery and to avoid the fast and deep discharging rate.

13.3.2.3 LIDAR

LIDAR stands for Light Detection and Ranging. Invented in the 1960s, these sensors are used to receive the light reflected from the source and compute the distance [58,59]. The basic principle behind the LIDAR is described as follows. Initially, a LIDAR sensor generates a pulsed light beam (waves) which travels in the surrounding environment. These environments have certain objects that act as the blocks where the generated pulsed light beam got reflected after hitting these objects. This back-scattered light beam from the objects is received by the LIDAR sensor which calculates the time of flight and the distance between the source and sink. From the calculated time, the exact distance between the object and the electric vehicle can be obtained which is used to notify the user about the obstructions in the surrounding environment. This process is repeated a million times which generates a 3D map. This 3D map is used by the vehicle's intelligent system for navigation purposes [60]. Figure 13.8 shows the working principle of LIDAR.

LIDAR adopts three ways to measure the TOF. They are

i. Pulsed TOF approach.
ii. Continuous wave amplitude modulation (CWAM).
iii. Continuous wave frequency modulation (CWFM).

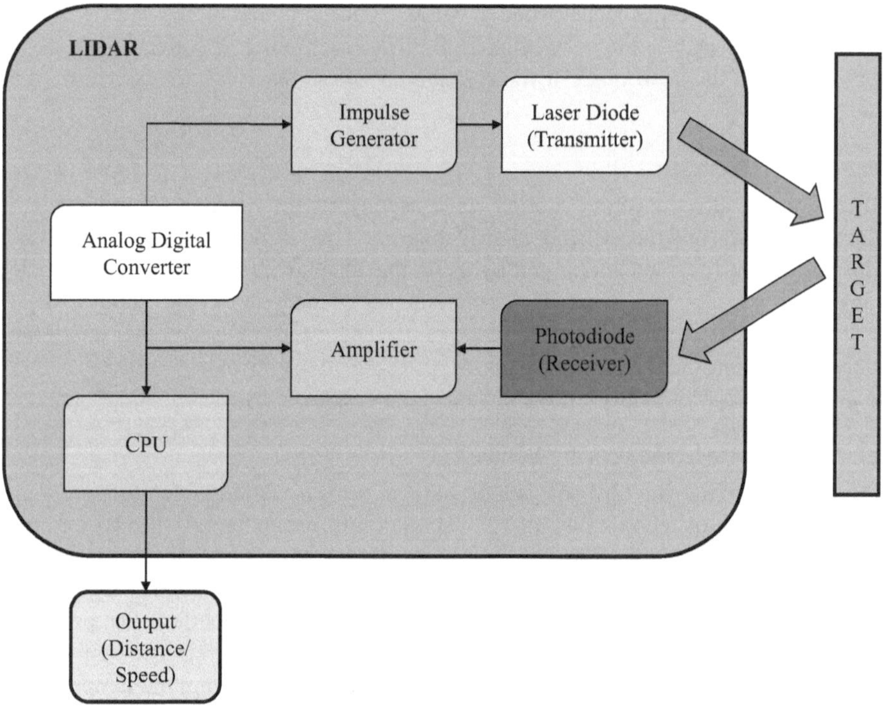

FIGURE 13.8 LIDAR block diagram.

The pulsed TOF approach uses a pulsed laser beam as the generation source. This pulsed beam is emitted from the source and travels into the surrounding environment. After being reflected from the object, this beam is received by the receptor which calculates the travel distance and time. Simplicity, fast and stable way based on the intensity detection is considered as the pros. But the low signal-to-noise ratio levels at long distances serve as the major drawback of this approach [61]. On the other hand, the CWAM approach uses the continuous light wave by incorporating a phase shift in the continuous wave signal. This signal travels between the source and receptor through target to calculate the distance range. This approach holds good for indoor environments [62]. The CWFM approach is reported to be the advanced approach that could be implemented in the future autonomous vehicles. This approach is entirely different from the pulsed TOF and CWAM approaches. It involves Fourier domain homodyne detection rather than the intensity detection. By implementing it in the Fourier domain, this approach overcomes the drawbacks of the previous two methods. It holds good for outdoor environments, has high resolution, and can be applied for long distances [63].

Since the LIDAR needs a source and detector to calculate the time and distance range, many devices can be used as the source and detector materials. Some of them are listed as follows:

i. *Sources*: Fibre lasers, microchip lasers and diode lasers.
ii. *Photodetectors*: PIN photodiodes, avalanche photodiodes, photomultiplier tubes, optical phased arrays, multipixel photon counters, single-photon avalanche photodiodes, and so on.

13.3.2.4 RADAR

Similar to LIDAR, the RADAR sensor operates by detecting radio waves. RADAR stands for Radio Detection and Ranging. This sensor generates the electromagnetic waves which are termed as radio waves. The emitted radio waves travel into the surrounding environment and get reflected after facing an obstruction. The reflected waves are sensed by the receptors and based on the generated and received waves, the RADAR calculates the distance, angle, velocity and position of the obstruction/object with respect to the vehicle's position (movement). It operates under the electromagnetic radiation principle which could adopt different frequency bands such as 24, 77 GHz, etc [64,65]. By adopting high-frequency ranges, this sensor produces high-resolution images that help the user to clearly distinguish various obstacles in the surrounding environment. These sensors are less expensive and easily available compared to the LIDAR sensors. Normally these sensors are short-range sensors which have the detectability of around 100 meters. But some RADAR sensors generate high-resolution images up to 150 meters. Figure 13.9 shows the block diagram of RADAR. Based on the obtained RADAR signals, three major signal processing strategies are implemented to detect the obstacle. The first strategy is called target or cluster RADAR which generates precise two-dimensional location of the obstacle with minimal ambiguity of phases. In order to cluster, several methods such as

1. long short-term memory (LSTM),
2. density-based spatial clustering of application with noise (DBSCAN),
3. random forest, etc were used [66]. The second category was termed as RADAR cube which employs the 3D data. Range, azimuth, and Doppler velocity were considered as the 3D data that could be useful in predicting the speed distribution at various 2D azimuthal locations. Since the RADAR cube uses pre-processed data, the ambiguity effect and resolution effect are

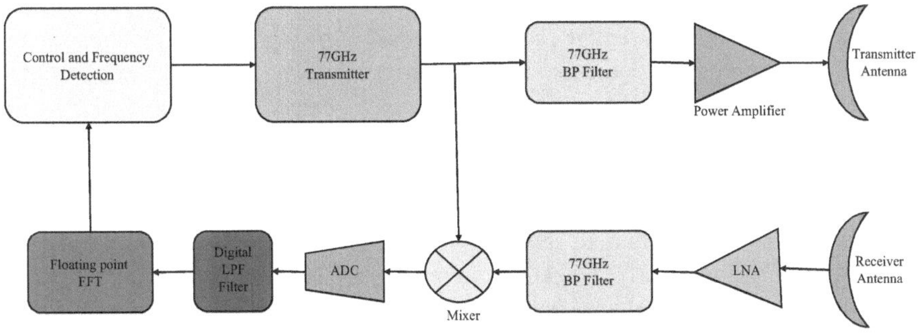

FIGURE 13.9 RADAR block diagram.

unavoidable [67]. In the near future, the third category is said to be implemented in the autonomous vehicle which combines the processed (RADAR cluster) and pre-processed (RADAR cube) for enhancing the detection range and resolution of the obstacle images.

13.3.2.5 Ultrasonic Sensors

As the name tells, the ultrasonic sensor uses sonic waves to detect the object/obstruction in the surrounding environment. This sensor is said to work under the principle of time of flight for detecting the obstruction nearby. These sensors transmit and receive ultrasonic waves. To compute the distance between the obstacle and the vehicle, these sensors emit ultrasonic waves that get reflected by the obstacle [68]. This reflected wave is often termed as the ultrasonic echo or backscattered echo and is received by the receiver and processed to compute the distance as given below:

$$d = 0.5 * TOF * v$$

where d is the distance in meters, TOF is the time of flight in seconds and v is the ultrasonic wave velocity in meters/second. Most of the ultrasonic sensors used in autonomous vehicle are piezoelectric sensor that operates on the piezoelectric effect which converts the electric charges applied into mechanical vibrations (ultrasonic waves) or vice versa [69]. The nominal frequency range used for detecting the obstacle is in the range of 40–50 kHz. Waves higher than this frequency range will lead to weaker ultrasonic echoes because of the attenuation effect [70].

13.3.2.6 Cameras

Cameras are predominantly used as environment recognition patterns. These cameras work under the principle of capturing the light that gets emitted from the environment through a lens (camera lens) and converting it into an image that is generated on a photosensitive surface. These are quite economical compared to the other sensors that are employed in detecting the obstacles. It is capable of generating images with higher quality and helps in identifying road signs, lane markings, road barriers and notifies the user about the obstacles existing in the surrounding environment [71,72]. Mostly the autonomous vehicle employs either monocular or binocular or a combination of both cameras for producing high-resolution images of the surrounding environment. The monocular category consists of a single camera that captures more pictures whereas the binocular has two lens setups. The major advantage of the cameras over the other sensors is that the camera can detect both static and moving objects and it can generate the colored images. But these cameras will get influenced if the light source's intensity is relatively low meaning that poor quality images are generated in low-intensity light conditions. By incorporating these images with certain algorithms, the autonomous vehicle system can take necessary actions if any malfunctions arise or mostly the generated images are used for traffic prediction in the surroundings or these images could be used by navigating devices that are incorporated in the autonomous vehicle. In most cases, visible and infrared cameras are employed in the electric vehicle to generate the images [73,74].

13.3.2.7 Global Navigation Satellite System (GNSS)

Global Navigation Satellite System is a predominantly used technology that gives the precise location of the object on the earth's surface. GPS belongs to the GNSS systems that employ satellites to track the objects located on the Earth's surface. The GNSS relies on the receiver's capacity to find at least four satellites, determine the distance to each of them, and utilize this data for trilateration (a process that determines the receiver's location) [75]. Some of the issues faced by the GNSS are as follows:

i. Timing errors.
ii. Delay in signals.
iii. Multipath effect.
iv. Satellite orbit's uncertainties.

To overcome these errors, it is better to merge the GNSS along with other sensors such as cameras, LIDAR, RADAR, and inertial measurement units.

13.3.2.8 Inertial Measurement Unit and Dead Reckoning

Dead reckoning is defined as the process which can be implemented to calculate the vehicle's position based on the previous measurements. Mostly encoders are employed as dead reckoning sensors [76]. On the other hand, the inertial measurement unit is reported as a device that is used to compute the angular rate, existing magnetic field, and force exerted. Accelerometers, gyroscopes, magnetometers, and other electronic devices which could calibrate the vehicle's position and other properties in three dimensions [77]. From the obtained parameters, the vehicle's system can able to make necessary action in the case of any malfunction activities.

13.3.3 MEMS-Based Sensors in Electric Vehicles

Micro electro-mechanical system (MEMS) based sensors play a major role in electric vehicles. Mostly these sensors were implemented as inclinometers, flow and pressure sensors, gyroscopes and accelerometers which ensure the passenger and vehicle safety. Some of the units that implement the MEMS-based sensors are described as follows:

i. *Passenger safety*: Most of the airbags used in all types of electric vehicles are made up of MEMS-based accelerometers. These accelerometers will monitor the acceleration of the vehicle and in deceleration mode, the variation in capacitance has been recorded and sent rapidly to the chip that sends the required signal to the airbag to operate thereby saving the life of the passengers. These accelerometers also avoid the need for the most expensive g-switches [78].
ii. *Tire pressure sensors*: These sensors provide information about the air pressure in tires and air leaks. A separate battery setup is required to energize these sensors. But nowadays, the MEMS technology has uplifted the issues

faced by the traditional tire pressure sensors. It changes the traditional one to a more compact, optimized one and most importantly the MEMS-based tire sensors are battery-less [79].

iii. *Electronic stability control*: This unit consists of a chip that alerts the drivers to control the stability in response to sudden jerks. Mostly gyroscopes and accelerometers are used for the electronic stability control unit. By incorporating the MEMS technology, this unit becomes more economical [80].

iv. *Engine management and rollover detection*: MEMS-based pressure sensors aid in optimizing the airflow in the engines. MEMS-Barometric pressure sensors detect the variations in ambient pressure [81]. Moreover, a MEMS-based gyroscope is used to calibrate the angular velocity which aids in rollover detection in most electric vehicles [82].

v. *Anti-theft electric vehicles*: MEMS-based 3D accelerometers were used for this purpose. Electric cars are quite expensive and there are more chances of theft of these cars. Before the MEMS technology, certain sensors were used but they will detect only major tilt or changes. After implementing the electric cars with these 3D accelerometers, a minute tilt can be noticed because this accelerometer will calibrate the inclination angle and alerts the driver in case of any theft activities [83].

vi. *MEMS-based navigation*: Most of the vehicle navigation system uses GPS for navigation purpose. But the information provided by a single GPS is reported to be inadequate considering the technological advancements. Hence, MEMS-based gyroscopes were implemented in the vehicle navigation unit to obtain more information about the minute movements. This information will help to know more about the navigation behaviors [84].

vii. *Road slope estimation*: The measurement of longitudinal velocity and slope angle estimation is a tedious and expensive process. Moreover, the sensors used in the slope angle estimation will not give accurate results. These drawbacks were overcome by implementing the MEMS-based technology where earlier detection with high-precision measurement can be obtained [85].

viii. *MEMS-based piezoresistive sensor*: MEMS-based technology is not new in the case of pressure sensors. The barometric MEMS sensor has been implemented in most electric vehicles. But in recent years, the MEMS-piezoresistive sensor has huge attention in the field of fuel cell electric vehicles. Because these sensors can detect the leakage from each unit cell of the fuel cell stacks [86].

ix. *MEMS-based flux sensing*: MEMS sensors are capable of detecting magnetic flux levels which remove the barrier to the commercial use of hybrid reluctance permanent magnet (HRPM) in electric vehicles [87].

x. *Self-balanced EVs*: University campuses, theme parks, and museums have large areas that cannot be accessed by cars. But now, these areas can be accessed by means of self-balanced EVs which contain parallely mounted wheels to maintain the equilibrium and position sensors made by using MEMS-based accelerometers and gyroscopes [88].

Thus, the MEMS technology has huge applications in the field of electric vehicles.

13.4 COMMUNICATION PROTOCOLS

Communication protocol plays a vital role in electric vehicles. These protocols describe the rules for transmitting or exchanging the data across networks. There are two major protocols used in the field of electric vehicles. They are described as follows:

13.4.1 Intra-Vehicle Communication Protocol

Most of the electric vehicles consist of electronic components like sensors, processors and sensing devices. By means of a feedback system, these components integrate to form a closed loop system. The controllers, sensing units and actuators contribute to the intra-vehicle communication system in electric vehicles. The sensing unit converts any physical disturbances to electric signals. The electronic control unit (controllers) contains processors, memory and other communication units where the data is processed. The actuators will take actions depending on the decisions made by the controllers [89]. Some of the intra-vehicle communication protocols used by the electric vehicles are as follows:

i. Media-oriented system transport (MOST).
ii. Flex ray.
iii. Local interconnect network (LIN).
iv. Control area network (CAN).

13.4.2 Inter-Vehicle Communication Protocol

As the name suggests, this communication protocol is used to transmit the data from one electric vehicle to the other vehicles or data transfer stations. This protocol aids in improving the safety and makes the electric vehicle more user-friendly [90]. Some of the inter-vehicle communication protocols used by the electric vehicles are listed below:

i. Bluetooth, Wi-Fi, ZigBee technology.
ii. Time-triggered protocol (TTP).
iii. Communication access for land mobiles (CALM).
iv. Wireless access for vehicular environment (WAVE).
v. Dedicated short-range communication (DSRC).

13.5 SENSOR FUSION CONCEPTS—FUTURE PROSPECTS

Electric vehicles consist of many sub-units such as electric control unit, vehicle control unit, battery management unit, and so on, and the probability of errors made by these units is comparatively high. Hence the researchers are focusing on the sensor fusion concepts [91,92]. Sensor fusion can be defined as combining the data of separate sub-units and processing it. By means of this concept, the probability of error

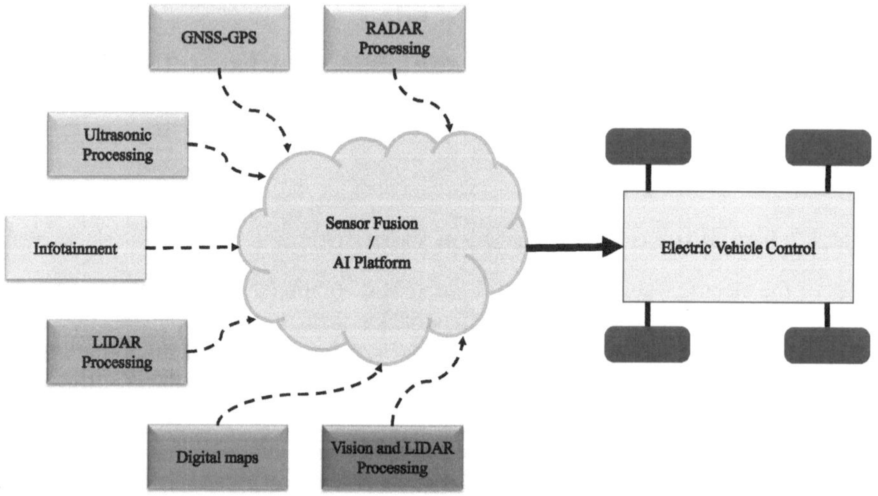

FIGURE 13.10 Sensor fusion concept.

can be minimized and more information from the different platforms is available which will pave the way for a better understanding of the procedures related to the automation [93–95]. Figure 13.10 represents the sensor fusion concept.

13.6 CONCLUSION

Electric vehicles are going to rule the entire transportation industry in the future due to their huge advantages over the traditional IC engine-propelled vehicles. Since these vehicles employ major units such as powertrain, energy storage unit, and electric motors, there are many factors which will influence the overall performance of these vehicles. To overcome this, there is a need for a system which monitors, diagnose, and notify the user about the service condition. This can be done by incorporating various sensor types in the electric vehicle. This chapter elaborates the types of sensors that could be used to monitor the overall performance and diagnose the malfunctions that happen in the electric vehicles. Further, this chapter gives details about the position sensors, encoders, LIDAR, RADAR, ultrasonic sensors, cameras and so on which helps in supervising each unit's performance in the electric vehicles. In addition, the MEMS-based sensor types were discussed along with the sensor fusion concepts that could be incorporated with the electric vehicles in the near future.

CONFLICTS OF INTEREST

The authors declare that there are no conflicts of interest.

ACKNOWLEDGMENT

The authors like to acknowledge the editors of this book for providing an opportunity to write this section.

REFERENCES

[1] Ashton TS. *The Industrial Revolution 1760–1830*. OUP Catalogue, 1997.
[2] Hudson P. *The Industrial Revolution*. Bloomsbury Publishing, 2014.
[3] More C. *Understanding the Industrial Revolution*. Routledge, 2002.
[4] Toynbee A. *The Industrial Revolution*. Read Books Ltd, 2013.
[5] Crafts NF. The first industrial revolution: A guided tour for growth economists. *The American Economic Review*. 1996:86(2):197–201.
[6] Loy AC, Chin BL, Sankaran R. Industrial revolution 5.0. In: *The Prospect of Industry 5.0 in Biomanufacturing*. Show PL, Chew KW, Ling TC, editors. CRC Press, 2021, 1st edn.
[7] van Agtmael A. Industrial revolution 2.0. *Foreign Policy*. 2007(158):40.
[8] Loy AC, Chin BL, Sankaran R. Industrial revolution 1.0 and 2.0. In: *The Prospect of Industry 5.0 in Biomanufacturing*. Show PL, Chew KW, Ling TC, editors. CRC Press, 2021, pp. 1–14.
[9] Greenwood J. *The Third Industrial Revolution: Technology, Productivity, and Income Inequality*. American Enterprise Institute, 1997.
[10] Janicke M, Jacob K. *A Third Industrial Revolution: Long-Term Governance for Social-Ecological Change*. Routeledge, 2013, pp. 47–71.
[11] Shahroom AA, Hussin N. Industrial revolution 4.0 and education. *International Journal of Academic Research in Business and Social Sciences*. 2018;8(9):314–9.
[12] Alaloul WS, Liew MS, Zawawi NA, Kennedy IB. Industrial revolution 4.0 in the construction industry: Challenges and opportunities for stakeholders. *Ain Shams Engineering Journal*. 2020;11(1):225–30.
[13] Harahap NJ, Rafika M. Industrial revolution 4.0: and the impact on human resources. *Ecobisma (Jurnal Ekonomi, Bisnis Dan Manajemen)*. 2020;7(1):89–96.
[14] Abas N, Kalair A, Khan N. Review of fossil fuels and future energy technologies. *Futures*. 2015;69:31–49.
[15] Doğan B, Erol D. The future of fossil and alternative fuels used in automotive industry. In: *2019 3rd International Symposium on Multidisciplinary Studies and Innovative Technologies (ISMSIT)*. IEEE, 2019, pp. 1–8.
[16] Höök M, Tang X. Depletion of fossil fuels and anthropogenic climate change: A review. *Energy Policy*. 2013;52:797–809.
[17] Barbir F, Veziroğlu TN, Plass Jr HJ. Environmental damage due to fossil fuels use. *International Journal of Hydrogen Energy*. 1990;15(10):739–49.
[18] Hoel M, Kverndokk S. Depletion of fossil fuels and the impacts of global warming. *Resource and Energy Economics*. 1996;18(2):115–36.
[19] https://www.ipcc.ch/2018/10/08/summary-for-policymakers-of-ipcc-special-report-on-global-warming-of-1-5c-approved-by-governments/.
[20] Li Z, Khajepour A, Song J. A comprehensive review of the key technologies for pure electric vehicles. *Energy*. 2019;182:824–39.
[21] Chau KT, Wong YS, Chan CC. An overview of energy sources for electric vehicles. *Energy Conversion and Management*. 1999;40(10):1021–39.
[22] Raslavičius L, Azzopardi B, Keršys A, Starevičius M, Bazaras Ž, Makaras R. Electric vehicles challenges and opportunities: Lithuanian review. *Renewable and Sustainable Energy Reviews*. 2015;42:786–800.

[23] Xiong R, Yu Q, Shen W. Review on sensors fault diagnosis and fault-tolerant techniques for lithium ion batteries in electric vehicles. In: *2018 13th IEEE Conference on Industrial Electronics and Applications (ICIEA)*. IEEE, 2018, pp. 406–410.

[24] Yao E, Shao C, Jin F, Pan L, Zhang R. Battery electric vehicles in China: ownership and usage. *Transport and Energy Research*. Elsevier, 2020, pp. 177–222.

[25] Nieuwenhuis P, Cipcigan L, Sonder HB. The electric vehicle revolution. In: *Future Energy*. Trevor M. Letcher, editor, Elsevier, 2020, pp. 227–243.

[26] Hannan MA, Azidin FA, Mohamed A. Hybrid electric vehicles and their challenges: A review. *Renewable and Sustainable Energy Reviews*. 2014;29:135–150.

[27] Zhang F, Wang L, Coskun S, Pang H, Cui Y, Xi J. Energy management strategies for hybrid electric vehicles: Review, classification, comparison, and outlook. *Energies*. 2020;13(13):3352.

[28] Zhuang W, Li S, Zhang X, Kum D, Song Z, Yin G, Ju F. A survey of powertrain configuration studies on hybrid electric vehicles. *Applied Energy*. 2020;262:114553.

[29] Alagarsamy T, Moulik B. A review on optimal design of hybrid electric vehicles and electric vehicles. In: *2018 3rd International Conference for Convergence in Technology (I2CT)*. IEEE, New York, 2018, pp. 1–5.

[30] Wirasingha SG, Emadi A. Classification and review of control strategies for plug-in hybrid electric vehicles. *IEEE Transactions on Vehicular Technology*. 2010;60(1):111–22.

[31] Green II RC, Wang L, Alam M. The impact of plug-in hybrid electric vehicles on distribution networks: A review and outlook. *Renewable and Sustainable Energy Reviews*. 2011;15(1):544–53.

[32] Amjad S, Neelakrishnan S, Rudramoorthy R. Review of design considerations and technological challenges for successful development and deployment of plug-in hybrid electric vehicles. *Renewable and Sustainable Energy Reviews*. 2010;14(3):1104–10.

[33] Muthukumar M, Rengarajan N, Velliyangiri B, Omprakas MA, Rohit CB, Raja UK. The development of fuel cell electric vehicles-A review. *Materials Today: Proceedings*. 2021;45:1181–7.

[34] Sorlei IS, Bizon N, Thounthong P, Varlam M, Carcadea E, Culcer M, Iliescu M, Raceanu M. Fuel cell electric vehicles-A brief review of current topologies and energy management strategies. *Energies*. 2021;14(1):252.

[35] İnci M, Büyük M, Demir MH, İlbey G. A review and research on fuel cell electric vehicles: Topologies, power electronic converters, energy management methods, technical challenges, marketing and future aspects. *Renewable and Sustainable Energy Reviews*. 2021;137:110648.

[36] Singla MK, Nijhawan P, Oberoi AS. Hydrogen fuel and fuel cell technology for cleaner future: A review. *Environmental Science and Pollution Research*. 2021;28(13):15607–26.

[37] Rahman KM, Jurkovic S, Stancu C, Morgante J, Savagian PJ. Design and performance of electrical propulsion system of extended range electric vehicle (EREV) chevrolet volt. *IEEE Transactions on Industry Applications*. 2014;51(3):2479–88.

[38] Ding N, Prasad K, Lie TT. The electric vehicle: A review. *International Journal of Electric and Hybrid Vehicles*. 2017;9(1):49–66.

[39] Dutta J, Mathew R. An overview of sensors in intelligent transportation systems and electric vehicles. In: *AI Enabled IoT for Electrification and Connected Transportation*. Marati N, Bhoi AK, Albuquerque D, Kalam A, editors. Springer,2022, 61–73.

[40] Djahel S, Doolan R, Muntean GM, Murphy J. A communications-oriented perspective on traffic management systems for smart cities: Challenges and innovative approaches. *IEEE Communications Surveys & Tutorials*. 2014;17(1):125–151.

[41] Damaj IW, Serhal DK, Hamandi LA, Zantout RN, Mouftah HT. Connected and autonomous electric vehicles: Quality of experience survey and taxonomy. *Vehicular Communications*. 2021;28:100312.

[42] Xiong R, Sun W, Yu Q, Sun F. Research progress, challenges and prospects of fault diagnosis on battery system of electric vehicles. *Applied Energy*. 2020;279:115855.
[43] Zhou N, He H, Liu Z, Zhang Z. UKF-based sensor fault diagnosis of PMSM drives in electric vehicles. *Energy Procedia*. 2017;142:2276–83.
[44] Alkheir AA, Aloqaily M, Mouftah HT. Connected and autonomous electric vehicles (caevs). *IT Professional*. 2018;20(6):54–61.
[45] Aujla GS, Kumar N, Singh M, Zomaya AY. Energy trading with dynamic pricing for electric vehicles in a smart city environment. *Journal of Parallel and Distributed Computing*. 2019;127:169–83.
[46] Sahayadhas A, Sundaraj K, Murugappan M. Detecting driver drowsiness based on sensors: a review. *Sensors*. 2012;12(12):16937–53.
[47] Phan AC, Nguyen NH, Trieu TN, Phan TC. An efficient approach for detecting driver drowsiness based on deep learning. *Applied Sciences*. 2021;11(18):8441.
[48] Hu J, Min J. Automated detection of driver fatigue based on EEG signals using gradient boosting decision tree model. *Cognitive Neurodynamics*. 2018;12(4):431–40.
[49] Finnefrock M, Jiang X, Motai Y. Visual-based assistance for electric vehicle driving. In: *IEEE Proceedings. Intelligent Vehicles Symposium*. IEEE, New York, 2005, pp. 656–661.
[50] Chen BC, Kuo CC. Electronic stability control for electric vehicle with four in-wheel motors. *International Journal of Automotive Technology*. 2014;15(4):573–80.
[51] Fleming WJ. New automotive sensors-A review. *IEEE Sensors Journal*. 2008;8(11):1900–21.
[52] Zhai L, Sun T, Wang J. Electronic stability control based on motor driving and braking torque distribution for a four in-wheel motor drive electric vehicle. *IEEE Transactions on Vehicular Technology*. 2016;65(6):4726–39.
[53] Ebbesson C. *Rotary Position Sensors Comparative Study of Different Rotary Position Sensors for Electrical Machines Used in an Hybrid Electric Vehicle Application*. CODEN: LUTEDX/TEIE, 2011.
[54] Kim KC, Hwang SJ, Sung KY, Kim Y. A study on the fault diagnosis analysis of variable reluctance resolver for electric vehicle. In: *SENSORS, 2010 IEEE*. IEEE, 2010, pp. 290–295.
[55] Yang H, Jianhua G, Xuhui W. The application of magnetic encoders in induction motor vector control. *Sixth International Conference on Electrical Machines and Systems: ICEMS*. 2003;2:544–547.
[56] Uzair M, Abbas G, Hosain S. Characteristics of battery management systems of electric vehicles with consideration of the active and passive cell balancing process. *World Electric Vehicle Journal*. 2021;12(3):120.
[57] Cherukuri M, Kanthi M. Battery management system design for electric vehicle. In: *2019 IEEE International Conference on Distributed Computing, VLSI, Electrical Circuits and Robotics (DISCOVER)*. IEEE, 2019, pp. 1–6.
[58] Mehendale N, Neoge S. Review *on LiDAR Technology*. Available at SSRN 3604309, 2020 May 18.
[59] Hsu CP, Li B, Solano-Rivas B, Gohil AR, Chan PH, Moore AD, Donzella V. A review and perspective on optical phased array for automotive LiDAR. *IEEE Journal of Selected Topics in Quantum Electronics*. 2020;27(1):1–6.
[60] Baxter JA, Merced DA, Costinett DJ, Tolbert LM, Ozpineci B. Review of electrical architectures and power requirements for automated vehicles. In: *2018 IEEE Transportation Electrification Conference and Expo (ITEC)*. IEEE, 2018, pp. 944–949.
[61] Sarbolandi H, Plack M, Kolb A. Pulse based time-of-flight range sensing. *Sensors*. 2018;18(6):1679.
[62] Koskinen M, Kostamovaara JT, Myllylae RA. Comparison of continuous-wave and pulsed time-of-flight laser range-finding techniques. *Optics, Illumination, and Image Sensing for Machine Vision VI*. 1992;1614:296–305.

[63] Jha A, Azcona FJ, Royo S. Frequency-modulated optical feedback interferometry for nanometric scale vibrometry. *IEEE Photonics Technology Letters*. 2016;28(11):1217–20.

[64] Varghese JZ, Boone RG. Overview of autonomous vehicle sensors and systems. In: *International Conference on Operations Excellence and Service Engineering*. 2015, pp. 178–191.

[65] Campbell S, O'Mahony N, Krpalcova L, Riordan D, Walsh J, Murphy A, Ryan C. Sensor technology in autonomous vehicles: A review. In: *2018 29th Irish Signals and Systems Conference (ISSC)*. IEEE, 2018, pp. 1–4.

[66] Schumann O, Wöhler C, Hahn M, Dickmann J. Comparison of random forest and long short-term memory network performances in classification tasks using radar. In: *2017 Sensor Data Fusion: Trends, Solutions, Applications (SDF)*. IEEE, 2017, pp. 1–6.

[67] Schubert E, Meinl F, Kunert M, Menzel W. High resolution automotive radar measurements of vulnerable road users-pedestrians and cyclists. In: *2015 IEEE MTT-S International Conference on Microwaves for Intelligent Mobility (ICMIM)*. IEEE, 2015, pp. 1–4.

[68] Vargas J, Alsweiss S, Toker O, Razdan R, Santos J. An overview of autonomous vehicles sensors and their vulnerability to weather conditions. *Sensors*. 2021;21(16):5397.

[69] Park WJ, Kim BS, Seo DE, Kim DS, Lee KH. Parking space detection using ultrasonic sensor in parking assistance system. In: *2008 IEEE Intelligent Vehicles Symposium*. IEEE, 2008, pp. 1039–1044.

[70] Nagaoka S, Raihanul IM, Okajima K, Ito R, Watanabe K, Ito T. Evaluation of attenuation of ultrasonic wave in air to measure concrete roughness using aerial ultrasonic sensor. *GEOMATE Journal*. 2018;14(42):158–63.

[71] Xiaoming L, Tian Q, Wanchun C, Xingliang Y. Real-time distance measurement using a modified camera. In: *2010 IEEE Sensors Applications Symposium (SAS)*. IEEE, 2010, pp. 54–58.

[72] Mehta S, Patel A, Mehta J. CCD or CMOS Image sensor for photography. In: *2015 International Conference on Communications and Signal Processing (ICCSP)*. IEEE, 2015, pp. 291–294.

[73] Gade R, Moeslund TB. Thermal cameras and applications: a survey. *Machine Vision and Applications*. 2014;25:245–62.

[74] Olmeda D, de la Escalera A, Armingol JM. Far infrared pedestrian detection and tracking for night driving. *Robotica*. 2011;29(4):495–505.

[75] Rosique F, Navarro PJ, Fernández C, Padilla A. A systematic review of perception system and simulators for autonomous vehicles research. *Sensors*. 2019;19(3):648.

[76] Pastore TH, Everett HR, Bonner K. *Mobile Robots for Outdoor Security Applications*. Space and Naval Warfare Systems Center, 1999.

[77] Bao Z, Hossain S, Lang H, Lin X. High-definition map generation technologies for autonomous driving: A review. *arXiv preprint arXiv:2206.05400*, 2022.

[78] Ernest P. *MEMS@ Bosch: Automotive Application and Beyond*. BOSCH Co, 2010. www.mstbw.de/imperia/md/content/mstbw/bestpractice/bosch_mems_12_micromachine_symposium_ernst. pdf.

[79] Zhang J, Zhang ZH, Chen T, Kong XM, Ren TL, Liu LT. A tire pressure monitoring system based on MEMS sensor. *Key Engineering Materials*. 2011;483:370–373.

[80] Reze M, Osajda M. MEMS sensors for automotive vehicle stability control applications. In: *Mems for Automotive and Aerospace Applications*. Kraft M, White NM, editors, Woodhead Publishing, 2013, pp. 29–53.

[81] Jin L. Applications and prospects of mems sensors in automotive. *Journal of Physics: Conference Series*. 2021;1884(1): 12010.

[82] Classen J, Frey J, Kuhlmann B, Ernst P, Bosch R. MEMS gyroscopes for automotive applications. In: *Advanced Microsystems for Automotive Applications*. Valldorf J, Gessner W, editor. Springer, Berlin, 2007, pp. 291–306.

[83] Babin A, Rizoug N, Mesbahi T, Boscher D, Hamdoun Z, Larouci C. Total cost of ownership improvement of commercial electric vehicles using battery sizing and intelligent charge method. *IEEE Transactions on Industry Applications*. 2017;54(2):1691–700.

[84] Alvarez AD, Garcia FS, Naranjo JE, Anaya JJ, Jimenez F. Modeling the driving behavior of electric vehicles using smartphones and neural networks. *IEEE Intelligent Transportation Systems Magazine*. 2014;6(3):44–53.

[85] Klomp M, Gao Y, Bruzelius F. Longitudinal velocity and road slope estimation in hybrid electric vehicles employing early detection of excessive wheel slip. *Vehicle System Dynamics*. 2014;52(sup1):172–88.

[86] Lee M, Nam K, Lee S, Kim H, Kim C, Park Y, Ahn B, Kim T, Seo H. Design and fabrication of low-pressure piezoresistive MEMS sensor for fuel cell electric vehicles. In: *2014 IEEE 64th Electronic Components and Technology Conference (ECTC)*. IEEE, 2014, pp. 1616–1621.

[87] McCann RA, Damugatla SR. Analysis of MEMS based rotor flux sensing in a hybrid reluctance permanent magnet motor. In: *2005 IEEE Vehicle Power and Propulsion Conference*. IEEE, 2005, p. 6.

[88] Sun H, Zhou H, Li X, Wei Y, Li X. Design of two-wheel self-balanced electric vehicle based on MEMS. In: *2009 4th IEEE International Conference on Nano/Micro Engineered and Molecular Systems*. IEEE, 2009, pp. 143–146.

[89] Abdelhamid S, Hassanein HS, Takahara G. Vehicle as a mobile sensor. *Procedia Computer Science*. 2014;34:286–295.

[90] Guerrero-Ibáñez J, Zeadally S, Contreras-Castillo J. Sensor technologies for intelligent transportation systems. *Sensors*. 2018;18(4):1212.

[91] Bhatti G, Mohan H, Singh RR. Towards the future of smart electric vehicles: Digital twin technology. *Renewable and Sustainable Energy Reviews*. 2021;141:110801.

[92] Yeong DJ, Velasco-Hernandez G, Barry J, Walsh J. Sensor and sensor fusion technology in autonomous vehicles: A review. *Sensors*. 2021;21(6):2140.

[93] Dai D, Chen Z, Bao P, Wang J. A review of 3D object detection for autonomous driving of electric vehicles. *World Electric Vehicle Journal*. 2021;12(3):139.

[94] Elvas LB, Ferreira JC. Intelligent transportation systems for electric vehicles. *Energies*. 2021;14(17):5550.

[95] Juang JC, Li TH, Chen SH, Li YC, Chen YJ. Development of a sensor fusion system for the navigation and control of an autonomous electric vehicle. In: *25th World Battery, Hybrid and Fuel Cell Electric Vehicle Symposium and Exhibition: Sustainable Mobility Revolution*. EVS, Shenzhen, China, 2010.

14 Trends in Electrical and Automated Vehicles

M. Baskaran, V. Sundararaju, and C. Vijayakumar
K.S. Rangasamy College of Technology

S. Senthilraja
SRM Institute of Science and Technology

14.1 INTRODUCTION

Transport vehicles, specifically two-wheelers, three-wheelers, four-wheelers, and buses, are referred to as electric vehicles (EVs) in general. Any vehicle that has electric combustion and operates with driver assistance or fully automatically falls under the category of an electric car. The purpose of transport automation is to eliminate the "human element" from transportation systems to avoid accidents.

All computer-controlled modes of land, air, and sea transportation would be included in a fully automated system. Nevertheless, there is a continued global drive towards electric and autonomous vehicles, supported by government incentives and policies. By 2025, pure EVs will compensate about 12% of all new vehicles sold worldwide. That percentage will surpass 50% by 2040. The first companies to produce an affordable, unsubsidised electric car will have a significant competitive edge. The battery, which makes up 30% of the production expenses, is one important component in terms of costs. Therefore, this step change will be made possible through advances in technology and scalable, streamlined production.

Globally, there were more than 500 electric car models on the market in 2020, a rise of more than 12% from the offerings in 2019 and more than twice as many as there were in 2017. Driving range continues to be prioritised by consumers. To boost sales, automakers generally seek for longer ranges. On the other side, increasing the driving range typically demands the use of larger batteries, additional materials, and more merchandise. When a market optimum vehicle range is established and quick charging becomes more accessible, it is projected that driving range will plateau over time.

Sales of EVs have grown exponentially in recent years, offering greater range, a wider choice of models and superior performance. If the growth of the past 2 years continues, CO_2 emissions from vehicles could follow a trajectory consistent with the net zero emissions scenario of 2050. We forecast that 17% of new cars sold in 2023 will be electric. However, EVs have not yet become a mainstream phenomenon. Sales have lagged in poor and developing countries due to higher supply costs and a lack of charging infrastructure.

DOI: 10.1201/9781003354901-14

Improved efficiency in that most EVs and AVs feature high-density batteries that have a greater capacity for energy storage than ordinary batteries of comparable size. The possibility of having a denser pack means that the battery design for autonomous vehicles might be streamlined, freeing up more room for the installation of other high-tech car parts. According to recent studies, autonomous vehicles convert around 59%–62% of electrical energy into power, whereas gasoline automobiles only transform about 17%–21%. The internal combustion engines' heat dissipation is the reason for this discrepancy in conversion rates. In addition, the electrification of automobiles makes it possible to use higher voltages while using less current to provide a certain amount of electricity. For instance, a 4× increase in the 12 V supply voltage, or 48 V, results in a 16× loss in power. Additionally, the integration of technology such as the global positioning system (GPS), vehicle sensors, such as inertial sense, position, direction, speed, and LiDAR scanning method, has brought the concept of autonomous cars to the complete satisfaction of all stakeholders.

14.2 SIGNIFICANCE OF ELECTRICAL VEHICLE TECHNOLOGY

The life cycle of an EV is fraught with several difficulties, as we can see by visualising the production process as follows.

14.2.1 Electric Vehicle Manufacture

EVs are now domestically built in every nation, but they lack the in-house capability to produce our personal lithium-ion batteries. Aside from the aforementioned problem, we do not strictly follow international best practices when it comes to the testing and certification of trade in lithium-ion batteries. There may be fluctuations in every aspect during the numerous means of travel for the lithium-ion batteries imported from abroad, leading to impacts, vibration, unusual temperature conditions, and wear and tear.

Since handling sensitive lithium-ion batteries improperly has the potential to affect their quality and dependability, the cells must undergo stringent inspection, testing, and certification procedures prior to importation in order to obtain the United Nations 38.3 Certification. The Government of India needs to step up and impose the mandatory battery management system (BMS) safety requirements in order to guarantee the dependability and quality of all imported cells without fail. The most demanding certification standard we now have is AIS 156, which includes a test for fire resistance in which the battery is exposed to both direct and indirect flame for more than 2 minutes. For the sake of safety, the fire resistance test ought to be performed at the assembly level with self-certification. No safety should be compromised while producing or assembling related EV components. Only then will we be able to fully overcome any potential flaws and other harm done to the imported battery cells.

14.2.2 Safe Design Aspects

A battery unit is made up of many single cells, a module is made up of many units, and a battery pack is made up of numerous modules. As previously said, a BMS becomes a crucial part of electric vehicle technology in order to keep track of the battery packs' state of charge (SoC) and state of health (SoH). Figure 14.1 shows typical parameters which significantly affect the battery performance. SOH is a combination of depth of discharge, SoC, and C-Rate (rate at which the battery is charged/discharged) (DoD). The amount of stored energy utilised during each cycle is gauged by the depth of discharge.

It is obvious that the battery sizes in EVs will grow as technology develops because the battery's storage capacity, driving range, and charging speed have all seen constant improvement. The battery storage cabinet in EVs has engineering issues as a result of this rapid technological improvement. Additionally, this improved performance energy storage device must be naturally resistant to all unfavourable circumstances, including crash/impact damage and unusual weather.

14.3 EV FIRE ACCIDENTS IN INDIA

The battery of a pure electric two-wheeler detonated in a home in Telangana's Nizamabad District, an old man was killed and a few other people were wounded. In another terrible event involving a single electric two-wheeler, a Boom Motors e-scooter exploded while being charged at home, killing a 40-year-old man in Vijayawada, Andhra Pradesh [1]. The same incident also severely burned his wife, and two kids. In the summer of March and April 2022, three pure EVs, one OLA, three Okinawa, and twenty JITENDRA EV scooters have already caught fire [2]. Most of the Southern States in India have experienced temperatures that are almost nearer to or beyond 40°C during summer. It is so obvious that the major opponent of lithium-ion battery (LiB) chemistry has caused numerous calamities in Andhra Pradesh, Telangana, Tamil Nadu, and Maharashtra.

In the most recent incident, which happened in Hosur, Tamil Nadu, an e-seat, scooter's which is where most Indian two-wheelers carry their lithium-ion batteries, caught fire. Frequently, there is limited ventilation and the battery is enclosed in heat-absorbing black plastic. The likelihood of an explosion increases when there is a concentration of chemical energy in a restricted area. In April 2022, the Indian government established a central panel to look into the causes of recent EV fires and explosions.

14.3.1 Causes of Fire in EV

Heat is the LiB's first adversary, according to a widely accepted theory about the chemistry of these batteries. When the rated power supply is at its best, the LiB should operate between 150°C and 300°C. The power capacity is determined by cold cracking when the temperature falls below 150°C. When the ambient temperature rises above 300°C, the traditional "thermal runaway" occurs, causing the system to collapse into fire or explosion. The BMS of today faces three research challenges: reliability, efficacy, and safety. Lithium-ion batteries are prone to "thermal runaway," a weak chemical retort brought on by a number of surroundings such as over-voltage and high temperatures [3].

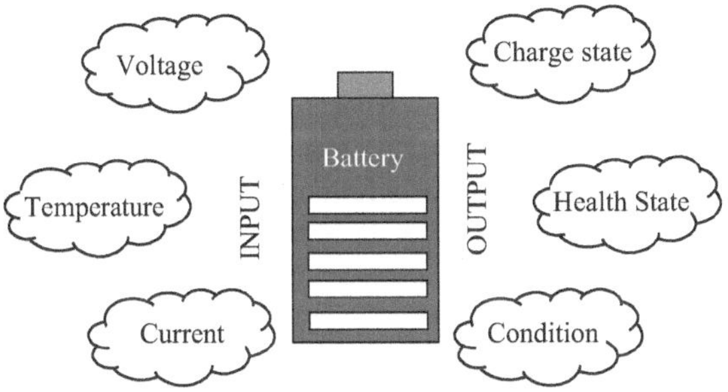

FIGURE 14.1 Battery management system.

The following are scenarios that could be vulnerable to risk as a result of the introduction of EVs:

- Battery fires while EVs are being charged on-board in a home/office,
- Battery fires are caused by a rise in the surrounding temperature or a fire nearby, battery fires following an accident colliding with another vehicle or structure,
- Release of flammable, explosive, or toxic gases that results in fire, explosion, or fatalities
- Thermal runaway is brought on by a core short-circuit caused by improper handling or powers during manufacturing or transportation activities.

EV fire suppression becomes more challenging when the on-board battery catches fire because the burning battery is unapproachable to heavily sprayed suppressant and can re-ignite without adequate cooling. As a result, a lot of suppressants is required to lower the heat in the battery, put off the fire, and completely prevent re-ignition [4]. The strength and durability of the material used to construct the battery storage cabinet that holds the LiB pack in EV must thus be improved through extensive research. Automated components based on metal matrix composites (MMCs) for the production of LiB storage cabinets can be viewed and perfected after careful evaluation focused on research and development [5]. The right combination of MMCs will help resist the heat and dust of the atmosphere as well as adverse conditions due to favourable parameters such as lightweight, long life and high resistance to corrosion, wear and shock collisions/collision situations.

14.3.2 Battery Charging Infrastructure

Since lithium-ion batteries are used in stationary storage applications and EVs, the higher number of energy densities increases the risk of potentially catastrophic events. The concept of safety for lithium-ion batteries is inversely related to the concept of misuse. The former falls to zero as the latter rises [6]. The instances that have happened around the world, but particularly in our nation, and have received a lot of attention from business, academia, and the general public. Fast battery charging, which is essential for EV applications, raises the risk of thermal runaway and other safety concerns.

An approach used to classify the safety of lithium-ion batteries is the hazard level recommended by the European Council for Automotive Research and Development (EUCAR), as shown in the accompanying Figure 14.2.

Recyclable batteries present a new problem for battery management systems (BMSs). Within the next 10 years, it is anticipated that sales of EVs would increase by about 500%. Modern BMS algorithms are only useful for first-time battery use because they rely so much on earlier characterisation performed in labs. Traditional BMS approaches that rely on empirical modelling and assume that batteries of the alike chemical composition and size have identical properties will not be sufficient to achieve used batteries because the initial usage of a battery affects its electrochemical properties in different ways.

The BMS technology available today is insufficient to correctly forecast a battery's SoH. Either premature battery replacement or waiting until a specific failure event occurs are options [7]. Both of these options have unfavourable outcomes: early replacement would increase costs for the consumer and generate excessive waste for the environment; waiting will reduce the consumer's safety and experience quality. For EV users to provide convenient charging services and to increase public acceptance of EVs, charging station placement is crucial. The placement of EV charging stations has received a lot of attention, but less attention has been paid to how EV drivers use their charging stations, which is crucial for determining how well they work. EV drivers choose between charging stations based on a variety of criteria, including as proximity, cost, and the length of the queue at each station, among others. In turn, EV drivers' decisions have a big impact on some issues like queuing.

FIGURE 14.2 Classification of lithium-ion battery hazard levels as per EUCAR.

Large queuing times should be anticipated as more EVs use the same station. To minimise the social cost associated with EV drivers' congestion at charging stations. This chapter offers a behavioural model as a starting point for representing how EV drivers choose charging sites. An ideal charging station placement model is then presented based on this concept.

Battery swapping, which is the act of exchanging a discharged battery for a fully charged battery at a battery swap station, represents another obstacle to EV safety (BSS). The BSS serves as a battery aggregator, providing the infrastructure for the charging of numerous batteries and making charged batteries available to aspirational EV drivers [8]. The adoption of the shared economy concept in the field of energy storage has been made possible by this approach. The Ministry of Road Transport and Highways broadcasted in 2020 that battery packs are not essential for the sale of electric two- and three-wheelers.

By removing the cost of the battery from the total cost of the car, this claim's main goal is to reduce the price of electric automobiles. The original equipment manufacturers (OEM) can supply the battery separately and function as an energy service provider. The ecosystem for battery swapping and EV adoption will benefit from the battery swapping station (BSS) [9]. The standardisation of battery packs, the dependability of the battery packs that are leased, and a commercially viable business model are all obstacles to this approach. Many businesses in our nation are stepping forward to commercialise switching technology, with Sun Mobility leading the way.

14.4 BUILDING ICT ARCHITECTURE IN EVs

Information and communication technology (ICT) in electric automobiles will become increasingly crucial as a result of electro mobility. This will have a significant impact on the automotive sector, particularly in light of prospective ICT architecture development to secure the predicted security matrix in EVs in line with social developments. ICT, for instance, can assist in lowering the required energy per kilometre because of the following circumstances:

- To lessen the total weight of the EVs, ICT can implement functions in software that were formerly realised mechanically or as distinct hardware pieces;
- Further, sophisticated predictive management tools can cut the amount of energy usage;
- ICT can assist reduce the incidence of accidents by incorporating proactive safety features: future cars must be able to see and comprehend the surrounding environment to react quickly in harmful situations [10].

14.5 MASS COMMUNICATION ON EV SAFETY

It is essential to enhance consumer knowledge regarding EVs safety so they are aware of the engineering difficulties, such as thermal runaway, which can result in disastrous circumstances and countless fatalities in the mobility industry. Mass communication is conceivable given the current trend of communication detonation through mass media platforms like LinkedIn, WhatsApp, Twitter, Instagram, Facebook, etc. It is important to develop a global emergency response protocol as well as strategies for accurate documentation and the impacted people's ongoing medical care [11]. Following the Bhopal Disaster of 1984, which left thousands of people dead and thousands more injured, we have already carried out a similar exercise. Thanks to the Government of India's tireless efforts, technical terms like "assembly point," "escape route," "emergency plan," "mock drill," and others are now constantly in the minds of everyday people thanks to the advanced knowledge, skill, and expertise that was once only available abroad and was later brought to India through constant education and training.

In a similar vein, we must accept the challenge and implement safety codes of practices for EV manufacturing, SoP for EVs charging, recognised internationally best practices for battery substitution, as well as public relations campaigns on the prevention of EV fires and explosions. Seeing the right safety posters that clearly depict the appropriate safety code of practice should provide consumers with information about safety messaging as shown in Figure 14.3 [12].

14.6 REVOLUTIONS IN AUTOMATED VEHICLES

Due to their many benefits, including environmental sustainability and improved efficiency, EVs and AVs have caught the attention of policymakers worldwide as the clean technology options. The by-products of burning fossil fuels for energy purposes include hazardous pollutants and greenhouse gases (GHGs). The primary factor contributing to rapid climate change, including polar ice melt and global warming, is the release of GHGs. The GHGs are mainly comprised of carbon-di-oxide (CO_2), carbon monoxide (CO), nitrous oxides (NO_x) and methane (CH_4). The aforementioned issues are overwhelmed with the usage of EVs. EVs come in a wide variety of forms, as listed in Table 14.1.

According to the level of intelligence and automation capabilities, the Society of Automotive Engineers (SAE) has defined six levels of driving automation, ranging from 0 (completely manual) to 5 (totally autonomous) as shown in Figure 14.4.

14.6.1 Level 0 (Conventional Driving)

Today's roads are filled with a lot of Level 0 automobiles. Manually operated. Despite the availability of assistance technologies, the driver is still responsible for

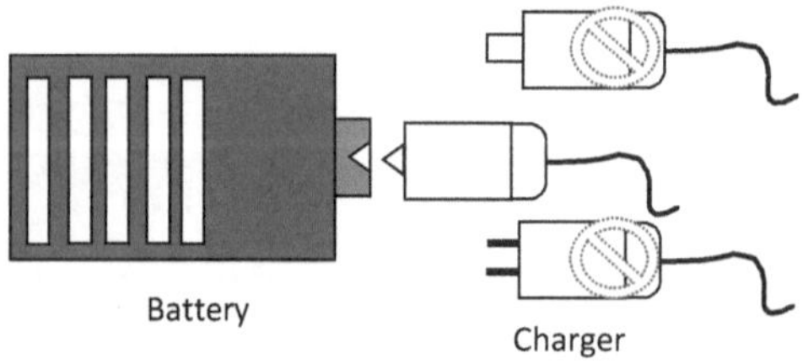

FIGURE 14.3 Appropriate charger selection.

TABLE 14.1
Types of Electric Vehicles [13]

Types of EVs	Description of EVs
BEV	A battery electric vehicle (BEV) is an all-electric: other synonyms include "pure electric vehicle" and "only-electric-vehicle". BEVs do solely rely on the power of an internal battery for energy.
HEV	Hybrid electric vehicles (HEVs) are cars that cannot be charged using an external power source. The vehicle combines a traditional Internal Combustion Engine with electric battery propulsion. They are also capable of converting other excess energy, such as kinetic energy from braking into electric power for internal electronics like Radio and Dash-board Light.
NEV	In China, NEL refers to a class of new electric vehicle encompassing both plug-in-electric -vehicle and hybrid electric vehicles. In the United States, however, NEV is the term given to the neighbourhood electric vehicle that have legal limitations to the speed of no more than 72 km/h (45 moles/h).
PHEV	An external source can be used to charge a PHEV, although they can also rely on an engine and fuel power. A PHEV's range, which relies solely on electric power, is often denoted by a number after PHEV acronym; for example, PHEV 32 KM.
REEV	PHEVs with a series of power train: the main components that generate and deliver power to the road surface, are sometimes referred to as Range-Extended-Electric-Vehicle (REEV).

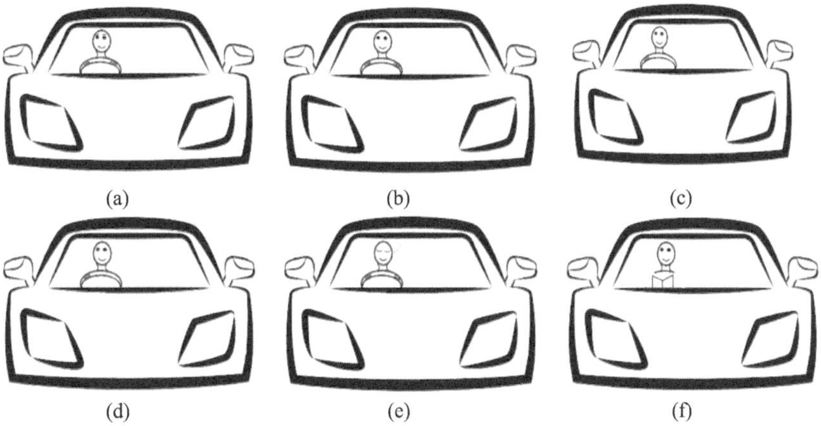

FIGURE 14.4 Levels of driving automation as per the Society of Automotive Engineers (SAE) (a) conventional driving, (b) driver assistance, (c) partly automated, (d) controlled automation, (e) provisional automation and (f) fully automated.

performing the "dynamic driving responsibility". As an example, consider the emergency braking system, which isn't considered automation because it doesn't "steer" the car.

14.6.2 Level 1 (Driver Aided)

This phase of automation is the lowest.

One autonomous driving assist system that may steer or accelerate is present in the vehicle (cruise control). Adaptive cruise control, which enables a vehicle to be maintained at a safe distance behind a vehicle in front of it, is categorised as Level 1 since the human driver is in charge of all other aspects of driving, such as steering and braking.

14.6.3 Level 2 (part Automation)

This refers to sophisticated driving assistance technologies (ADAS). The car has steering, acceleration, and deceleration controls. As a human is still in the driver's seat and has the ability to take over at any time, this automation falls short of self-driving in this instance.

14.6.4 Level 3 (Control Automation)

The capability of Level 4 vehicles to interpose when something goes incorrect or a system miscarries is the primary distinction between Level 3 and Level 4 automation. These cars normally don't need to engage with people in this way. However, a person can still manually override.

14.6.5 Level 4 (Provisional Automation)

The capability of Level 4 vehicles to interpose when something goes incorrect or a system miscarries is the primary distinction between Level 3 and Level 4 automation. These cars normally don't need to engage with people in this way. However, a person can still manually override.

14.6.6 Level 5 (Fully Automation)

With Level 5 cars, the "dynamic driving challenge" is no longer necessary. Vehicles at Level 5 won't even have steering wheels or pedals for stopping or acceleration. They will be able to go wherever, be unrestricted by geofencing, and perform any duty that a qualified human driver can. Fully autonomous vehicles are being tested in several parts of the world, but no autonomous vehicle is yet running for public use.

There are currently two revolutions revolving side by side in the global automotive industry:

- *Number-1*: Transition to electric power and
- *Number-2*: The rise of electric and automated vehicles

The promotion of EVs and AVs has led to the development of yet another invention: the Software-as-a-Service (SaaS) distribution paradigm for cloud computing. Similar to the SaaS model, which provides economical services without the need to install the necessary Licensed Software, this one-click access to the specific App is advantageous in the EV charging industry. With a quicker and more effective payment process, SaaS is supporting both the plug-and-charge and the scan-and-charge methods. Using mobile devices on the scene, the car could be charged right away. Furthermore, once transportation transitions to a pay-as-you-go model entirely, advanced driver assistant system (ADAS) and the trend of shared mobility and autonomous driving together have the potential to upend vehicle ownership and insurance. Additionally, battery swapping stations are set up at predetermined intervals to allow for the replacement of the vehicles partially or entirely depleted battery with a fully charged battery.

14.7 NEED OF AUTOMATION IN TRANSPORTATION SYSTEM

The Henrich's Domino Model of Accident Causation states that risky activities, often known as human error, are to blame for 80% of accidents; however, this percentage may as well be 100% given that all decisions that lead to accidents are made by people. Regardless of whether it is a work-related accident, a maritime accident, a building site accident, or a transportation accident, someone makes a decision, or omits something, which causes the accident. This is merely to acknowledge that things are as they are, not to place blame. We naturally concentrate on people, creating innovative techniques for enhancing the safety culture or constructing mechanisms that exclude individuals from the system. Electric and automated vehicles were created as a result of one such revolutionary design method. Nevertheless, it is crucial to understand that we do not support the doctrine of complete disengagement from the innovation. Because we are fully aware that the human brain is the only thing that can serve as the ultimate foundation for all future advancements in automated and EVs.

The environmental sustainability and additional benefits that could be attained through electrification are yet another good argument for the automation of the transportation system. Additionally, autonomous vehicles might be able to maximise eco-motorisation for single vehicles and increase their energy effectiveness within connected vehicle groups. It may also be possible to lessen the traffic congestion that increases vehicle emissions by developing eco-motorisation programming for a wide range of prospective cars. The deployment of autonomous vehicles has the potential to further the sustainability of the environment through their use.

With the development of lithium-ion batteries, our obligation to preserve the environment has led us to choose EV and automated vehicle (AV) technologies (LIBs). One such wonder that doesn't emit any greenhouse gases (GHG) is EV technology. We'll be astonished to learn that 23% of the world's emission comes from the transportation sector. Many nations, including India, have pledged to address the growing problem of air pollution through standard laws and goods. Numerous illnesses, including lung cancer and acute respiratory distress syndrome, are associated with air pollution, which also contributes to the ongoing global mortality rate. The electrification of transportation vehicles is thought to be the most viable and workable solution for keeping the air quality on track in time, out of many different ways and means to address the aforementioned issue. The NITI Aayog, is the top policy think tank of the Government of India and is responsible for developing the Electric Vehicle Mobility Vision 2030 [14]. NITI Aayog offers incentives to investors, such as cash subsidies based on the total value-added Kilowatt Hour of the nation.

Only 10.5 lakhs EVs were sold in 2022 in India, which is around 5% of automobile sales in India, according to the recently released e-Mobility Revolution for India's Transportation Portal, and 5,100 charging stations were installed in public areas in India. The India Energy Storage Alliance (IESA) has predicted in a different analysis that the market for EVs in India will expand at a compound annual growth rate (CACR) of 36% through 2026. The rationale for the anticipated expansion of the EV industry is that, despite the higher initial investment cost, EVs compete favourably with petrol and diesel vehicles when it comes to formal decisions regarding fuel and maintenance costs. Thus, EVs have established themselves as "environmentally benign" and "cheap" alternatives to gasoline and diesel vehicles.

The vehicle population, particularly those with internal combustion engines (ICEs), had significantly contributed to the deterioration of the air quality during the Corona crisis with a national and state-level lockdown, and in 2020, 36 out of 50 cities had dangerously high levels of air pollution. As a result, the Indian government set a goal to replace all new ICE (petrol/diesel) car sales with plug-in EVs by 2030. Since the start of 2019, the faster adoption of manufacturing hybrid and electric vehicles (FAME) II Scheme has gained popularity. It is critical to keep in mind that demand-side incentives were taken into consideration when the FAME II Policy, India's general policy, was being developed. A total of 100% of the available funds will be used to create the infrastructure for EV charging, with 86% going towards consumer incentives for EV purchases. NITI Aayog and Rocky Mountain Institute have examined the oil, net energy, and net CO_2 release savings related with these vehicles. To clarify how FAME II and other supportive measures would affect India's current vehicle market and how it will affect oil usage. The results of the analysis are shown in Figure 14.5.

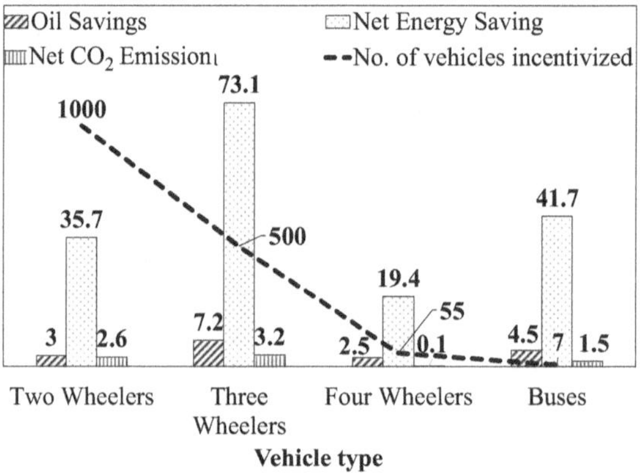

FIGURE 14.5 Faster adoption of manufacturing hybrid & electric vehicles (FAME) II effect.

14.8 IMPORTANCE OF BMS

An e-vehicle's battery pack is made up of several modules of cells, each of which is made up of different individual cells. The fact that each cell in each module can be charged and drained at a different pace makes it challenging to control the performance of a battery pack. Additionally, due to variations in temperature, health, and charge, each cell has a unique operational state. Therefore, it is necessary to monitor each battery cell separately to ensure efficient and secure operation. To track the charge rate across the entire pack down to the cell level, a BMS is needed. This monitoring assures secure and dependable battery operation, leading to the development of safer and more effective EVs.

14.8.1 Lithium Ion Battery Technology

The terms "electric vehicles" (EVs) and "automated vehicles" (AVs) are only used in this chapter to refer to automobiles that are either completely or partially propelled by a LiB pack that supplies all of the electrical energy. The propulsion system, which is referred to as a technology differentiator, is the only vehicle component that differs significantly amongst all other vehicle components.

Propulsion systems that employ conventional technology have been popular for many years. Concerns about climate change, however, have led to a paradigm shift towards the creation and usage of low-carbon or electricity-based cars. Currently, the only connection between autonomous vehicles (AVs) and electric cars (EVs) is that they both share a letter in their acronym. In general, the level of human involvement distinguishes automated from autonomous systems. An autonomous vehicle is more intelligent and independent than an AV. Therefore, the terms "driverless," "autonomous," "self-driving," and "automated" are closer synonyms.

EVs use rechargeable battery packs comprised of many cell modules organised in parallel and series. Simple fuel tanks for internal combustion (IC) engines and complicated fuel tanks for EVs can be compared. LiB packs, the most popular technology for EVs, benefit from features like unlimited energy and power density, extended cycle life, lightweight, and small size.

The BMS, an implanted system that keeps an eye on the parts closer to the battery cell, is necessary for LiB technology because the battery cell needs to be closely watched to ensure that there are no voltage variations or imbalances in the voltage circumstances. Additionally, BMS controls cell optimisation through cell balancing, which prolongs the battery's lifespan. At no point in time can the lithium-ion batteries be overcharged or permitted to reach a deep discharge state. When a battery is being charged or overcharged, the current running through it can result in a phenomenon called thermal runaway, which raises the temperature of the cell. Such circumstances can reduce the battery's lifespan or capacity [15]. It becomes imperative to consistently check the voltage and current flowing through the BMS to make sure the aforementioned condition is not present. The BMS ensures that these particular functions work as intended. For example, just as all vehicles have a gasoline gauge, EVs should also have a battery SoC gauge to establish an efficient management of the discharging current over time.

14.9 CASE STUDIES IN AUTONOMOUS VEHICLES

Nearly every day, stories about autonomous vehicles, driverless cars, and autopilot are reported. Every business in this industry is skilled at using the numerous buzzwords that are associated, and a lot of them are investing a lot of money to keep their business safe. The general public won't be able to buy a fully self-driving car for many more years, despite the huge advancements in the technology that underpins them. It may not be possible to give an exact year at this time. According to some predictions, fully autonomous vehicles may be widely used by 2035. To assure the technology's smooth development, the challenges must be overcome. Driverless cars, along with drones and the Internet of Things, have been named as one of the primary disruptors in the impending technology revolution and as a crucial topic for future research since the Internet and smartphone revolutions.

Tesla's autopilot may be a sophisticated driver-assistance technology that combines the facility to call the car to and from a garage or parking space with lane centering, an adaptive controller, self-parking, and the ability to physically change lanes. The company's stated intention is to eventually offer full self-driving (FSD) as an upgrade to the standard autopilot's capabilities, while noting that there are still technical, legal, and regulatory obstacles to be cleared.

Human errors while driving lead to 90% of vehicle crashes in the world. The tireless efforts in the development of safety systems such as airbags, child seating, steering wheel with energy absorption and so on reduces the accidents [16]. The technological development in the automation and automobile industries made the development of autonomous vehicles so-called self-driving cars. The level of automation has been categorised in five levels by the SAE [17]. In that Level 5, express the full level of automation where there is no manual assistance in the operations. Though self-driving cars have the potential advantages, the safety of the passenger as well as the public is most considerable and essential one. Before implementing self-driving cars on road, they need to be tested and certify its performance, reliability and safety. Some of the major companies involved in the development of self-driving cars are Tesla, Google, Waymo, Delphi, and Audi.

In 2009 Google started its development of self-driving cars in a confidential way, and in 2010, it announced it in its official blog [18]. From the report of the *New York Times*, it was known that the Toyota Prius is used for self-driving cars which were integrated with the GPS. Google started the testing of self-driving cars on road under the manual supervision in 2009 [19]. A report on the on road crashes during the testing of self-driving cars in the United States shows 11 crashes in autonomous mode and five crashes in conventional mode by the Google self-driving car; 0 crash in autonomous mode and one crash in conventional mode by the Delphi self-driving car; 0 crashes in both autonomous mode and conventional mode by the Audi self-driving car [20]. Out of 10 Google car crashes in the Mountain View, United States only three were found to be reportable by the police. Besides, the Google disengagement report 2015 shows that the test driver was indicated by the simulation as the vehicle is likely to be in contact with an object or a vehicle at 13 instances [21]. Of that 13 instances, 10 instances caused the contact, and of that two were been in the traffic cones. In many instances, the crashes which need to be reported to police are not often reported owing to several reasons. From the analysis by Second Strategic Highway Research Program (SHRP2) the crash rate by the conventional vehicles is higher than that of Google cars.

Coming to the passenger vehicles Autopilot from Tesla, Super cruise from General Motors, Pro pilot from Nissan is in the development of self-driving cars at Level 2 automation [22]. Besides some of the companies like Mercedes Benz, Embark, Kodiak Robotics, and TuSimple are been on the development of autonomous trucks. The trucks are the one which rides on the road for a longer time at higher speed. TuSimple used the cameras to sense and stop the trucks instead of LIDAR technology. Uber tested an autonomous vehicle in 2018, and the unfortunately it had a hit on the pedestrian. The incident shows that the autonomous vehicle is not able to detect the objects and people around them.

Due to the vehicle collision of 554 road users were killed in Germany in 2020 and a high number of accidents were reported in the curves [23]. Tesla's Autopilot and VM's Travel Assist are developed with advanced driver assistance system and are experimented on road. The test showed no deviation in the track at the curve radius of ≥370 m, whereas at the curve radius of 225–265 m, Tesla autopilot showed no deviation and VM Travel assist showed deviation in three out of four curves. Tesla Autopilot had a collision with the Michigan state police trooper on 17th March 2021 and the police reports that two such collisions happened in the same week [24]. Tesla had a crash with the highway patrol cruiser on 28 August 2021 which caused minor injuries to both the vehicle drivers. The report shows the Tesla model S and Tesla Model 3 underwent a crash on 14 July 2020 and 7 December 2019, respectively. Again the Tesla model S underwent a crash in January 2018 and 11 May 2018. In the above incidents, the investigation says the drivers are observed to be over-relied on the autopilot and was careless.

Waymo started self-driving car project in 2005 and landed its first self-driving car on road in October 2015 [25]. With the objective of reduction of injuries, death from public accidents and public user mobility enhancement Waymo initiated the self-driving cars. It was shown that the 47 crashes occurred in the automated driving of 6.1 million miles. Besides, 16 crashes were reported in rear-end collision with the stationary vehicle and at slowing down for traffic management. Of the 47 crashes, 18 were actual incidents and 29 were been simulated events where there are no severe injuries [26].

14.10 FUTURES OF EV

Customers are choosing low-emission transportation over environmentally harmful fossil fuel-powered vehicles, and governments throughout the world are supporting and regulating the EV business. The business did not take off as a result of the original EVs' extremely high initial cost, constrained battery range, sluggish speed, and minimal environmental considerations. Nonetheless, interest from OEMs, consumers, and governments has significantly increased during the past 10 years. As a result, substantial investments in EVs manufacturing and battery technology have been made, and millions of cars have been sold in numerous nations.

The whole cost of owning a vehicle includes significant expenses for operation and upkeep. It is now well known that EV purchasers spend significantly less on fuel/energy and maintenance because EVs have fewer moving parts than a gasoline engine and are hence simpler to operate. In EVs, the components that could break down and require work are rather small. As a result, replacing the engine oil is not a concern, making it easier and less expensive to maintain electric cars. Oil causes a range of problems because it is still a key source of energy and is necessary for transportation. With an EV, however, everyone can travel more safely and sustainably. The need for millions of barrels of oil will likely shrink as there are more EVs on the road.

14.11 CONCLUSION

The major issue in automotive vehicles is pollution which induces the global Green House effect. This shifted the focus to creating and releasing EVs with little emissions. The performance of the EV depends on the battery technology and the BMS. The recent trend is the development of self-driving vehicles i.e., passenger cars and trucks. This chapter disclosed the case studies on the recent development and its limitations on the deployment of self-driving cars on road. From the case studies it is perceived that the crashes induced during the testing of self-driving cars are majorly due to the drivers overreliance on the autopilot and their carelessness. Therefore it concludes the usage of self-driving cars will reduce the accidents and still there is a need for the enhancement of self-driving cars to be made completely on road.

REFERENCES

[1] *Times of India Newspaper*, April 15, 2022
[2] *Deccan Herald Newspaper*, May 1, 2022
[3] B. Balasingam, M. Ahmad and K. Pattipati, Battery management systems, challenges and some applications. *Energies*, 13, 2825, 2020.
[4] P. Sun, R. Bisschop, H. Niu and X. Huang, A review of battery fires in electric vehicles. *Fire Technology*, 56, 1361–1410, 2020.
[5] H. Singh, G. Singh Brar, H. Kumar and V. Aggarwal, A review of metal matrix composites for automobile applications. *Materials Today Proceedings*, 43, 1, 2021.
[6] E. Cabrera-Castillo, F. Nidermer and A. Jossen, Calculation of the state of safety (SOS) for lithium-ion battery. *Journal of Power Sources*, 324, 509–520, 2016.
[7] National Science Foundation, *Lithium Battery Safety, Handling and Storage* (STEPS-SOP-Version 1). 2015, United States Antarctic Program.
[8] A. Lam, Y.-W. Leung and X. Chu, Electric vehicle charging station placement: calculation of the state of safety (SOS) for lithium-ion battery. *EEE Transactions on Smart Grid*. doi:10.48550/arXiv.1310.692
[9] EV Reporter, Battery swapping: A paramatic complement to EV charging. *EV Reporter*, September 23, 2019.
[10] C. Buck, A. Comex and C. Simon, *The Software Car: Building ICT Architecture for Future Electric Vehicles*. IEEE
[11] N. D. Jayaprakash, The path to sustainable development: lessons from Bhopal disaster. *People's Democracy*, December 14, 2014.

[12] NITI Aayog, *A Handbook of Select Global and Indian Best Practices in Mobility: NITI Aayog-Global Mobility Summit*, September 7–9, 2018. Govt of India.
[13] USEPA, *Electric Vehicles Trends and Projections: US-EPA: State & Local Energy and Environment Programs, Webinar Series.* USEPA, 2019
[14] NITI Aayog, *NITI Aayog Electric Vehicle Mobility Vision 2030 Report.* Govt of India.
[15] A. Rameshkumar and S. Kumar Padmanabhan, Electric vehicles for india- overview and challenges. *IEEE*, 14, 2, 2019
[16] B.W. Smith, Human error as a cause of vehicle crashes. *The Centre for Internet and Society (CIS)*, 12, 9–20, 2013.
[17] T. Related, D.A. Systems, *Vehicles, Taxonomy and Definitions for Terms Related to Driving Automation Systems for On-Road Motor Vehicles.* SAE International, 2016.
[18] X. Zhu, K. Du, Incumbent business ecosystems in the face of new entry: An event study of Google's autonomous car announcement. *Information & Management*, 60, 2023. doi:10.1016/j.im.2023.103760.
[19] E.R. Teoh, D.G. Kidd, Rage against the machine ? Google's self-driving cars versus human drivers. *Journal of Safety Research*, 63, 57–60, 2017. doi:1016/j.jsr.2017.08.008.
[20] B. Schoettle, M. Sivak, *A Preliminary Analysis of Real-World Crashes Involving Self -Driving Vehicles.* University of Michigan Transportation Research, 2015, p. 20. https://www.umich.edu/~umtriswt/PDF/UMTRI-2015-34_Abstract_English.pdf%5Cnhttps://www.umich.edu/~umtriswt/publications.html.
[21] Google, *A Summary: Google Self-Driving Car Project Monthly Report Google Self-Driving Car Project Monthly Report.* Google, 2015.
[22] G. Spiteri, *The Future of Truck Driving in America.* 2023. https://aegai.nd.edu/latest/the-future-of-truck-driving-in-america/.
[23] D. Paula, T. König, M. Bauder, F. Petermeier, T. Kubjatko, H. Schweiger, Transport means. In: Part II: *Proceedings of the 26th International Scientific Conference, Transp. Means.* 2022. doi:10.5755/e01.2351-7034.2022.p2.
[24] Car, *Tesla Autopilot and Collisions with Stationary Emergency Vehicles.* 2023, https://aegai.nd.edu/latest/tesla-autopilot-and-collisions-with-stationary-emergency-vehicles/.
[25] IvyPanda, *Waymo Case Studies and Future Efficiency Report*, July 28, 2022, https://ivypanda.com/essays/waymo-case-studies-and.
[26] M. Schwall, T. Daniel, T. Victor, F. Favaro, and H. Hohnhold, 2020. Waymo public road safety performance data. arXiv preprint arXiv:2011.00038.

Index